LASER

MATERIALS

LASER

MATERIALS

Fuxi Gan

Shanghai Inst. Optics & Fine Mechanics
Academia Sinica

World Scientific
Singapore•New Jersey•London•Hong Kong

Published by

World Scientific Publishing Co. Pte. Ltd.

P O Box 128, Farrer Road, Singapore 9128

USA office: Suite 1B, 1060 Main Street, River Edge, NJ 07661

UK office: 57 Shelton Street, Covent Garden, London WC2H 9HE

ISBN 981-02-1580-0

Printed in Singapore.

Preface

Laser material means the laser active medium, which is the heart part of the laser device. The active media for gas lasers are more simple, such as inert and diatomic molecules, it is unnecessary to discuss them specially. Therefore, the laser materials are mainly condensed matters represented by solid state materials. There are two kinds of solid state lasers: light-pumped and electrically excited, the latter ones are semiconductor lasers. As the preparation of semiconductor laser materials—semiconductor thin films always ties with laser devices, the semiconductor materials are always discussed in device monographs. Hence, in this book we introduce mainly the light-pumped and ion-activated solid state materials, in which the authors have long been engaged.

Since emergence of the first laser device—ruby laser, solid state lasers continue to occupy a prominent place among the different kinds of laser devices, which are widely used in industrial, medical, military and scientific fields. The laser materials are the research keystones for development of new solid state lasers.

In 1970s several monographs concerning laser materials have been published, such as "Laser Crystals" written by A. A. Kaminskii in 1979 (in Russian and English), "Phosphate Laser Glasses" edited by M. E. Zabotinskii in 1980 (in Russian), but the most introduction of laser materials were compiled in some handbooks, such as "Laser Handbook" edited by M. Bass and M. L. Stitch in 1985. No comprehensive monographs concerning solid state laser materials—crystals and glasses have been published up to now.

The crystal and glass lasers have been developed in China at the beginning of 1960s, just 2–3 years later since the emergence of the first solid state laser in the world. All the laser materials were produced in China. The research and development of laser materials have been paid great attention in China in the past thirty years. Authors of this book have been actively engaged in this field and made notable contributions to the above mentioned subjects and published many papers in journals both at home and abroad. An attempt was made in this monograph to summarize the research and development results of laser materials, which have been achieved by the authors' research groups. The emphasis has been put on the optical spectroscopy, laser physics, structure and defects, as well as new laser crystals and glasses. Therefore, the main effort has been directed towards a generalization of the above mentioned subjects and compilation of all experimental information from journals and proceedings.

It is not quite enough to select those chapters in this monograph for discussing the above mentioned subjects due to the imbalance of our research work. We tried our best to choose those subjects that contain the authors' own research achievements (ideas, data and results). It might be possible to omit some important subjects or contents due to the limit of the authors' knowledge and incomplete access to foreign materials. We hope that our readers, especially the laser materials experts, will give us their valuable suggestions in order to be able to make correction and complements for us in the next edition.

Acknowledgements are expressed to those journals and proceedings from which some figures have been reproduced. The authors of this book are grateful to Prof. Kewu Wang for his help in English, and particularly thank my colleagues Mr. Donghong Gu, Mr. Xiaodong Tang and Mr. Qiying Chen for their assistance in preparing the manuscript, without their help this book would not come into public.

Fuxi Gan

October, 1994

Contents

1. Fundamentals of Optical Spectroscopy of Paramagnetic Centers in Dielectric Solids

The optical spectroscopy of laser materials is mainly concerned with the interaction of electronic centers (i.e. paramagnetic centers, including the ions of transition metals and rare-earth metals) with optical radiation in insulating solids (crystals and glasses). In this part the electronic spectroscopic fundamentals are introduced and the spectroscopic properties of transition metal ions ($3d$ ions) and rare earth metal ions ($4f$ ions) are described in detail.

1.1 Energy Levels

The energy level splitting of electronic states of ions in solids are discussed in this section. The energy levels are mainly associated with $3d^n$ and $4f^n$ electronic configuration of ions, which are the most important active ions for laser materials.

1.1.1 Eigen States

The energy E and eigen wave function ψ of paramagnetic ions in solid states can be obtained by solving Schrödinger equation:

$$H\psi = E\psi, \tag{1.1}$$

where the total Hamiltonian of paramagnetic ions is

$$H = H_{ic} + H_{esl} + H_{edl} + H_{dl}, \tag{1.2}$$

where H_{ic} is the Hamiltonian of the isolated center, while in the ion-activated glass and crystal, it is the free ion Hamiltonian H_{fi}. H_{esl} is the static interaction of ligand field with the active ions, H_{dl} the vibrational energy of ligand field represented by phonon mode, H_{edl} the dynamic interaction of ligand field with active center. The Hamiltonian of free ions, H_{fi}, can be expressed as

$$H_{fi} = H_0 + H_e + H_{so}, \tag{1.3}$$

1

where H_0 is the interaction of spheric symmetric ion nuclei with an optical active electron, H_e the static electric (coulomb) interaction between electrons, which configurates the electrons into L and S terms, $H_{s,o}$ the interaction of spin with orbital, which forms J and M terms.

1.1.2   Energy Level of Free Ions

Energy level of paramagnetic ions is denoted with spectral term of $^{2S+1}L_J$. In the cases of multi-electron configurations, the spectral term is represented by the total angle quantum number L produced by multi-electronic interaction. When $L =0, 1, 3...$, it is denoted by S, P, D, F.... The upper-left hand number indicates the multi-state, i.e. the number of spectral sub-terms in the same spectral term (namely the number of different J values). When $L>S$, the multi-state is dominated by $(2S+1)$, when $L<S$, it is dominated by $2L+1$. The lower-right-hand number indicates the J value, which is produced by the interaction of L with $S(H_{s,o})$ formed spectral sub-terms.

The energy value in the energy levels for each spectral term can be obtained from calculation of the interaction of multi-electrons theoretically. Slater, Condon and Shortley *et al.* used three parameters of F_0, F_2 and F_4 (radical integration of static interaction) to denote the level position of d^n electron configuration, known as the Slater parameters. Based on group theory, Racah used A, B, C to denote them, we call the Racah parameter. The relation between these two parameters is $A = F_0 - 49F_4$, $B = F_2 - 5F_4$, $C = 35F_4$. In the Racah parameter, A is the same for all the energy levels and only contributes shift.

Racah used E_0, E_1, E_2 and E_3 to represent the energies of all the spectral terms formed by f^n electron configuration. Slater added a parameter F_6 to it. The relations between them are

$$
\begin{aligned}
E_0 &= F_0 - 10F_2 - 33F_4 - 286F_6, \\
E_1 &= (70F_2 + 231F_4 + 2002F_6)/9, \\
E_2 &= (F_2 - 3F_4 + 7F_6)/9, \\
E_3 &= (5F_2 + 6F_4 - 91F_6)/3.
\end{aligned}
$$

Table 1.1 gives the energy of all the spectral terms by taking $3d^3(\mathrm{Cr}^{3+})$ and $4f^3(\mathrm{Nd}^{3+})$ electron configuration as an example.

The parameters of A, B, C and E_1, E_2, E_3 for all the electron configurations can be found in the literatures [1, 2].

The energy in the level of spectral subterm is determined by multi-state splitting values (known as the Lande parameter) of the interaction between electrons in multi-electron configuration and orbital ($H_{s,o}$),

$$
\xi_{\mathrm{nL}} = 5.8 \int_0^\infty \frac{R^2 du(r)}{r\,dr} dr. \tag{1.4}
$$

Table 1.1. Term spacing energy of electron configuration $3d^3(\text{Cr}^{3+})$ and $4f^3(\text{Nd}^{3+})$.

Electronic configuration	Spectral term	Energy of level	Electronic configuration	Spectral term	Energy of level
$3d^3$	F	0	$4f^3$	^4I	0
	^4P	$15B$		$^4F, ^4S$	$21E_3$
	^2G	$4B+3C$		^4G	$33E_3$
	^2H	$9B+3C$		^4D	$54E_3$
	^2F	$24B+3C$		^2H	$3E_1+21E_2-3E_3$
	^2D	$18B+3C$		^2K	$3E_1-135E_2+10E_3$
				^2P	$3E_1+10E_3$

Generally, ξ is directly proportional to $Z^2\alpha^2$, where Z is the atomic number and α the electric valence of ion plus 1. The multi-state splitting value is an important parameter to $5f^n$ and $4f^n$ electron configuration. The value of ξ_{nL} generally ranges from 10^3 to 10^4 cm^{-1}. It is small in d^n electron configuration. For example, $\xi_{3d}\approx0.1\sim0.9\times10^3$cm^{-1}, $\xi_{4d}=0.3\sim2.3\times10^3$cm^{-1}, $\xi_{5d}\approx6\times10^3$cm^{-1}.

1.1.3 *Ligand Field Action*

The ligand theory is evolved from the crystal lattice theory. Bethe and Van Vleck have studied the crystal lattice theory theoretically [3, 4]. It takes account of static electric interaction exclusively, no account of electron exchange, and completely neglects the action of covalence bonding. For materials of covalence, e.g. the orbital energy in organic materials, the molecular orbital theory is generally used to deal with it. Most inorganic glasses and crystals have polarized covalent bonds. The ligand field theory incorporates both theories, and considers the total energy splitting caused by static electric interaction and the σ and π bonding. The ligand field theory can be found in the work by Orgel, Jorgensen and Schlafer *et al.* [5–7] in detail.

All the electron orbitals consists of an equivalent orbital, i.e. the orbital is degenerated. The electron cloud of the S orbital has completely spherical symmetry, i.e. the change of wave function with nuclei spacing is independent of the direction, the equivalent orbital is 1. Orbital P consists of 3 equivalent orbitals, which are composed of two spherical orbitals along a certain section and alternated. The sections are xy, yz and zx separately. As the electrons at S and P orbitals are of bonding electrons, they have no contribution to the formation of energy levels. When the ionic bonding is formed, S and P are all of full shells.

The contribution of the crystal lattice field is to d and f electrons of the inner shells which have not associated with the bonding. There are 5 equal valence orbitals in d orbital, namely d_{xy}, d_{yz}, d_{zx}, d_{z^2}, $d_{x^2-y^2}$, and 7 in f orbital as xyz, x^3, y^3, z^3, $z(x^2-y^2)$, $y(x^2-z^2)$, $x(y^2-z^2)$. Due to s, p, d outer electron shielding outside of f orbital in f^n electron configuration, the contribution of crystal lattice field is less,

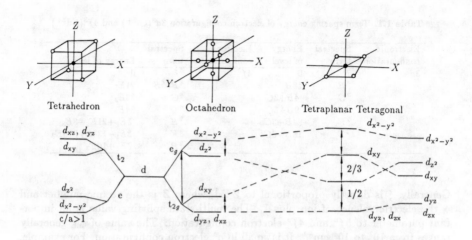

Fig. 1.1. Energy level splitting of the d orbitals in different crystal fields.

and ligand field is the main contributor to d^n electron configuration.

With the interaction of ligand field, 5 degenerated equivalent d orbitals are splitting, which are dominated by symmetry of the ligand. Because the d electrons without bonding repel the ligand with negative charge, and they are difficult to enter the orbital along the direction of ligand linking with metal ion, so the orbital energy at these directions are higher than that in other orbitals. Figure 1.1 shows the energy level splitting of the d orbital at different symmetric conditions. Its value $(t_{2g}-e_g)$ is symbolized with Δ or 10 D_q. Two kinds of symbols for energy level of ligand field splitting are of Bethe and Mullikin. Their relation is as follows:

Bethe	Γ_1	Γ_2	Γ_3	Γ_4	Γ_5
Mullikin	A_1	A_2	E	T_1	T_2
	A_{1g}	A_{2g}	E_g	T_{1g}	T_{2g}

Table 1.2 gives the splitting number of spectral terms in different symmetric ligand field and the spectral sub-terms in cubic symmetry for multi-electron configuration d^n.

The Stark splitting number of spectral sub-terms with different J value formed by L and S coupling is also related to symmetry, shown in Table 1.3. The energy splitting value Δ of the d orbital is dependent on the properties of central metal ions and ligands. By using static electric theory or disturbance theory of quantum mechanics, the energy splitting value can be estimated approximately.

$$\Delta = \frac{5eq\bar{r}^4}{3R^5} \quad \text{or} \quad \Delta = \frac{5e\bar{\mu}\bar{r}^4}{R^6}, \tag{1.5}$$

Table 1.2. Splitting number of spectral terms in different symmetry ligand field and the spectral sub-terms in cubic symmetry.

	Symmetry				
Spectral term	C_2	C_{4v}, D_{4v}	D_{6h}	O_h, T_d	
S	1	1	1	1	A_{1g}
P	3	2	2	1	T_{1g}
D	5	4	3	2	$E_g + T_{2g}$
F	7	5	5	3	$A_{2g} + T_{1g} + T_{2g}$
G	9	6	6	4	$A_{1g} + E_g + T_{1g} + T_{2g}$

where $\bar{r}$ is the distance of metal ion from nucleus to the electron at d shell, R the distance of metal ion to the center of ligand, e the electric charge, $\bar{\mu}$ the dipole vector, and q the ligand charge. So the characteristics of metal ions contribute to the following points:

a) Similar Δ value for the ions with same electric valence in the same period;

b) Δ values increase with increasing electric valence of ions;

c) Δ values increase with the increasing period.

The contributions of ligand to the splitting value are generally increasing with the decreasing radii of central ions of ligand. The interaction sequence for often used organic and inorganic ligands are

$$I^- < Br^- < Cl^- \sim SCN^- \sim (NH_2)CO \sim OH^- \sim NO_2^- \sim HCOO^-$$
$$< H_2O < SCN^- < EDTA < NH_3 < SO_3^= < NO_2^- \ll CN^-.$$

It should be noted that the static electric theory can not explain the above sequence because from the view of static electric interaction, halides must have higher field intensity than those of H_2O and NH_3. Both charges and polarization of OH^- are higher than those of H_2O. However, for the Δ value, the former is less than the latter. So the interaction of π bonding of ligand should be taken into account. The interaction of metal ions with π orbital of ligand needs a supplementary explanation of molecular orbital theory. So in case of a mixing orbital wavefunction which is not so strong, covalent bonding factor β could be used to describe such an effect.

Because the increase of the covalent bond makes the repulsion force between inner electrons decrease, both Racah and Landé parameters are less than those of the corresponding free ions, β value can be expressed as

$$\beta \approx B/B_{fi} \approx C/C_{fi} \approx \xi_{nl}/\xi_{nl(fi)}. \tag{1.6}$$

One may find that the stronger the covalent bond, the less the β value.

Hence, it can be concluded from empiric analysis of the experimental data that

$$\Delta = f(\text{ligand}) \times g(\text{central ion}), \tag{1.7}$$

Table 1.3. Ligand field symmetry and Stark splitting.

Symmetry		Stark splitting number								
Cubic: O_h,O,T_d	i	1	1	2	3	4	4	6	6	7
T_h,T	h	1	1	2	3	3	4	5	5	6
Hexagonal D_{6h},D_6,C_{6v},C_{6h}	i	1	2	3	5	6	7	9	10	11
C_6,D_{3h},C_{3h},D_{3d}	h	1	2	3	4	5	6	7	8	9
D_3,C_{3v},S_6,C_3										
Tetragonal D_{4h},D_4,C_{4v}	i	1	2	4	5	7	8	10	11	13
C_4,D_{2d},S_4	h	1	2	3	4	5	6	7	8	9
Low D_{2h},D_2,C_w	i	1	3	5	7	9	11	13	15	17
Symmetric D_{2h},C_2,I	h	1	2	3	4	5	6	7	8	9
S_2,C_1										
J value	i	0	1	2	3	4	5	6	7	8
	h	1/2	3/2	5/2	7/2	9/2	11/2	13/2	15/2	17/2

Table 1.4. Parameters for calculated Δ and β values.

Ligand	f	h	Metal ions	$g(1000\ \text{cm}^{-1})$	k
$6F^-$	0.9	0.8	V^{2+}	12.3	0.08
$6H_2O$	1.0	1.0	Cr^{3+}	17.4	0.21
$6(NH_2)CO$	0.91	1.2	Mn^{2+}	8.0	0.07
$6NH_3$	1.25	1.4	Fe^{3+}	14	0.24
$3OX^2$	0.98	1.5	Ni^{2+}	8.9	0.12
$6Cl^-$	0.80	2.0	Mo^{3+}	24	0.15
$6CN^-$	1.70	2.0	Rh^{3+}	27	0.30
$6Br^-$	0.76	2.3	Pt^{4+}	36	0.50
3dtp	0.86	3.8			

$$(1 - \beta) = h_{(\text{ligand})} \times k_{(\text{central ion})}. \tag{1.8}$$

Table 1.4 lists some of f, h values of ligands and g, k values of central ions.

1.1.4 Energy Level of Paramagnetic Ions in the Ligand Field

The energy level splitting is different for various electronic configurations at different intensities of the ligand field. It can be divided into several cases.

a) Interaction between the same electrons $(\Delta L)>$coupling of spin with orbital $(L+S)>$ligand field function (Δ). This case corresponds to the energy level splitting of rare-earth ions $(4f^n)$. Figure 1.2 is the energy level diagram of Nd^{3+}.

b) $\Delta L >\Delta>L+S$. This case corresponds to the energy level splitting of transition element ions $(3d^n)$ at weak ligand field, or rare-earth ions at very high intensity

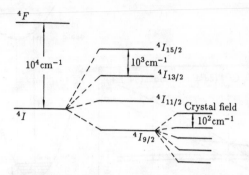

Fig. 1.2. Energy level splitting of $4f^3$ electronic configuration.

of the ligand field. Figure 1.3 is the energy level diagram of Cr^{3+} ($3d^3$) and Mo^{3+} ($4d^3$).

c) $\Delta > \Delta L > L + S$. This case corresponds to the energy level splitting of transition element ions at very high ligand field.

d) $L + S > (L + L)S + S > \Delta$. This case corresponds to the energy level splitting of Ac family ions ($5f^n$) at weak ligand field.

e) $L + S > \Delta > S + S(L + L)$. This case corresponds to the energy level splitting of Ac family ions or transition elements ($4d^n$, $5d^n$) at very weak ligand field.

f) $\Delta > L + S > S + S(L + L)$. This case corresponds to transition elements ($4d^n$, $5d^n$) at high ligand field.

The energy levels of $3d^n$ ions in the ligand field were calculated and calibrated by Sugano *et al.* [8, 9]. As the splitting terms have large spaces in the intense field, it does not take account of interaction between the splitting terms, so the energy changes of various levels with the Δ value has a linear relationship. But at the weak field, it becomes curves because of the interaction of splitting terms (see Fig. 1.3).

1.1.5 *Action of the Vibration State*

Above were discussed all of the static interactions of ligands with paramagnetic ions. Now let us discuss the dynamic effect of the ligand, i.e. the interaction of vibration modes with energy level and electron transition (H_{edl}).

The configurational coordinate is always used here [10, 11], shown in Fig. 1.4. Vibration of ligand is parabolic simple harmonic motion expressed as

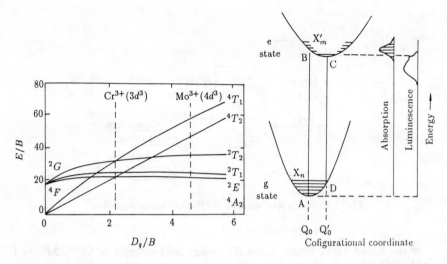

Fig. 1.3. Energy levels of the d^3 electronic configuration in an octahedral field.

Fig. 1.4. Configurational coordinate diagram for analysis of optical transitions between e and g.

$\frac{1}{2}M\omega^2(Q-Q_0)^2$, where, M is the mass of atoms involved in the vibration, and ω the vibrational frequency. Several vibration states of the ground state g are denoted with x_n, and m kinds of vibration states of excited state are denoted with x'_m; the average configurational coordinate of the ground state, and the excited state are Q_0 and Q'_0 respectively. The difference of configurational coordinates $Q_0-Q'_0$ means the sensitivity difference of the two electronic levels to the host environment, to be expressed by the Huang-Rhys parameter S,

$$\frac{1}{2}M\omega^2(Q_0-Q'_0)^2 = Sh\omega. \tag{1.9}$$

Hence, absorption $(A-B)$ and emission $(C-D)$ both become broad-band taking an account of vibration state interaction (shown in Fig. 1.4), i.e. many phonon side-bands around the spectral lines of electron transition are produced, each vibration mode produces a phonon side-band. Sharp zero-phonon transition lines appears only at low temperature.

1.2 Electron Transition Process

In this section the electron transition process is mainly concerned with the electronic transition between the energy levels within an active ion, including selection rules of the electronic transition, radiative transition and nonradiative transition.

1.2.1 *Selection Rules of Electronic Transition*

Based on the calculation with quantum mechanics, the selection rules of electronic transition of electric dipole radiation for free ions can be summarized as follows:

a) For main quantum number n, Δn can be arbitrary during transition.

$$\Delta n = 0, 1, 2, 3, \ldots$$

b) Selection rules of angular moment L is

$$\Delta L = 0, \pm 1,$$

when $\Delta L=0$, the transition is very weak, and ΔL of the transition electron must be ± 1 (called Laporte selection rules), namely, transition between the same shells is forbidden.

c) The selection of total angle moment J is

$$\Delta J = 0, \pm 1,$$

additional limitation is that $J =0$ can not be configurated with another $J =0$.

d) The selection of spin quantum number is

$$\Delta S = 0,$$

usually called the Russel-Sounder rules.

Violation to the above selection rules is called forbidden transition. The forbidden transitions are usually weak in intensity.

Violation of the above selection rules is caused by:

a) Selection rule $\Delta S =0$ is completely valid when $L + S$ interaction is weak. With the coupling of $L + S$ becomes strong, it will not be so strict. The higher the atomic number, the larger the multi-state splitting and the higher the possibility of violation of the selection rules.

b) There may be quadri-pole radiation ($\Delta J =0$, $+1$, ± 2) and magnetic dipole radiation ($\Delta L =0$, ± 1, ± 2) when the electric dipole radiation is forbidden, the intensity of them is very low.

c) The selection rules can be changed at high electric field or high magnetic field. Thus the forced electric dipole radiation and the magnetic dipole radiation are generated.

The above selection rules are further broken down in the ligand field. The causes for such violations are mainly:

a) If the ligand distribution is not completely symmetrical, the orbital may be partially mixed, e.g. d and p, f and d orbitals mixed. So some of the transitions become non-forbidden.

b) Due to the vibration of the crystal lattice, the central ions shift from their equilibrium position and lose the temporary symmetry, consequently produce mixed orbitals. When the lattice vibration causes electron-vibration spectra, not only new spectral lines are produced, but also the intensity of the spectral lines is increased.

1.2.2   *Radiative Transition*

When the light radiation interacts with the electron centers, radiative transition (absorption and emission) is induced, as shown in Fig. 1.5. The probability W_r of radiative transition between the two levels a and b is proportional to the second order of matrix element. Transition intensity S can be expressed as [10]

$$S_{(ab)} = \sum_{a,b} |<b|D|a>|^2, \tag{1.10}$$

when it is an electric dipole transition, D is replaced by electric dipole vector P; when it is a magnetic dipole transition, D is replaced by magnetic dipole vector m. The oscillator strength f always used in absorption spectra can be expressed as

$$f(ab) = \frac{1}{g_a} \cdot \frac{8\pi^2 m\nu}{3he^2} S(ab), \tag{1.11}$$

where ν is the transition frequency, $h\nu = |E_b - E_a|$. m, e, is the mass and the charge of the electron, h the Planck constant, the allowed electric dipole transition $f_{ED} \approx 1$, and the allowed magnetic dipole transition $f_{MD} \approx 10^{-6}$.

If the absorption occurs by a dipole process, the relation of oscillator strength with absorption coefficient $\sigma(\nu)$ for cgs units becomes

$$\int \sigma(\nu)d\nu = \frac{\pi e^2}{mc} [(\frac{n^2+2}{3})^2 \cdot \frac{1}{n}] f_{ED}(ab). \tag{1.12}$$

The probability of Einstein spontaneous transition $A(ba)$ can be expressed as

$$A_{ED}(\nu) = \frac{8\pi \nu^2 e^2}{mc^3} [(\frac{n^2+2}{3})^2 \cdot n] f_{ED}(ab). \tag{1.13}$$

The decay time of radiative transition $\tau_R = A_{ED}^{-1}(ba)$. The probability relation between the stimulated transition B and the spontaneous transition A is

$$\frac{A}{B} = \frac{8\pi h\nu^3}{c^3} n^3. \tag{1.14}$$

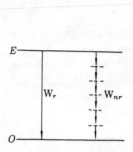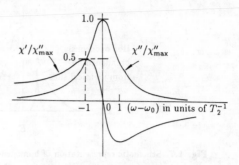

Fig. 1.5. Schematic representation of the radiative and the nonradiative processes.

Fig. 1.6. A plot of the real(χ') and imaginary(χ'') parts of the susceptibility for negligible saturation ($\mu^2 E_0^2 T_2 \tau / \hbar^2 \ll 1$).

1.2.3 *Non-radiative Transition*

As shown in Fig. 1.5, when the excited state changes to the ground state, several phonons can be generated instead of photon radiated, and non-radiative transition is induced by loosing its energy. The larger energy spacing between the two levels, the higher probability of the radiative transition ($\bar{W}_{nr}$), and the lower probability of the non-radiative transition because it needs more phonons associated with the multi-phonon process,

$$\bar{W}_{nr} = \bar{W}_0 \varepsilon^{p_i}, \tag{1.15}$$

where $\bar{W}_0$ is the relaxation rate from energy level b and a at the absolute temperature of zero, ε the attenuation probability of multi-phonon (<1), p_i the phonon order number, $p_i = \Delta E / \hbar \omega$, ΔE the energy spacing between level a and b, $\hbar \omega$ the phonon energy.

1.3 Spectral Line Profile and Broadening

The atomic polarizability χ consists of real part χ' and imaginary part χ'' and can be expressed by

$$\chi = \chi' - i\chi''. \tag{1.16}$$

The Lorentz line profile function $g(\nu)$ is

$$g(\nu) = \frac{\frac{\Delta \nu}{2\pi}}{(\nu - \nu_0)^2 + (\frac{\Delta \nu}{2})^2}, \tag{1.17}$$

where ν_0 is the central wavelength, $\Delta \nu$ halfwidth of the line profile, it should be

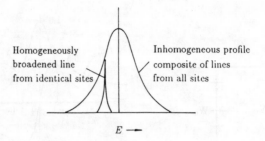

Homogeneously
broadened line
from identical sites

Inhomogeneous profile
composite of lines
from all sites

$E \longrightarrow$

Fig. 1.7. Schematic representation of homogeneous and inhomogeneous line broadening.

$$\int_{-\infty}^{+\infty} g(\nu)d\nu = 1. \tag{1.18}$$

Figure 1.6 shows the Lorentz absorption (χ'') and the dispersion curves.

The width of spectral lines of paramagnetic ions in solid state is constituted by homogeneous and inhomogeneous broadening. The relaxation rate of energy level dominates the natural broadening of energy level ΔE (energy uncertainty). Its dependence on the life-time of energy level is $\Delta E \cdot \tau = h$, where h is the Planck constant. The energy uncertainty of energy level ΔE causes the homogeneous broadening of ion line δ, $\delta = h\sum_i W_i$, where $\sum_i W_i$ is the sum of transition probability. Due to the effect of the vibration state, the interaction of electron-phonon is induced, and homogeneous broadening of spectral lines is increased. In addition to the interaction of electron-phonon caused by the vibration state mentioned above, the width of zero-phonon line varies with temperature owing to the different thermal distribution of phonons. The relaxation rate introduced by phonons is given by $W = W_0'(1+\bar{n}_0)$, where W_0 is the relaxation rate at absolute temperature, n_0 the thermal distribution of phonon.

The spectral line of paramagnetic ions in solids is wider than that of the homogeneous broadening mentioned above, because the different sites of paramagnetic ions induce inhomogeneous spectral broadening. At low temperature, homogeneous broadening of zero-phonon line is only 1 cm^{-1}, the homogeneous broadening induced by phonons is usually $10 \sim 10^2$ cm^{-1}, while the inhomogeneous broadening reaches $10^2 \sim 10^3$ cm^{-1}. As shown in Fig. 1.7, the ions on each side have a homogeneous broadening line, while the spectral lines from different sites overlap to form a very wide spectral band. Since the long range disorder of glassy materials produces serious site disorder, so the inhomogeneous line-broadening of paramagnetic ions in glass is more serious than that in crystals.

In the case of existing site difference and phonon interaction, the absorption line (S_{ab}) and the emission line (S_{em}) become

$$S_{ab} = \alpha Z_{ab}(\nu - \nu_i) + (1 - \alpha)V_{ab}(\nu - \nu_i), \tag{1.19}$$

$$S_{em} = \alpha Z_{em}(\nu_i - \nu) + (1 - \alpha)V_{em}(\nu_i - \nu), \tag{1.20}$$

where Z and V are the zero-phonon line and the phonon side-band respectively, ν_i the central frequency of zero-phonon line for i type site, α the ratio of zero-phonon lines to the total lines. For rare-earth ions, α approaches 1, and for transition metal ions, α approaches 0.

Let $D(\nu_i)$ represent the central frequency distribution of ions at different site in solids. If a narrow monochromatic light with frequency ν_L is used for excitation, and the distribution of excited state as $D(\nu_i) \times S_{ab}(\nu_L - \nu_i)$, then the fluorescence spectra formed are given by

$$F(\nu) = \int S_{em}(\nu_i - \nu)D(\nu_i)S_{ab}(\nu_L - \nu_i)d\nu_i. \tag{1.21}$$

We now substitute the Eqs. 1.19, 1.20 into it, 4 terms are obtained: $Z_{em}Z_{ab}$, $V_{em}Z_{ab}$, $Z_m V_{ab}$ and $V_{em}V_{ab}$. The first and the third terms are the zero-phonon line or phonon side band excitation, which emits at zero phonon line, the second and the fourth terms are the phonon side band emission. We may find from them that the fluorescence line becomes narrow when the monochromatic light with the narrow line is used to excite the ions at certain site. This is called the Fluorescence Line Narrowing (FLN), an important means for studying the site condition and energy transfer between the sites of paramagnetic ions in solids.

1.4 Energy Transfer Process

There exist two types of energy transfer process: one is the ion-ion interaction, the other is that between the ion and the host.

1.4.1 Ion-ion Interactions

Energy transfer between like and unlike paramagnetic ions occurs at concentrations where the ion separations become small ($\widetilde{<}1$–2 nm). The energy transfer may lead to energy migration, fluorescence quenching and sensitization. Depending on the concentration the energy transfer process can be taken by dipole-dipole, dipole-quadrupole and quadrupole-quadrupole. There are two types of energy transfer: resonant and non-resonant. We define non-resonant energy transfer as that occurring when the energy mismatch between an optically excited "donor" (D) ion and an unexcited "acceptor" (A) to which the excitation will be transferred is much than kT and/or the inhomogeneous linewidth.

Fig. 1.8 shows the energy migration and the self-quenching for Nd^{3+} by resonant energy transfer. The theoretical calculation of Dexter [12] indicated that resonant

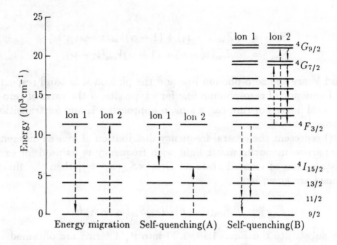

Fig. 1.8. Pairs of energy-conserving transitions for Nd^{3+} arising from ion-ion interactions.

energy transfer probability is inversely proportional to the sixth power of d—the distance between two centers if the two centers belong to the dipolar transition, and the energy transfer probability will be inversely proportional to the higher magnitude of d($>$6th power) in case of quadripole transition. The non-resonant energy transfer is always followed by multiphonon relaxation. Fig. 1.9 shows some pairs of sensitization process by non-resonant energy transfer. The energy transfer from D to A can be shown in expression as:

$$\eta_t = 1 - \frac{\eta_l}{\eta_d^0}, \qquad (1.22)$$

where η_d and η_d^0 denote respectively the luminescence quantum efficiency of donor ions with and without existence of acceptor ions. Quantum efficiency η_d/η_d^0 can be replaced with luminescence ratio of I_d/I_d^0 in the case of non-excitation of acceptor ions or very weak absorption. Within a certain excitation wavelength range of donor ions (I_d and I_d^0 indicate respectively the luminescence intensity of donor ions with and without existence of acceptor ions), Eq. 1.22 can be written as follows:

$$\eta_t = 1 - \frac{I_d}{I_d^0}, \qquad (1.23)$$

which is applicable to the case of absorbing same number of photons at the same excitation wavelength. The energy transfer probability is:

$$P_{da} = \frac{1}{\tau d}(\frac{\eta_d^0}{\eta_d} - 1). \qquad (1.24)$$

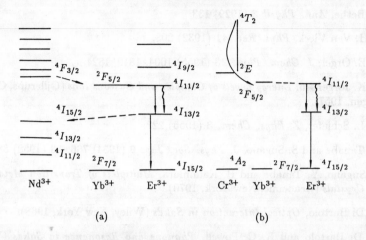

Fig. 1.9. Ion sensitizations (a) Nd^{3+}-Yb^{3+}-Er^{3+} (b)Cr^{3+}-Yb^{3+}-Er^{3+}.

1.4.2 *Interaction between Ion and Host*

The energy transfer between ion and host is always considered as a multiphonon relaxation process. This is a nonradiative transition. The probability of nonradiative energy transfer can be described as that in [13],

$$\bar{W}_{nr} = \bar{W}_0 \cdot \varepsilon_p, \tag{1.25}$$

$$\bar{W}_{nr} = C \cdot \exp(-\alpha \Delta E) \cdot (\eta(T) + 1)^{p_i}, \tag{1.26}$$

where ε represents the decay probability of multi-phonon, $\varepsilon = \bar{w}_p / \bar{w}_{p-1} < 1$; p_i is the order of phonon, $p_i = \Delta E / h\omega$; ΔE is the energy gap of the nearest energy levels, $h\omega$ the phonon energy, $\bar{n}(T)$ the phonon mode, $\bar{n}(T) = [\exp(h\omega/kT) - 1]^{-1}$, C, α, constants relevant to the hosts.

Therefore, non-radiative transition probability of multi-phonon process relies first of all on the phonon order, i.e. the energy gap of energy levels, and also the phonon energy. The former determines the energy level structure of paramagnetic ions, and the latter is dependent on the structure of hosts.

References

1. E. U. Condon and G. H. Shortley, *Theory of Atomic Spectra*, (Cambridge, 1953).

2. G. Racah, *Phys. Rev.* **62** (1942) 438; **63** (1943) 367.

3. H. Bethe, *Ann. Physik* **3** (1929) 133.

4. J. H. Van Vleck, *Phys. Rev.* **41** (1932) 208.

5. L. E. Orgel, *J. Chem. Phys.* **23** (1955) 1004, 1819, 1824.

6. C. K. Jorgensen, *Energy Levels of Complex and Gaseous Ions* (Gjllerups, Copenhagen, 1957).

7. H. L. Schlafer, *Z. Phys. Chem.* **3** (1955) 222.

8. Y. Tanabe and S. Sugano, *J. Phys. Soc. Jap.* **9** (1954) 766; **13** (1958) 880.

9. S. Sugano, Y. Tanabe and H. Kamimura, *Multiplets of Transition Metal Ions in Crystals* (Academic, New York, 1970).

10. B. Di Bartolo, *Optical Interaction in Solids* (Wiley, New York, 1968).

11. B. Di Bartolo and R. C. Powell, *Phonons and Resonance in Solids* (Wiley-Interscience, New York, 1976).

12. P. L. Dexter, *J. Chem. Phys.* **21** (1963) 876.

13. L. A. Risebery and M. J. Weber, *Prog. Opt.* **14** (1976) 91.

2. Spectroscopy of Transition Metal Ions (3d, 4d) in Dielectric Solids

For transition metal (TM) ions the electron-static lattice interaction H_{cf} is comparable with the Coulomb interaction of electrons, hence, it is customary to adopt the strong field scheme, in which one first considers how the d orbitals are affected by the electrostatic field of the surrounding ions which is approximately of either octahedral or tetrahedral symmetry, and then takes account of the interactions between the electrons and the configurations. The energy separation between the twofold degenerate e state and threefold degenerate t_2 is labelled with $10D_q$, where D_q is a parameter which measures the strength of the field. The spin-orbit interaction H_{so} is relatively weak in the domain of optical measurement. Hence for each ion the energy separations between various levels depend on two parameters D_q and B, which can be calculated by the absorption spectra and Tanabe and Sugano diagrams for the $(3d)$ ions [1]. The Racah parameter B of TM ions in solids is less than that in free ion state, which indicates the influence of covalence of chemical bond of hosts on the TM ions.

The spectroscopic behaviours of transition metal ions (3d, 4d) in solid state media are greatly dependent on the structure and the chemical composition of the host materials due to the ligand field interaction. Therefore, we discuss here the spectroscopy of transition metal ions in inorganic glasses and crystals separately.

2.1 Spectroscopic Properties of Transition Metal Ions in Inorganic Crystals

Since the emergence of the first solid state laser (Cr^{3+} : Al_2O_3 ruby laser) the spectroscopic properties of transition metal ions in crystals have been studied in detail.

The valence state of transition metal ions is changeable, Table 2.1 shows the number of valence electrons for the common valence states of all the ions in the $3d$ group.

Table 2.1. Number of $3d$ electrons for $3d$ ions.

Number of electrons	0	1	2	3	4	5	6	7	8	9	10
Ions	Ti^{4+}	Ti^{3+}	Cr^{4+}	Cr^{3+}	Mn^{3+}	Fe^{3+}	Fe^{2+}	Co^{2+}	Ni^{2+}	Cu^{2+}	Cu^{+}
			V^{3+}	V^{2+}	Cr^{2+}	Mn^{2+}	Co^{3+}	Ni^{3+}			
			Ti^{2+}								
			Mn^{5+}								

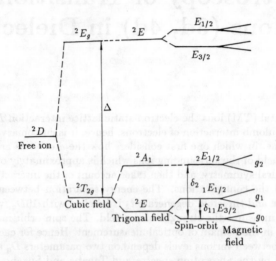

Fig. 2.1. Schematic energy-level diagram for the Ti^{3+} ion in $Ti:Al_2O_3$.

A lot of TM ions have been found as laser active centers in crystals, among them the most important ones are Ti^{3+}, Cr^{4+}, Cr^{3+}, Co^{2+}, Ni^{2+}.

2.1.1 $3d^1$ Ti^{3+}

Trivalent titanium ions (Ti^{3+}) are the active ions in tunable lasers. Since the electronic structure of Ti^{3+} ion is a closed shell plus a single $3d$ electron, the energy level diagram and spectral properties should be simple. The most famous tunable laser crystal is titanium doped sapphire ($Ti^{3+}:Al_2O_3$).

In the α-Al_2O_3 lattice Ti^{3+} ($3d^1$) substitutes Al^{3+} and is surrounded by an octahedron of oxygens with a slight trigonal distortion. A schematic energy-level diagram of $Ti^{3+}:Al_2O_3$ system shows in Fig. 2.1. The d-electron levels are split by the crystal field of α-Al_2O_3. In the cubic field the unique ^2D term of the $3d^1$ con-

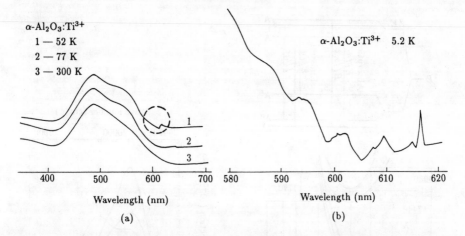

Fig. 2.2. (a) Temperature dependence of the visible absorption spectrum of Ti^{3+}:α-Al_2O_3 [3]; (b) enlarged view of the absorption spectrum (at 5.2 K) in the 580–620 nm region [3].

figuration is split in a ground $^2T_2$ triplet and an excited ^2E doublet. The trigonal field splits the ground $^2T_2$ state into two levels, and the lower of these two levels is split further into two sublevels by the spin-orbit interaction and Jahn-Teller effect, but the cubic splitting and electron-couplings determine the range of laser emission of Ti^{3+}:Al_2O_3 single crystal. Since both initial and terminal emission levels have the same spin parity, the emission is characterized by a large cross section $(4.5\times10^{-19}cm^2)$ [2]. There is no excited state absorption because no other d^1 levels exist above the emitting level.

According to the measurement of absorption spectra of Ti^{3+}:Al_2O_3 by A. Lupei *et al.* [3], a broad absorption band was observed in the visible region with a peak at 490 nm and a shoulder at 520 nm. The temperature dependence of the visible absorption spectrum and the enlarged view of the absorption spectrum (at 5.2 K) in the 580–620 nm region are shown in Fig. 2.2. The broad band feature corresponds to the phonon sideband of the transition from the lowest level $^2T_2$ to the two splitting levels of ^2E. The visible part of absorption spectrum shows a very interesting feature, that is for 60° Z-cut sample the peak of absorption band is observed at 489.1 nm with e-light excited, and at 484.0 nm with o-light excited. If using 90° Z-cut sample the peak of absorption band is shifted to 491.6 nm with e-light excited, and to 485.6 nm with o-light excited.

At 5 K, the visible absorption spectra show on the lower wavelength side a very sharp line at 616.6 nm shifting toward higher energies by a progression of at least five narrow lines with an average splitting of about 194 cm^{-1}. Each of these lines has a weaker satellite at about 46 cm^{-1} (Fig. 2.2b). The sharp line at 616.6 nm can be assigned to the zero-phonon transition between the lowest electronic

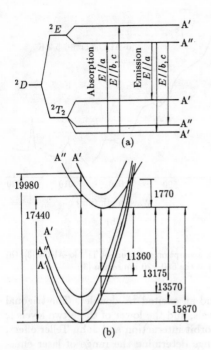

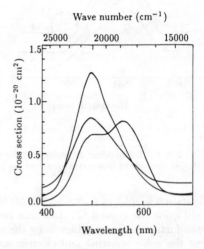

Wave number (cm^{-1})

Fig. 2.3. (a) Term splitting of the $(3d)^1$ state. From the left: free ion, splitting by a cubic field, and splitting by an orthorhombic field; (b) schematic diagram of the potential curves for the ^2E and $^2T_2$ states of Ti^{3+} in chrysoberyl [4].

Fig. 2.4. Absorption spectra of $Ti:BeAl_2O_4$ at room temperature for various directions of polarization. Band maxima: 500 nm (19980 cm^{-1}) and 573 nm (17440 cm^{-1}) for $E\|a$ and 500 nm (19980 cm^{-1}) for $E\|b$ and $E\|c$ [4].

components of the $^2T_2$ and ^2E levels. The progression observed in low temperature spectra is due to electron-phonon interaction in the excited state. The value of 194 cm^{-1} determined here directly corresponds to the coupling with the $_1E$ phonon of $\alpha\text{-}Al_2O_3$. The broad band feature observed in the visible part of the absorption spectrum (Fig. 2.2, 5.2 K) corresponds to the phonon sideband of the transition from the lowest component of the $^2T_2$ level to the intense sidebands indicate a strong electron-phonon coupling, which in turn explains the large Jahn-Teller splitting of ^2E level.

Ti^{3+} in chrysoberyl ($BeAl_2O_4$) occupies the C_s site, its possible energy diagram is shown in Fig. 2.3. From the absorption spectra (Fig. 2.4) the absorption peak near 500 nm that appears when the polarization is $E\|b$ and $E\|c$ can be attributed to the $^2T_2(A')\text{-}^2E(A')$ transition. The lower energy sideband near 580 nm observed when $E\|a$ is probably due to the $^2T_2(A')\text{-}^2E(A'')$ transition [4].

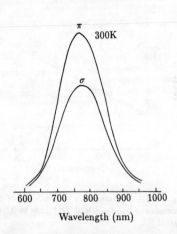

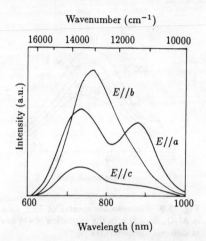

Fig. 2.5. Polarized fluorescence spectra of Ti^{3+} ions in Al_2O_3 crystal at 300 K [5].

Fig. 2.6. Luminescence spectra of $Ti:BeAl_2O_4$ at room temperature with various directions of observed polarization [4].

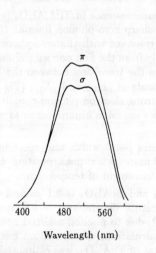

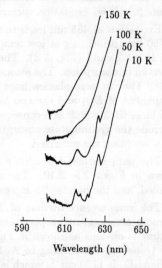

Fig. 2.7. Corrected excitation spectra of $Ti:Al_2O_3$ for emitting wavelength of 770 nm at 300 K [5].

Fig. 2.8. Part of fluorescence spectra of Ti^{3+} ions in Al_2O_3 crystal with the excitation of 488 nm at lower temperature (10–100 K) [5].

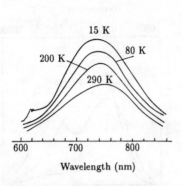

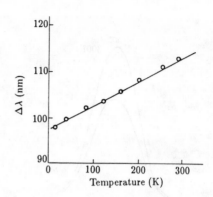

Fig. 2.9. Fluorescence spectra of Ti^{3+} ions in Al_2O_3 crystal with the excitation of 488 nm at different temperatures [5].

Fig. 2.10. Temperature dependence of the fluorescence bandwidth of Ti^{3+} ions in Al_2O_3 crystal [5].

Figures 2.5, 2.6 show the luminescence spectra of Ti^{3+} in Al_2O_3 and in $BeAl_2O_4$ at room temperature, strong polarization characteristics of the luminescence can also be observed. In $Ti^{3+}:Al_2O_3$ the ratio of emission intensity I_π/I_σ is equal to 1.5 (λ_{ex}=488 nm) and ratio of emission bandwidth $\Delta\lambda_\pi/\Delta\lambda_\sigma \approx 1$ can be obtained [5]. Figure 2.7 shows excitation spectra for emitting wavelength of 770 nm at 300 K.

Excitation at 488 nm leads to a broad red luminescence in $Ti^{3+}:Al_2O_3$ peaking at 750 nm, presenting at low temperature two sharp zero-phonon lines at 16215.3 and 16181.2 cm^{-1} (Fig. 2.8). The first line is resonant with the zero-phonon line observed in absorption. The second is split partly from the first one with about 34.1 cm^{-1}. The two zero-phonon lines correspond to the transition between the lowest component of $E_{3/2}$ and the two lowest components of $_1E_{3/2}$ and $_1E_{1/2}$ (Fig. 2.1). The large Stokes shift is a consequence of the strong electron-phonon coupling and preclude the ground state absorption of the laser emission luminescence as well as the absorption are polarized.

The temperature dependence of luminescence band width and spectrum are shown in Figs. 2.9, 2.10. The red shift of luminescence peak position can be notified, and the bandwidth increases with the increment of temperature.

The luminescence lifetime of Ti^{3+} in Al_2O_3 and $BeAl_2O_4$ is 3.1 μs and 4.7 μs at room temperature respectively, the temperature dependence of luminescence lifetime is obvious as shown in Figs. 2.11, 2.12, due to phonon-assisted nonradiative transition. According to A. Sugimoto's calculation, the activation energy of $Ti:BeAl_2O_4$ is 1770 cm^{-1}, which is similar to that of $Ti:Al_2O_3$, was estimated to be 1860 cm^{-1} [4].

We have studied the concentration effect of luminescence of $Ti^{3+}:Al_2O_3$ crystals. The experimental results are listed in Table 2.2. No obvious concentration

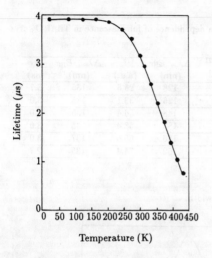

Fig. 2.11. Fluorescence lifetime versus temperature for the $^2E \rightarrow {}^2T_2$ transition in Ti:Al$_2$O$_3$ [2].

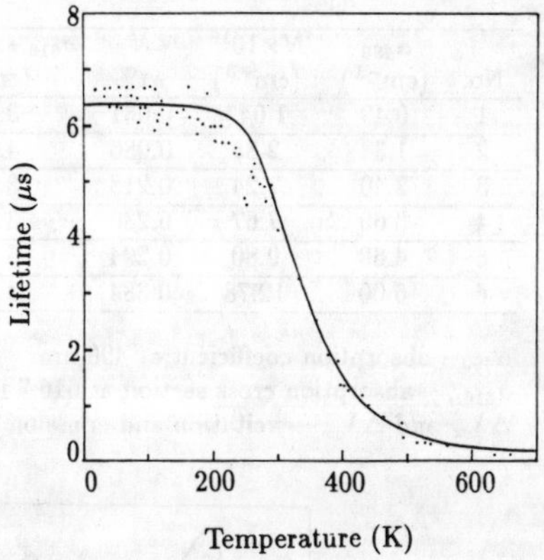

Fig. 2.12. Luminescence lifetime versus temperature for $^2E \rightarrow {}^2T_2$ transition in Ti:BeAl$_2$O$_4$ [4].

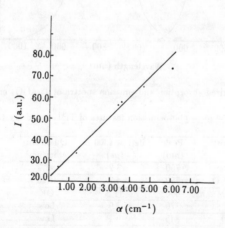

Fig. 2.13. Corrected fluorescence intensities vs Ti^{3+} concentration for six Ti:Al$_2$O$_3$ crystal samples [6].

Table 2.2. Spectral characteristics of concentration dependence of luminescence in Ti:Al$_2$O$_3$ crystals [6].

No.	α_{490} (cm^{-1})	$N \times 10^{19}$ (cm^{-3})	N wt%	$\sigma_{616.7} \times 10^{-21}$ (cm^2)	$\Delta\lambda_{ex}$ $E\|c$ (nm)	I (a.u.)	$\Delta\lambda_{em}$ $E\|c$ (nm)	τ (μs)
1	0.49	1.043	0.031	3.86	128	26.5	135	3.1
2	1.34	2.85	0.086	4.40	126	33.1	135	3.0
3	3.40	7.24	0.218	3.99	148	56.4	135	3.0
4	3.60	7.67	0.230	3.98	152	58.4	135	3.0
5	4.60	9.80	0.294	5.22	156	65.6	137	3.0
6	6.00	12.78	0.384	4.56	162	74.4	135	2.7

α_{490}—absorption coefficient at 490 nm.
$\sigma_{616.7}$—absorption cross section at 616.7 nm.
$\Delta\lambda_{ex}$ and $\Delta\lambda_{em}$—excitation and emission bandwidth.

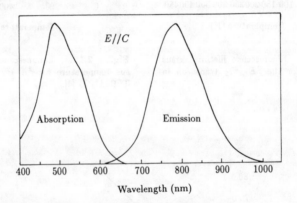

Fig. 2.14. Polarized absorption and emission spectra of Ti:Al$_2$O$_3$ crystal.

Table 2.3. The main photoemission features of Ti^{3+}-doped crystals.

Host	Peak (nm)	Lifetime (300 K) (μs)	Quantum efficiency
Al$_2$O$_3$	750	3.2	80%
BeAl$_2$O$_4$	750	4.7	High
YAlO$_3$	600	11.5	OK
YAG	770	1.0	Low
ScBO$_3$	845	0.1	Low
Rb$_2$NaScF$_6$	859	—	Low

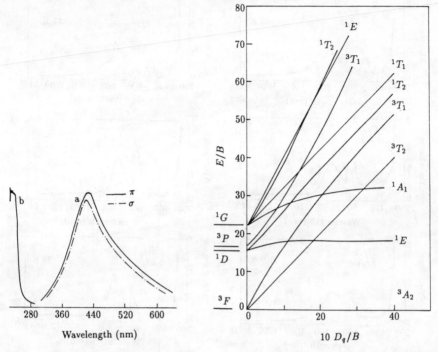

Fig. 2.15. Polarized luminescence of Ti:Al$_2$O$_3$ excited by 236 nm wavelength laser light (a) and excitation spectrum at emission wavelength 480 nm (b) both at 300 K.

Fig. 2.16. Tanabe-Sugano diagram for Cr^{4+} (d^2) in tetrahedral symmetry.

quenching can be observed, even though the Ti^{3+} ion concentration is up to 1.278×10^{20} cm^{-3} (or 0.384 wt%). The luminescence intensity is linear with absorption coefficient of Ti^{3+} ion at 490 nm (Fig. 2.13). According to Dexter's resonant energy transfer theory, the dipole-dipole resonant energy transfer probability between donor-acceptor ions depends on the overlapping area of absorption and luminescence curves. As shown in Fig. 2.14, the overlapping area is very small. This is the advantage for Ti^{3+}:Al$_2$O$_3$ crystal working at high titanium concentration.

Excitation in UV region leads to strong and broad polarized luminescence with a maximum around 425 nm (Fig. 2.15). The lifetime of this luminescence is about 35 μs, but a deviation from the exponential is observed, suggesting a superposition with a short-lived emission. The origin of the blue emission is not clear, it may be due to aggregate of color center-impurity. Luminescence emission in blue region (420–430 nm) promoted by diferent impurities has been observed recently in corundum [7].

The main photoemission features of Ti^{3+}-doped crystals are listed in Table 2.3.

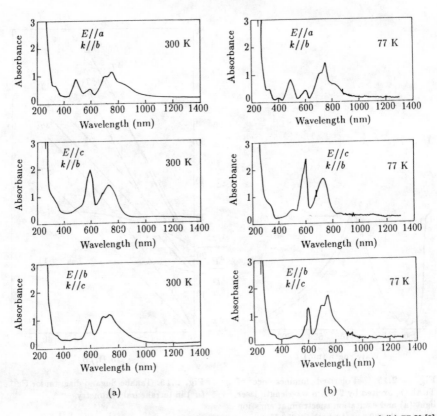

Fig. 2.17. Polarized absorption spectra of Cr^{4+}:Y_2SiO_5 at (a) room temperature and (b) 77 K [8].

2.1.2 d^2 $(Cr^{4+}, V^{3+}, Ti^{2+}, Mn^{5+})$

Cr^{4+}, V^{3+}, Ti^{2+}, Mn^{5+} belong to d^2 electron configuration, but valence state of V^{3+}, Mn^{5+} and Ti^{2+} are more unstable than that of Cr^{4+}. Laser action of Cr^{4+} in Mg_2SiO_4, Ca,Cr:YAG and Y_2SiO_5 crystals has been demonstrated quite recently. In this section we will pay more attention on spectroscopic properties of Cr^{4+} in crystalline state.

According to ionic radius Cr^{4+} ion is stable in tetrahedral site. Figure 2.16 shows Tanabe-Sugano diagram of energy levels for Cr^{4+} (d^2) in tetrahedral symmetry. Cr^{4+} in Cr^{4+}:Y_2SiO_5 is a purely Cr^{4+} doped crystal, allowed a better understanding of the spectroscopic properties of Cr^{4+} in tetrahedral sites, the site symmetry of Cr^{4+} in Y_2SiO_5 is $C_{3\nu}$. The absorption and fluorescence spectra of Cr^{4+}:Y_2SiO_5 at 77 K are shown in Figs. 2.17, 2.18. It has shown that the fluorescence intensity depends on the excitation wavelength and the absorption is polarized [8,9]. C. Deka

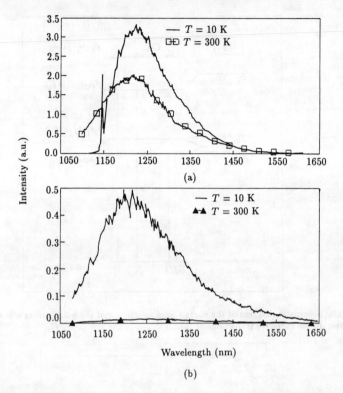

Fig. 2.18. Fluorescence spectra of Cr^{4+}:Y_2SiO_5 at different temperatures due to excitation at (a) 1064 nm and (b) 532 nm [8].

deduced the Racah parameter B, C and the crystal field splitting parameter D_q, assuming the Cr^{4+} site in terms of ideal symmetry group T_d, the obtained values are as follows [10]:

$$D_q = 831.1 \text{ cm}^{-1}, B = 468.44 \text{ cm}^{-1}, C = 2500 \text{ cm}^{-1}.$$

In the symmetry $C_{3\nu}$, the $^3T_1(F)$ will split into two components, namely ^3E [$^3T_1$ (^3F)] and $^3A_2$ [$^3T_1$ (^3F)] (see Fig. 2.19), the absorption peaks at 602 nm and 801.9 nm are corresponding to the transitions from the ground state to the above mentioned levels respectively. Because $10D_q/B$=18.42, the Cr^{4+} site is in a strong crystal field.

The sharp line at 1146 nm in low temperature specturm is attributed to the spin-forbidden transition from the ^1E level to the ground state $^3A_2$, and the broad-band luminescence is attributed to the ^3E ($^3T_2$)–$^3A_2$ transition, and the zero-phonon line for this transition is at 1158 nm.

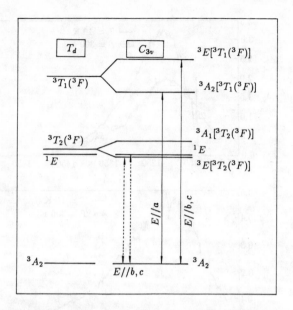

Fig. 2.19. Schematic diagram of the energy level splitting and the polarization selection rules for the electric dipole radiative transitions of the Cr^{4+} ion in the symmetry C_{3v}.

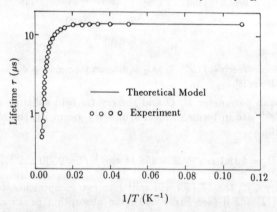

Fig. 2.20. Temperature dependence of fluorescence lifetime τ of $Cr^{4+}:Y_2SiO_5$. The fluorescence was excited by 1064 nm radiation [9].

The decay of the fluorescence resulting from the excitation by 1064 nm radiation is nearly exponential. The energy transfer between ions is possessed of dipole-dipole donor-acceptor nature. The lifetime of the fluorescence excited by 1064 nm pumping varies with temperature from 14.1 μs at 10 K to 593 ns at 300 K, as shown in

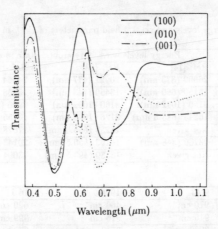

Fig. 2.21. Transmission spectra of the Cr-doped forsterite at room temperature. (100)—Transmission spectrum along a-axis; (010)—transmission spectrum along b-axis; (001)—transmission spectrum along c-axis.

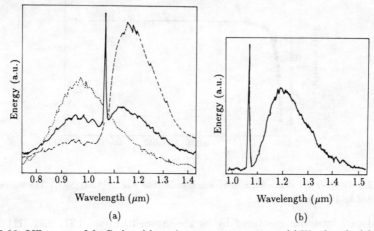

(a)   (b)

Fig. 2.22. LIF spectra of the Cr-doped forsterite at room temperature. (a) Wavelength of the pump laser is 0.532 μm —: $E\|a$, pumping direction $\|c$ and detecting direction $\|b$; - - -: $E\|c$, pumping direction $\|b$ and detecting direction $\|a$; ⋯⋯: $E\|b$, pumping direction $\|c$ and detecting direction $\|a$. (b) Wavelength of the pump laser is 1.064 μm.

Fig. 2.20.

Cr:Mg$_2$SiO$_4$ (forsterite) is a complex system. Its spectroscopic properties depend on the growth environment and pumping wavelength because of presence of Cr^{3+} and Cr^{4+} in crystal. The transmission and laser induced fluorescence (LIF) spectra

Table 2.4. Energy levels and crystal field parameters of Cr^{4+} in Cr:forsterite.

Energy states	Energy levels w.r.t. $^3A_2$ ground state in cm^{-1}		
	Ref. [14]	Ref. [13]	Ref. [12]
$^3T_1(^3P)$	26800 (373 nm)	20921 (477 nm)	24064 (415 nm)
$^3T_1(^3F)$	15150 (660 nm)	13459 (743 nm)	15358 (651 nm)
$^3T_2(^3F)$	9150 (1092 nm)	9160 (1092 nm)	9990 (1001 nm)
^1E	16245 (615.5 nm)	9126 (1095 nm)	10255 (975 nm)
$^1A_1$	Not given	15255 (655.5 nm)	17628 (567.3 nm)
$^1T_2$	28700 (348 nm)	18163 (550.57 nm)	20143 (496 nm)
$^1T_1$	Not given	16580 (493.5 nm)	23094 (433 nm)
Crystal field parameters			
$10D_q/B$	9.43	19.91	16.11
B	970 cm^{-1}	460 cm^{-1}	620 cm^{-1}
D_q	915 cm^{-1}	916 cm^{-1}	999 cm^{-1}
C	3980 cm^{-1}	2790 cm^{-1}	2728 cm^{-1}

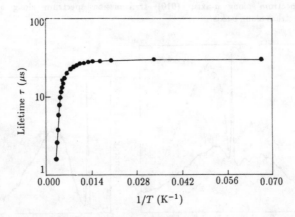

Fig. 2.23. Temperature dependence of fluorescence lifetime τ of Cr^{4+}:Mg_2SiO_4. The fluorescence was excited by 1064 nm radiation.

of Cr-doped forsterite measured by us are shown in Figs. 2.21, 2.22 [11]. The absorption bands at 460 nm and 700 nm as well as emission band at 960 nm belong to Cr^{3+} intra-ionic transition. New absorption bands at 530 nm, 570 nm and 800–1000 nm, as well as fluorescence band at 1150 nm are assumed due to Cr^{4+} intra-ionic transition.

Detailed analysis of the absorption spectrum of Cr^{4+}:Mg_2SiO_4 have been given by Verdun *et al.* [12], Moncorge *et al.* [13] and V. Petricevic *et al.* [14]. Table 2.4 summarizes the results of electronic energy levels of Cr^{4+} ion in Cr:forsterite. The difference arises because of a much weaker crystal field strength assumed for calculation by V. Petricevic *et al.*

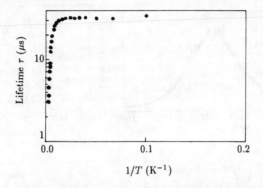

Fig. 2.24. Temperature dependence of fluorescence lifetime of Ca,Cr:YAG.

Table 2.5. Fluorescence intensity of Cr^{4+} in different hosts in the 1.1–1.5 μm band.

Crystal	Intensity at room temperature	Intensity at liquid nitrogen temperature
Cr^{4+}:Y_2SiO_5	Moderate	Strong
Cr^{4+}:$Ca_2Al_2SiO_7$	Weak	Weak
Cr^{4+}:$Ca_2Ga_2SiO_7$	Very weak	Very weak
Cr:Mg_2SiO_4	Strong	Strong
Ca,Cr:$Y_3Al_5O_{12}$	Weak	Weak
Be,Cr:$Ca_2Al_4O_7$	No fluorescence	No fluorescence
Cr^{4+}:$Ca_5(PO_4)_3F$	Weak	Weak

The temperature dependence of fluorescence lifetime τ of Cr^{4+}:Mg_2SiO_4 and Ca,Cr:YAG is shown in Figs. 2.23, 2.24. More intense luminescence of Cr^{4+} ion at room temperature can be observed in Cr^{4+}:Mg_2SiO_4.

We summarized the emission behaviours of Cr^{4+} ion in different crystalline hosts, as listed in Table 2.5. The luminescence spectra of Cr^{4+} ion in $Ca_2Al_2SiO_7$ [15], $Ca_2Ga_2SiO_7$ [15], $Ca_5(PO_4)_3F$ [10], $Y_3Al_5O_{12}$ [16] are shown in Fig. 2.25 for comparison with Cr^{4+} in Mg_2SiO_4.

The fluorescence line narrowing (FLN) of Cr^{4+} in Mg_2SiO_4 has been studied by S.T. Jacobsen *et al.* recently. As shown in Fig. 2.26, the emission spectrum of zero-phonon line, obtained with a conventional experiment, is shown by the top trace, the line width is around 5 cm^{-1}. The FLN spectrum is given by lower trace, it consists of a resonant FLN signal shifted by 15 cm^{-1} toward lower energy and two nonresonant signals separated by 2 cm^{-1}. The FLN results indicate that at 2 K the center relaxes nonradiatively by 15 cm^{-1} and then radiatively to the ground state and the ground state is characterized by three electronic levels separated by 2 cm^{-1}.

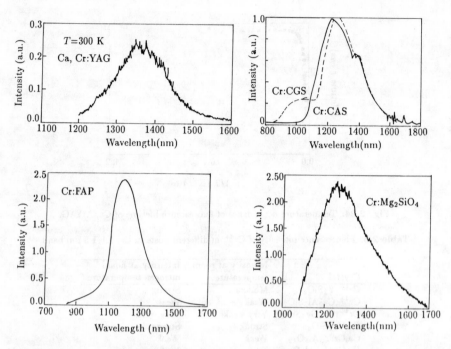

Fig. 2.25. Luminescence spectra of Cr^{4+} ion in $Ca_2Al_2SiO_7$ [15], $Ca_2Ga_2SiO_7$ [15], $Ca_5(PO_4)_3F$ [10], $Y_3Al_5O_{12}$ [16] as compared with Cr^{4+} in Mg_2SiO_4.

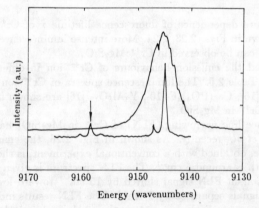

Fig. 2.26. Top: broad-band excited emission of the NIR center in Cr:forsterite in zero-phonon region at 2 K. Bottom: 2 K resonant and nonresonant FLN spectra with arrow denoting position of laser.

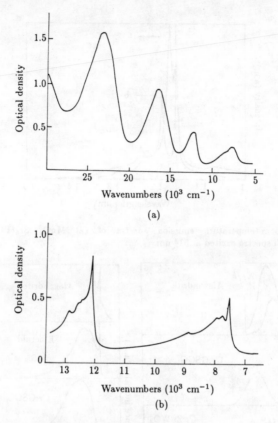

Fig. 2.27. Absorption spectrum (a) and near infrared absorption bands (b) of a 0.3 cm thick V:YAG single crystal recorded at 295 K.

Only optical absorption of V^{3+} ions in YAG crystal has been studied in refs. [16–19]. The absorption spectrum of V^{3+}:YAG at 295 K and 80 K are shown in Fig. 2.27. The two strong broad bands in the visible region were assigned to V^{3+}-substituted Al^{3+} in octahedral lattice site, fitted with calculation using $D_q = 1700$ cm^{-1} and $B = 600$ cm^{-1}, the weak absorption observed near 1250 nm (800 cm^{-1}) in YAG:Ca, V was assumed to be associated with tetrahedral sites of V^{3+} [16]. Using Td approximation the best fit to the experimental data is observed with $D_q = 800$ cm^{-1}, $B = 480$ cm^{-1} and $C = 2500$ cm^{-1}, these values are very close to that of Cr^{4+}:Y_2SiO_5, as mentioned above.

The spectral properties of Mn^{5+} in Ca_2PO_4Cl and $Sr_5(PO_4)_3Cl$ have been investigated by J.A. Capobiano *et al.* recently [20–22]. The emission spectra of Mn^{5+}:$Sr_5(PO_4)_3Cl$ and Ca_2PO_4Cl are shown in Fig. 2.28 [21]. The absorption

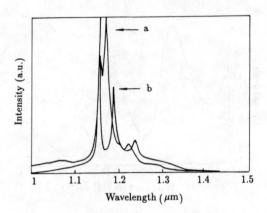

Fig. 2.28. Room-temperature emission spectra of (a) MnO_4^{3-}:$Sr_5(PO_4)_3Cl$ and (b) MnO_4^{3-}:Ca_2PO_4Cl spectra excited at 514 nm.

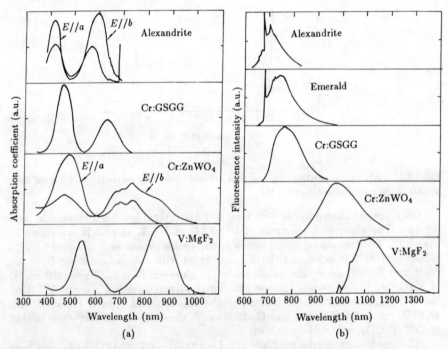

Fig. 2.29. Absorption and fluorescence spectra for a variety of d^3 systems at 80 K [23]. (a) Absorption; (b) fluorescence.

Table 2.6. Spectral characteristics of Cr^{3+}-doped crystals.

Host	Site symmetry	ΔE_s $(^4T_2 \to {^2E})$	Wavelength of R-line $^2E \to {^4A_2}$ (nm)	Central wavelength of broad band emission $(^4T_2 \to {^4A_2})$	D_q (cm^{-1})	B (cm^{-1})	C (cm^{-1})	τ (20°C) (μs)
Al$_2$O$_3$	C_{3v}	2300	694.3, 692.4	—	1800	640, 720	3300, 3000	2×10^4(R)
MgAl$_2$O$_4$	C_{3v}	3500	682	—	1825	700	3200	1.5×10^4(R)
SrTiO$_3$	D_{4h}		794	—	1626	470.6	2991.5	5×10^3(R)
BeAl$_2$O$_4$	C_s	800	680.4, 678.3	750	1630	780	2960	260
BeAl$_6$O$_{10}$	O_n	589	693, 690, 687.5	760	1663	643	3180	$2\sim4 \times 10^3$(R), 30
Be$_3$Al$_2$(SiO$_3$)$_6$		400						20
Y$_3$Al$_5$O$_{12}$	S_4	400	680	720	1730	630	3235	240
Y$_3$Ga$_5$O$_{12}$	S_6	600	693	720	1630	639	3235	160
Gd$_3$Ga$_5$O$_{12}$	S_6	300		750	1597	626	3236	145
Y$_3$(Sc, Ga)$_2$Ga$_3$O$_{12}$	S_6	250	696.3, 694.9	740	1613	630	3253	120
Gd$_3$(Sc, Ga)$_2$Ga$_3$O$_{12}$	S_6	50	690	760	1580	638	3237	
La$_3$La$_2$Ga$_3$O$_{12}$	S_6	<0			1480	619	3309	
MgO	O_n	−39	698.5	780				180
KZnF$_3$		−500	—	800				
Na$_3$AlF$_3$				720				40 (310μs, 77K)
Na$_2$MgAlF$_3$				773				40 (180)
MgF$_2$	D_{4h}	−2500	—	913	1410	601	2404	
ZnWO$_4$	$P2/c$	−3000	—	760				56
Ca$_3$(PO$_4$)$_3$F				745	1613	665	3325	
Y$_3$(In, Ga)$_2$Ga$_3$O$_{12}$	O_n							
Na$_4$Ge$_9$O$_{20}$			701.5		1710, 1460	549, 750	3400, 2750	10 (1 ms 77 K)

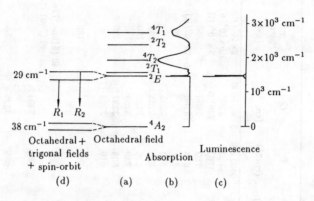

Fig. 2.30. (a) Energy level diagram of Cr^{3+} in an octahedral field; (b) absorption spectrum of ruby at low temperature [24]; (c) luminescence spectrum of ruby at 77 K; (d) fine structure in the ^2E and $^4A_2$ levels.

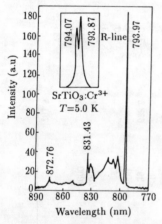

Fig. 2.31. ^2E–$^4A_2$ fluorescence spectrum of Cr^{3+}:SrTiO$_3$ crystal at 5.0 K [25].

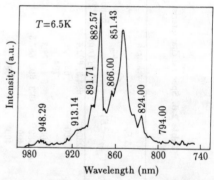

Fig. 2.32. Fluorescence spectrum of heavily doped SrTiO$_3$ crystal at 6.5 K [26].

Table 2.7. Ratio of integral intensity of R-line to total fluorescence, wavelength and linewidth of R-line of Cr^{3+}:SrTiO$_3$ crystal at various temperatures.

T (K)	5	20	40	60	80	100	120	140	160	180
β (%)	9.5	9.5	9.5	9.3	8.7	8.0	7.3	6.6	5.0	3.6
λ (nm)	793.97	793.91	793.77	793.65	793.54	793.20	792.77	792.24	791.84	
$\Delta\lambda$ (nm)	0.22	0.24	0.27	0.30	0.33	0.50	0.65	0.78		

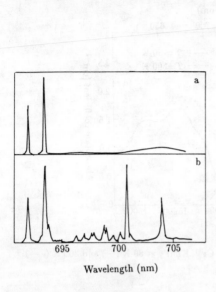

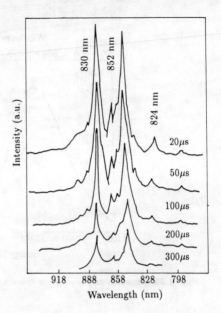

Fig. 2.33. A comparison of the spectrum of (a) dilute ruby and (b) heavily doped ruby taken at 77 K [30].

Fig. 2.34. Time-resolved fluorescence spectra of Cr^{3+}:$SrTiO_3$ (0.5 wt%) at 6.8 K. The excitation wavelength is 532 nm.

spectra taken at low temperature shows two intense bands at 10600 and 11250 cm^{-1} [20], assigned to $^3A_2$ to $^3T_2$ and ^3E transitions respectively. Crystal-field analysis of Mn^{5+}:$Sr_5(PO_4)_3Cl$ was performed using a C_4 site symmetry [22], the parameters used in the calculation were $B=491.58$ cm^{-1}, $C=2497$ cm^{-1}, $D_q=1100$ cm^{-1}.

2.1.3 d^3 (Cr^{3+}, V^{2+})

Cr^{3+} is the most well-known $3d$ ion for laser crystals, the combination Cr^{3+}:Al_2O_3, ruby was the first crystal to exhibit laser action, the combination Cr^{3+}:$BeAl_2O_4$ alexandrite was the first tunable laser crystal operated at room temperature. A series of Cr^{3+} doped laser crystals have been emerged recently.

Energy level diagram showing the variation of energy E with the ratio of D_q/B have been given by Tanabe and Sugano [1] for all TM ions. As shown in Fig. 1.3, depending on D_q value the emission can be taken place from excited state ^2E or $^4T_2$, the transition from $^2E \rightarrow {}^4A_2$ or "R-line" is the narrow spin-forbidden line and transition from $^4T_2 \rightarrow {}^4A_2$ is spin-allowed with vibronic in nature. One quantitative measure of the crystal-field strength is the separation in energy between the $^4T_2$ and ^2E levels, ΔE_s. This parameter determines the nature of emission of

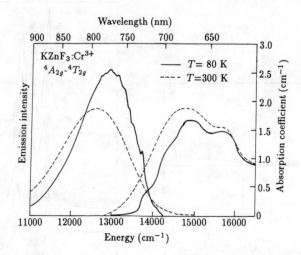

Fig. 2.35. Emission and absorption spectra of Cr^{3+}:$KZnF_3$ at 80 K (—) and 300 K (- - -).

Cr^{3+} ion. Table 2.6 lists the ΔE_s and D_q/B values, wavelengths of R emission line and broad band emission of different Cr^{3+}-based crystals. It is shown that the R-line luminescence is dominant when $D_q/B > 2.3$, $\Delta E_s > 0$ (strong field), and broad band emission occurs, when $D_q/B < 2.3$, $\Delta E_s < 0$. Figure 2.29 shows the absorption and luminescence spectra of Cr^{3+} in different crystal hosts [23]. The two absorption bands move toward longer wavelength with decrease of D_q/B value. In luminescence spectra the shift from a mixture of R-line and vibronic emission to purely vibronic emission with increase of ΔE_s value.

Figure 2.30 shows the energy level diagram for Cr^{3+} in ruby, where the low symmetry trigonal field and spin-orbit interaction are also taken into account. The absorption and luminescence spectra are measured at low temperature. The 29 cm^{-1} splitting in the ^2E state results in two sharp fluorescence lines. The 0.38 cm^{-1} splitting in $^4A_2$ state can be resolved in good quality ruby crystal. An associated vibrational sideband from the ^2E state can also be observed [24].

We studied the spectral properties of Cr^{3+}:$SrTiO_3$, which belongs to strong field ($D_q/B \sim 3.46$) too [25–27]. Figure 2.31 shows the emission spectrum of lightly doped Cr^{3+}:$SrTiO_3$ crystal at 5 K. The sample concentration is 0.01 wt%. The zero-phonon line (R line) is located at 794 nm and R line splitting is 0.2 nm due to tetragonal symmetry distortion in Cr^{3+}:$SrTiO_3$, which is lower than that of ruby. The red shift of R-line is caused by low B and C values due to high covalency in chemical bond of Cr^{3+}:$SrTiO_3$ in comparison with other strong field Cr-based crystals, as shown in Table 2.6. The vibronic emission band is at longer wavelength side of R-

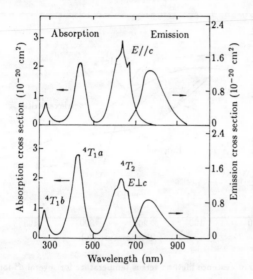

Fig. 2.36. Absorption and emission spectra of Cr:LiCAF.

line, the wavelength gap between vibronic emission lines 831.43 nm and 872.76 nm is 568 cm^{-1} and 1136 cm^2 respectively, which can be recognized as single-phonon and two-phonon lines. At elevated temperature except Stockes phonon emission the anti-Stock's emission appears at lower wavelength side of R-line, and the ratio of integral intensity of R-line to total fluorescence decreases. Table 2.7 lists the temperature dependence of fluorescence lifetime, wavelength and linewidth of R-line in Cr^{3+}:SrTiO$_3$ crystal at various temperatures. The violet shift of R-line as the rise of temperature was discovered first, it can be explained by crystal-field and covalent bond theories reasonably, the Racah parameters B and C increase at elevated temperature and cause the violet shift of R-line [25, 26]. The defects (oxygen vacancy) of charge compensation and nonradiative energy transfer by single phonon absorption in Cr^{3+}:SrTiO$_3$ crystal were considered as the main reason for quenching the fluorescence at high temperature.

At high Cr^{3+} concentration in Cr^{3+}:SrTiO$_3$ R-line diminished gradually and new luminescence lines (882.57 nm and 851.43 nm) appeared at longer wavelength, as shown in Fig. 2.32, the Cr-concentration is 0.5 wt%. The luminescence lifetime decreased also from 20 ms to 5 ms, as Cr-concentration increased. The similar behaviour has been observed in Cr-doped Al$_2$O$_3$, ZnGa$_2$O$_4$ [28, 29], it can be concluded to formation of Cr-pairs. Two ions may be near enough to interact strongly and form a distinct center. These pairs act as distinct luminescence centers. Figure

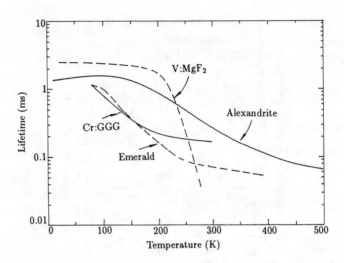

Fig. 2.37. Fluorescence lifetime versus temperature for several d^3-ion systems [23].

2.33 shows a comparison of the spectrum of (a) dilute ruby, and (b) heavily doped ruby taken at 77 K [30]. The additional lines in heavily doped ruby originate from exchange-coupled pairs of Cr ions. The strong pair line at 700.9 nm is caused by energy transfer excitation from single Cr ions. Figure 2.34 shows the time-resolved fluorescence spectra of highly doped Cr^{3+}:$SrTiO_3$ (0.5 wt%) at 6.8 K. The excitation wavelength is 532 nm. It can be seen that the decay time is less than 200 μs, which is much shorter than that of R-line. Due to dipole cross-relaxation, the neighbouring Cr^{3+} ions in pair are closer, the lifetime of luminescence will be more shorter.

The spectroscopic properties of Cr^{3+} doped new host crystals, such as $BeAl_6O_{10}$ [31–33], $Y_3Ga_5O_{12}$ [34–36], Na_2MgAlF_7 [37], $Y_3(In, Ga)_2Ga_3O_{12}$ [38, 39], $Na_4Ge_9O_{12}$ [40] and $ZnWO_4$ [41], have been studied by Chinese groups, the experimental results were summarized in Table 2.6. The Cr^{3+} doped $Na_4Ge_9O_{12}$, $BeAl_6O_{10}$ and $Y_3Ga_5O_{12}$, Na_2MgAlF_7 and $ZnWO_4$ belong to strong, middle and weak crystal fields respectively.

The broad band emission of transition $^4T_2$–$^4A_2$ is useful for tunable laser materials. The Cr^{3+} doped $KZnF_3$ is the typical low crystal field one. Figure 2.35 shows the emission and absorption spectra of Cr^{3+}:$KZnF_3$ at 80 and 300 K [42]. A series of Cr^{3+}-doped fluoride crystals have been reported recently [37, 43, 44], the spectroscopic behaviours are rather similar. Figure 2.36 shows the absorption and emission spectra of Cr^{3+}:LiCAF. The emission lifetime of $^4T_2$–$^4A_2$ is small due to spin allowed transition, the temperature dependence of fluorescence lifetime is

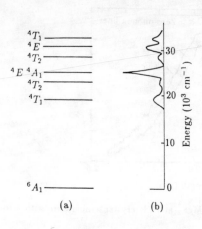

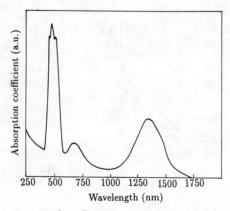

Fig. 2.38. (a) Energy level diagram of Mn^{2+} in an octahedral field; (b) room temperature absorption spectrum of MnF_2.

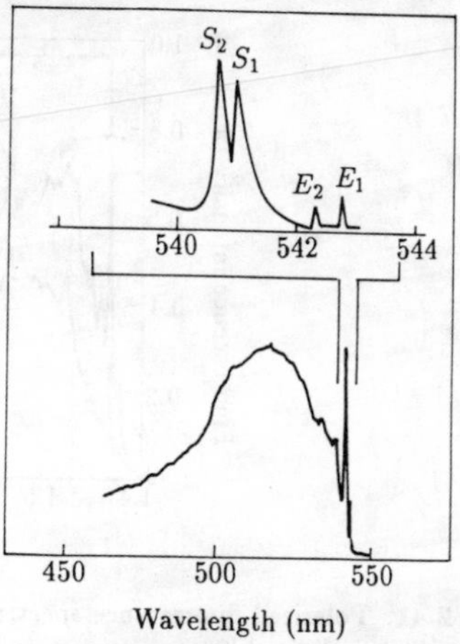

Fig. 2.39. Low trace shows the $^6A_1 \rightarrow {}^4T_1$ absorption transition in MnF_2 at 2 K. The upper trace shows in greater detail the sharp features on the low energy side of the broad band.

Fig. 2.40. Absorption spectrum for a $Co:MgF_2$ at 300 K temperature for unpolarized light in the c-axis direction.

important for making choice of tunable laser crystals, Fig. 2.37 shows the fluorescence lifetime versus temperature of several d^3-ion systems. Herein the alexandrite ($Cr^{3+}:BeAl_2O_4$) seems much better than others.

The V^{2+} ion behaves very similar spectral characteristics to Cr^{3+}. The first explored combination $V^{2+}:MgF_2$ belongs to low field interaction. The absorption

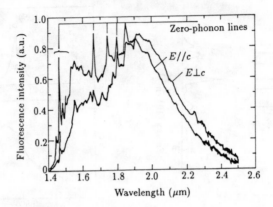

Fig. 2.41. Polarized fluorescence spectra from Co:MgF$_2$ at 77 K crystal temperature with zero-phonon lines indicated [23].

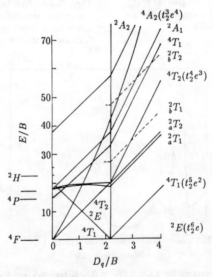

Fig. 2.42. Tanabe-Sugano diagram for Co^{2+} ion.

and luminescence spectra, as well as temperature dependence of lifetime are shown in Figs. 2.29, 2.37 respectively [45].

2.1.4 d^5, Mn^{2+}

Figure 2.38 shows the energy level diagram for Mn^{2+} in an octahedral field and the

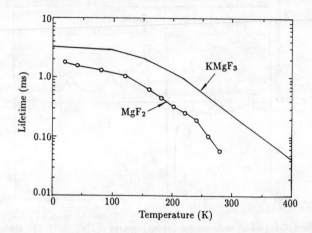

Fig. 2.43. Fluorescence lifetime versus temperature for Co:MgF_2 and Co:$KMgF_3$ crystals [47].

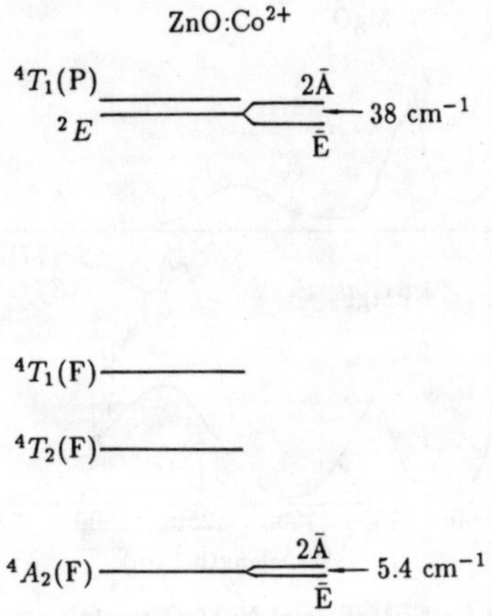

ZnO:Co^{2+}

$^4T_1(P)$ ───────
^2E ─────── $2\bar{A}$
           $\bar{\bar{E}}$ ← 38 cm^{-1}

$^4T_1(F)$ ───────

$^4T_2(F)$ ───────

$^4A_2(F)$ ─────── $2\bar{A}$
                  $\bar{\bar{E}}$ ← 5.4 cm^{-1}

Fig. 2.44. Energy level diagram for Co^{2+}:ZnO showing splitting of $^4A_2$ and ^2E levels [48].

absorption spectrum of MnF_2 at room temperature. The fluorescence occurs on-lyfrom lowest excited $^4T_1$ to ground state $^6A_1$ with a large energy gap around 18000 cm^{-1}. Therefore, the multiphonon nonradiative transition is unefficient, so the ions in $^4T_1$ level decay radiatively to the $^6A_1$ state. It has a long decay time—about 30 ms due to spin-forbidden transition.

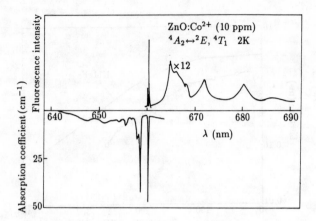

Fig. 2.45. Absorption (lower) and emission (upper) spectra of Co^{2+}:ZnO. The absorption is to ^2E and $^4T_1(P)$ levels [48].

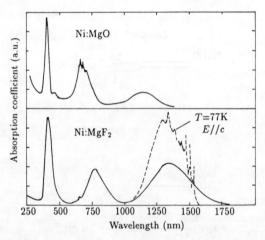

Fig. 2.46. Absorption spectra for Ni:MgF_2 and Ni:MgO crystals at 300 K, unless noted. Data for MgF_2 system are for unpolarized light in the c-axis direction, except 77 K data which is for $E\|c$ polarization [23].

The $^6A_1$–$^4T_1$ transition of absorption and emission at 2 K is shown in Fig. 2.39 [46]. Zero-phonon line with broad sideband in absorption spectrum can be clearly seen from Fig. 2.39, and because of fine structure in the $^4T_1$ level two no-phonon lines, E_1 and E_2, and two magnon sidebands at S_1 and S_2 occur, which is caused by lower symmetry crystal field, exchange and spin-orbit coupling.

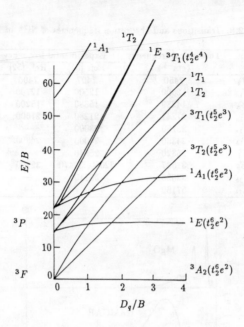

Fig. 2.47. Tanabe-Sugano diagram for Ni^{2+} ion.

2.1.5 d^7, Co^{2+}

Figures 2.40, 2.41 show the absorption and fluorescence spectra of Co^{2+} in MgF_2 [23]. As fluoride crystals (MgF_2, $KMgF_3$) are weak field hosts ($D_q/B<1$), the absorption bands around 1320 nm, 680 nm and 490 nm correspond to transitions from the ground state $^4T_1$ to $^4T_2$, $^2E_1$ and $^4A_2$ (Fig. 2.42). The fluorescence was resonant with the transition $^4T_2$-$^4T_1$. Due to first order spin-orbit splitting of $^4T_1$ ground state, 6 distinct zero-phonon lines can be observed in fluorescence spectra at low temperature originated from cubic site symmetry for the Co^{2+} ion. The highest splitting is 1500 cm^{-1} in MgF_2 and 950 cm^{-1} in $KMgF_3$.

The temperature dependence of fluorescence lifetime of Co^{2+} in host crystal MgF_2 and $KMgF_3$ is shown in Fig. 2.43. The decrease of lifetime is attributed to the non-radiative transition.

Visible emission is often observed from Co^{2+} doped oxide hosts, such as ZnO [48], $ZnAl_2O_4$ [49] and $LiGa_5O_8$ [50], originating from ^2E and $^4T_1$ excited states respectively. Taking an example of Co^{2+}:ZnO, Co^{2+} can be substituted for Zn in the distorted tetrahedral sites of $C_{3\nu}$ symmetry, the energy level diagram for Co^{2+}:ZnO is shown in Fig. 2.44. Three spin allowed quartet-quartet transitions were observed around 4000 cm^{-1} ($^4A_2$-$^4T_2$), 6500 cm^{-1} ($^4A_2$-$^4T_1$(F)) and 15500

Table 2.8. Transitions and absorption frequencies of Ni^{2+} in $KMgF_3$.

Transition	Theoretical frequency (cm^{-1})	Experimental frequency (cm^{-1})		
		Ref. [51]	Ref. [52]	Ref. [53]
$^3A_2 \rightarrow {}^3T_2$	7450	7407	7400	7250
$^3A_2 \rightarrow {}^3T_{1a}$	12600	12500	12700	12530
$^3A_2 \rightarrow {}^1E_a$	14289	15380	15200	15440
$^3A_2 \rightarrow {}^1T_{2a}$	21241	21280	21600	—
$^3A_2 \rightarrow {}^1A_{1a}$	23180	23000	—	—
$^3A_2 \rightarrow {}^3T_{1b}$	24360	24000	24300	23810
$^3A_2 \rightarrow {}^1T_1$	26320	25640	—	—
$^3A_2 \rightarrow {}^1E_b$	31537	32050, 32470	32700	31000
$^3A_2 \rightarrow {}^1T_{2b}$	32035	33110	—	32000
$^3A_2 \rightarrow {}^1A_{1b}$	57160	—	—	—

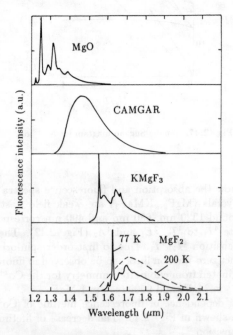

Fig. 2.48. Fluorescence spectra for several Ni^{2+}-doped crystals at 77 K, unless noted. Zero-phonon-line intensities are off the scale for Ni:MgO. Polarization is $E\|c$ for MgF_2 [23].

cm^{-1} ($^4A_2$-$^4T_1(P)$) in absorption spectrum. Resonant fluorescence with zero-phonon lines at 660.23 nm and 660.53 nm together with a single phonon sideband could be seen in Fig. 2.45. A fluorescence lifetime of 15 ns was measured [48].

Table 2.9. Linewidth and fluorescence peaks of $^3T_2 \to {}^3A_2$ transition of Ni^{2+}:$BeAl_2O_4$.

Temperature (K)	Linewidth (cm^{-1})	Peak 1 (cm^{-1})	Peak 2 (cm^{-1})	Peak 3 (cm^{-1})
10	1462	6990	7172	7335
20	1487	6969	7122	7340
30	1503	6920	7086	7336
50	1503	6881	7051	7315
69	1478	6864	7031	7294
80	1458	6859	7015	7279
100	1435	6857	7010	7250
115	1427	6858	7008	7245
130	1430	6860	7007	7247
150	1443	6862	7006	7260
168	1459	6863	7006	7280
180	1465	6862	7006	7294
200	1477	6861	7010	7319
220	1478	6859	7017	7339
240	1480	6857	7026	7351
260	1482	6858	7032	7355
270	1484	6859	7028	7353
288	1497	6861	7020	7351

2.1.6 d^8, Ni^{2+}

Ni^{2+} in MgF_2 is the first tunable TM-ion laser, the laser transition is from $^3T_2$ to $^3A_2$. Figure 2.46 shows the absorption spectra of Ni^{2+} in MgF_2 and MgO hosts, according to the Tanabe-Sugano diagram as shown in Fig. 2.47, the three absorption bands of Ni^{2+} ion correspond to the transitions from ground state $^3A_2$ to the $^3T_2$, $^3T_1(^3F)$ and $^3T_1(^3P)$ levels with values of $D_q/B<1$, where B is 1030 cm^{-1}. The absorption spectrum of Ni^{2+} in $KMgF_3$ has been analysed in detail [51]. The electronic transitions and corresponding absorption lines are listed in Table 2.8, the literature data are also shown in comparison. The crystal field and Racah parameters were calculated: D_q=745 cm^{-1}, C=3592 cm^{-1}.

Figure 2.48 presents Ni^{2+} fluorescence spectra in different host crystals, the zero-phonon line and phonon-side band are also shown at low temperature [23]. The infrared fluorescence spectra and thermal shift of energy levels of Ni^{2+}:$BeAl_2O_4$ crystal have been studied by G. Wang and S. Liu [54]. The infrared fluorescence is originated from transition $^3T_2 \to {}^3A_2$ with C_s symmetry, when Ni^{2+} ions are in distorted octahedral sites, the $^3T_2$ level is splitted into three. The experimental data of temperature dependence of bandwidth and three fluorescence peaks are listed in Table 2.9. The temperature dependence can be given by the following equation [54]:

$$\Delta\nu(t) = \Delta\nu(0)[\tanh(h\omega/2kt)]^{\frac{1}{2}}. \tag{2.1}$$

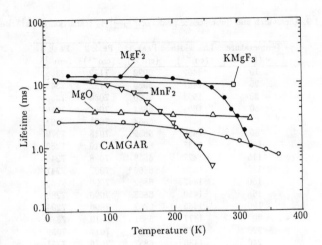

Fig. 2.49. Fluorescence lifetime versus temperature for several Ni^{2+}-doped crystals [23].

Here $\Delta\nu(0)$ is the linewidth at absolute zero temperature, $\hbar\omega$ the average phonon energy. By experimental data fitting we obtained: $\Delta\nu(0)=1476$ cm^{-1}, $\hbar\omega=671$ cm^{-1}.

The transition of $^3T_2$–$^3A_2$ belongs to the resonant ones, the large Stokes shift indicates the strong electron-phonon coupling, as shown that the central positions of absorption band and fluorescence bands of Ni^{2+} were 1410 cm^{-1} and 1710 cm^{-1} in MgF_2 and 1070 cm^{-1} and 1430 cm^{-1} in $BeAl_2O_4$ respectively.

The emission lifetime of Ni^{2+} ions in different host crystals is shown in Fig. 2.49, the temperature quenching effect is not obvious and in some case is nearly constant from low temperature to 300 K. Analysis of the polarization properties of absorption and fluorescence indicates that vibrationally induced electric-dipole transitions are not significant, but primarily magneto-dipole transition which accounts for the long (>1 μs) fluorescence lifetime.

The FLN experiment on Ni^{2+}:MgF_2 has been done by Tonucci *et al.* [55]. The broad-band emission and FLN spectrum of Ni^{2+}:MgF_2 at 2 K are shown in Fig. 2.50. The zero-phonon transitions are centered around 6500 cm^{-1} ($\sim$1.5 μm). From FLN spectrum it can be seen that resonant and non-resonant transitions consist of a total of five distinct lines, and this can be considered as follows: three components come from the splitting of ground state $^3A_2$ (in D_{2h} symmetry) and others are due to different subsets of Ni^{2+} ions.

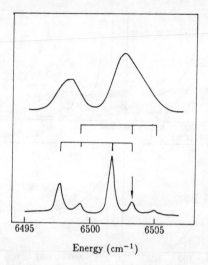

Fig. 2.50. Broad-band excited emission and FLN spectrum of Ni^{2+}:MgF_2 at 2 K. Top: broad-band excited NIR emission in zero-phonon region of $^3T_2 \rightarrow {}^3A_2$ transition. Bottom: 2 K FLN spectrum.

2.2   Spectroscopic Properties of Transition Metal Ions in Inorganic Glasses

Transition metal ions ($3d$, $4d$) have been used as color agents in glass industry for many years. A lot of work have been done concerning their absorption spectra [56, 57]. Due to strong and disorder interaction of glass host with transition metal (TM) ion, the non-radiative energy transition from activated TM ions to the host is great, up to now the laser action of TM ion in glasses has never been observed yet. Little research work has been carried out on luminescence and relaxation spectra so far. In this section, we put more emphasis on the influence of chemical bond characteristics of the host on spectroscopic properties and site structure of TM ion and to present some new results of luminescence spectra of low valent TM ions, such as Ti^{3+}, Cr^{3+}, Mn^{2+}, Cu^+ and Mo^{3+}. Based on spectroscopic experiments the possibility of laser action of TM ions in glass hosts has been discussed.

2.2.1   *Absorption Spectra*

Figure 2.51 shows the absorption spectra of $3d^n$ transition metal ions in the wavelength range from near UV to IR. The absorption peaks and corresponding transition energy levels in silicate glasses are listed in Table 2.10.

The principal features of absorption spectra of transition metal ions in silicate glasses have been interpreted in terms of ligand field theory by Bates [58]. We paid

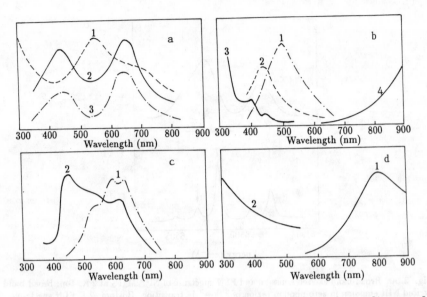

Fig. 2.51. Absorption spectra of $3d^n$ ions in silicate glasses. **a** 1 Ti^{3+}, 2 V^{3+}, 3 Cr^{3+}; **b** 1 Mn^{3+}, 2 Mn^{2+}, 3 Fe^{3+}, 4 Fe^{2+}; **c** 1 Co^{1+}, 2 Ni^{2+}; **d** 1 Cu^{2+}, 2 Cu^{+}.

more attention to the influence of glass hosts on the ligand field parameters. Absorption spectra of TM ion doped glasses changed with its valence and symmetry states. Cr^{3+} doped glasses melted in reducing atmosphere are more stable in valency and symmetry; Cr^{3+} ions are located in an octahedron. As an example, the measured peak wavelengths of absorption bands, their optical transitions and calculated values of D_q and B of the Cr^{3+} ion in different glass hosts are listed in Table 2.11. It can be found that in the order of phosphate-fluorophosphate-fluoride glasses, the values of D_q and B increase gradually [59]. That means there is an increase of the ligand field strength and the ionic contribution of the chemical bond by substitution of fluorine for oxygen anions. In the order of borate-silicate-phosphate, the D_q value decreases and Racah parameter B increases. Therefore, in oxide glasses the action of oxygen ligands on the activated ions (TM) is different and depends mainly on the nature of the polyhedron-formed glass structural network. From borate to phosphate glasses the ligand field strength and nephelauxetic effect all decrease.

We have summarized the calculated values of D_q and B of TM ions in glasses from experimental results, which are shown in Table 2.12. Analogous regularity is also found in other TM ion doped inorganic glasses. When the composition of the host glass is modified, from fluoride to phosphate and from borate to phosphate, the D_q values decrease due to the weakening electro-static interaction of the TM ion with ligand anion. For every TM ion, the B values decrease gradually from

Table 2.10. Absorption peaks and corresponding transition energy level of $3d^n$ transition metal ions in silicate glasses.

Electron configuration	Ion	Δ (cm^{-1})	B (cm^{-1})	Transition energy level		Absorption peaks (cm^{-1})
$3d^2$	Ti^{3+}	18100		$^2T_1 \rightarrow$	^2E	18100, 14300 (Jahn-Teller effect)
$3d^2$	V^{3+}	16600	644	$^3T_1 \rightarrow$	$^3T_2$	15500
					$^3T_1(P)$	23500
					$^3A_2$	32000
$3d^3$	Cr^{3+}	15300	760	$^4A_2 \rightarrow$	$^4T_2$	15300
	Mn^{2+}				$^4T_1(F)$	22200
					$^4T_1(P)$	34600
					^2E	14600
					$^2T_1$	15750
					T_2	21200
$3d^4$	Mn^{3+}	20000	(850)	$^5E \rightarrow$	$^5T_2$	20000
$3d^5$	Fe^{2+}	12300	720	$^6A_1 \rightarrow$	$^4A_1$	23400
					$^4E(G)$	23800
					$^4T_2(D)$	26300
					$^4T_1(G)$	13600
					$^4T_2(G)$	17600
$3d^6$	Fe^{3+}	9100	(950)	$^5T_2 \rightarrow$	^5E	9100
$3d^7$	Co^{2+}	7700	(920)	$^4T_1 \rightarrow$	$^4T_2$	6700
					^2E	12400
					$^4A_2$	14400
					$^4T_1(P)$	20000
$3d^8$	Ni^{2+}	7500	900	$^3A_2 \rightarrow$	$^3T_2$	7500
					$^5T_1$	13200
					^1E	15000
					$^3T_1(P)$	24000
$3d^9$	Cu^{2+}	12700		$^2E \rightarrow$	$^2T_2$	12700

Table 2.11. Absorption peak position, corresponding transition and parameters D_q and B of Cr^{3+} ions in glasses [59] (in cm^{-1}).

Glass	$^4A_{2g}, {}^2E(G)$	$^4T_{2g}$	$^2T_{1g}$	$^4T_{1g}$	$^4T_{1g}(P)$	D_q	B
Borate (B–La–Ba)	14600	15900	16100	22800		15900	710
Silicate (Si–Na)	14500	15300	15800	22200		15300	760
Phosphate (P–Ba)	14600	15100	15750	21900	33200	15100	750
Fluorophosphate (P–Al–Ca–Na)	14700	15400	15700	22600	33400	15400	700
Fluoride (Pb–Zn–F)	14920	15550	16210	22720	34480	15550	780

Table 2.12. D_q and B values of transition metal ions in different glasses.

		Fluoride		Fluorophosphate		Phosphate		Silicate		Borate	
3d	ion	D_q (cm^{-1})	B (cm^{-1})	D_q (cm^{-1})	B (cm^{-1})	D_q (cm^{-1})	B (cm^{-1})	D_q (cm^{-1})	B (cm^{-1})	D_q (cm^{-1})	B (cm^{-1})
d^1	Ti^{3+}			1890		1810[60]		1818		2050	
d^2	V^{3+}					1600	680	1660	640[61]		
d^3	Cr^{3+}	1555	800	1540	770	1510	750	1530	760	1590	710
d^4	Mn^{3+}							1390			
d^5	Mn^{2+}		770	780	760	765	750	680	740[61]		
								300	740		
	Fe^{3+}			1265	800			1230	720		
d^6	Fe^{2+}							910			
d^7	Co^{2+}	780	860	775	840	770	810			770	800[61]
		380		375		370		365	805	370	
d^8	Ni^{2+}	700	970	710	950	710	910			750	900[61]
										470	860
d^9	Cu^{2+}					1270[62]				1450[62]	

fluoride, fluorophosphate, phosphate, silicate to borate, which indicates that the covalent content of the chemical bond increases.

2.2.2 Luminescence Spectra

The 3d electrons are on the outer shell of the TM ion and very sensitive to the effects of the environment, i.e. dynamic lattice. The configuration coordinate model is successful in explaining the shape of the broadband transition. We can take trivalent chromium as an example, which has been extensively studied for spectroscopy in crystals because of a narrow $^2E-^4A_2$ transition, but it is much less useful in glasses. The lowest excited state in glasses is usually the bottom of the $^4T_2$ band and broadband, Stokes-shifted $^4T-^4A_2$ emission rather than $^2E-^4A_2$ emission is observed. This feature arises from the smaller value of the crystal-field parameter D_q in glasses, and the spin-allowed resonance transition is characterized by the broader and vibronic nature, shorter lifetime ($\sim\mu s$) and only small line narrowing effects at temperatures under 77 K [63].

Mo^{3+} has $4d^3$ electron configuration and a higher D_q value. The radiative transition $^2E-^4A_2$ is a spin-forbidden one. The luminescence bandwidth of Mo^{3+} is narrower than that of Cr^{3+} and its lifetime is longer ($\sim\mu m$). Similar behaviour of luminescence can be found in Mn^{2+} ion-doped glasses [63]. The radiative transition $^4T_1(G)-A_1(S)$ of Mn^{2+} is also spin-forbidden and its lifetime is in ms range.

Low valence transition metal ions, such as Cu^+ amd Ti^{3+}, can exist mainly in the phosphate and fluorophosphate glasses under strong reducing melting conditions [64, 65]. A strong UV absorption band in the vicinity of 40000 cm^{-1} and a fluorescent band around 22000 cm^{-1} of Cu^+ ions in glasses are attributed to the

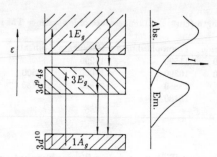

Fig. 2.52. Pattern of energy eigenvalue disorder and phonon-assisted emission of Cu^+ sites having extended octahedral symmetry. It is the cause of the broad-band structure appearing in the absorption spectra.

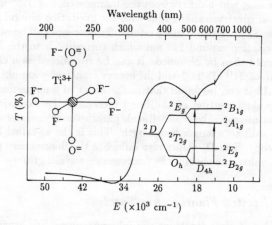

Fig. 2.53. Absorption spectra of titanium-doped fluorophosphate glass with the coordination and energy diagrams.

transition between the electron configurations $3d^{10}$ and $3d^9 4s^1$. Figure 2.52 shows the pattern of eigenvalue of Cu^+ sites having fuzzy octahedral symmetry. The absorption and emission are disordered and phonon-assisted with broadband structure. There is a large Stokes shift, which can be explained not only by the configurational diagram mentioned above; the site-inhomogeneity of ions in glasses and the effect of the lattice relaxation also contribute greatly to the shift, which will be discussed later.

The absorption spectra of Ti^{3+} have been studied in several glass systems [58], but the luminescence due to titanium ions in glasses has so far never been reported to our knowledge. We have observed the fluorescence with peak position at 550 nm as well as the bandwidth of 180 nm in Ti^{3+} doped fluorophosphate glasses [66].

Table 2.13. Peak wavelength (cm^{-1}) of luminescence of low valence TM ions in different glasses.

Ion	Fluoride	Fluorophoshpate	Phosphate	Silicate	Borate
Cr^{3+}	11680	11520	11290	11950[67]	12250[67]
Mn^{2+}	17570	16390	16000	16500	
Cu^+		23810	21830	18180 [64]	20800[64]
Ti^+		18180			
Mo^{3+}			9766		

The absorption spectrum of Ti^{3+} containing fluorophosphate glass with the coordination and energy diagram is shown in Fig. 2.53. The fluorescence is then to be assigned to the $^2B_{1g}$–$^2B_{2g}$ transition of Ti^{3+} ions due to this location in disordered octahedra. Its lifetime is about 2 μs. The luminescence of titanium doped glasses is not a resonance one, which differs from the luminescence of Ti^{3+} ions in crystals [23]. The excitation spectrum shows that with the excitation around 300 nm which corresponds to the $Ti^{4+}CT$ band, more intense fluorescence can be obtained. However, with the excitation around 240 nm which corresponds to the $Ti^{3+}CT$ band, no fluorescence could then be obtained. It can be considered as a charge transfer process in the form of Ti^{4+}–F–Ti^{3+} and the energy transfer by ion-phonon coupling.

From Table 2.13 it can be found that in the order of fluoride-fluorophosphate-phosphate the covalent portion of the bond increases; the exclusion among the electrons inside an atom weakens and Racah parameter B decreases. This makes the spectral lines shift to a longer wavelength. This is the so-called Nephelauxetic effect in spectroscopy. The D_q values also influence the fluorescence peak position; the larger the D_q value, the shorter the fluorescence wavelength.

2.2.3 *Laser Excited Fluorescence Spectra*

Due to the disorder of glass structure the inhomogeneous line-broadening of TM ions is caused by site multiplicity. A portion of the ion sites in glasses can be selectively excited by a narrow laser line, so that the spectral behaviour of the ions in different sites could be displayed in the fluorescence spectra by changing the excitation wavelengths.

Figures 2.54 and 2.55 show the fluorescence spectra of Cr^{3+} in fluorophosphate glasses at different temperatures for different excitations [68, 69]. From the figures it can be seen that the fluorescence spectra differ in the peak position λ_p and spectral linewidth $\Delta\nu$(FWHM) as the excitation wavelengths are different. From 578.2 nm to broadband excitation the spectra move toward the longer wavelength, as shown by curves 2∼6 in the figure.

Based upon the dynamic lattice model and the temporal disorder of the lattice, we suggest an extended Tanabe-Sugano diagram to explain the above distinctions in the spectra as follows (Fig. 2.56): the sites excited by 578.2 nm correspond to

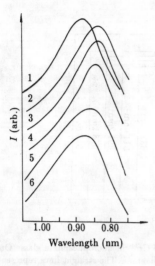

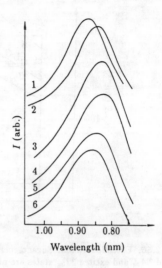

Fig. 2.54.   **Fig. 2.55.**

Fig. 2.54. Fluorescence spectra of Cr^{3+} at 77 K in fluorophosphate glass for different excitations.

1, 510.6 nm, λ_p=0.880 μm, $\Delta\lambda$=172 nm;   4, 611.4 nm, λ_p=0.844 μm, $\Delta\lambda$=180 nm;

2, 578.2 nm, λ_p=0.840 μm, $\Delta\lambda$=176 nm;   5, 632.8 nm, λ_p=0.856 μm, $\Delta\lambda$=207 nm;

3, 591.2 nm, λ_p=0.840 μm, $\Delta\lambda$=178 nm;   6, broadband, λ_p=0.868 μm, $\Delta\lambda$=221 nm.

Fig. 2.55. Fluorescence spectra of Cr^{3+} at room temperature in fluorophosphate glass for different excitations.

1, 510.6 nm, λ_p=0.870 μm, $\Delta\lambda$=164 nm;   4, 611.4 nm, λ_p=0.830 μm, $\Delta\lambda$=180 nm;

2, 578.2 nm, λ_p=0.830 μm, $\Delta\lambda$=170 nm;   5, 632.8 nm, λ_p=0.834 μm, $\Delta\lambda$=185 nm;

3, 591.2 nm, λ_p=0.830 μm, $\Delta\lambda$=184 nm;   6, broadband, λ_p=0.855 μm, $\Delta\lambda$=194 nm.

the ions with larger D_q values and can be represented by A-B, and similarly, C-D; E-F; G-H; correspond to 591.2 nm, 611.4 nm, 632.8 nm excitations respectively. Thus the distribution of sites extends by degrees. The 510.6 nm line locates the long wavelength edge of the $^4T_{1g}$ band, and can selectively excite the ions which possess smaller D_q values and can rapidly be relaxed by nonradiative transition to the lower $^4T_{2g}$ level, so that these ions will correspond to the most narrow fluorescence spectrum and the lowest emission frequency.

Comparing Figs. 2.54 and 2.55, we found that for excitation at the same wavelength the peak position at low temperature is located at longer wavelength and the spectrum is broader. We regard it as a result of the change in population of the vibronic states of $^4A_{2g}$ when the temperature increases. Therefore, A'-B' would be used to represent the site distribution at room temperature for 578.2 nm excitation. As compared with that at low temperature, it seems to be narrower and its energy

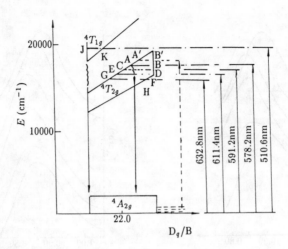

Fig. 2.56. Extended Tanabe-Sugano energy diagram of Cr^{3+} in fluorophosphate glass. Only the ground $^4A_{2g}$ and excited $^4T_{2g}$ states are plotted in the diagram. The straight lines represent excitation and emissions respectively, the undulating lines nonradiative relaxation. At low temperature the transitions occur mainly from the ground vibronic state of $^4A_{2g}$ to $^4T_{2g}$ for excitation and from the ground vibronic state of $^4T_{2g}$ to $^4A_{2g}$ for emission, which results in a Stokes shift. At room temperature the excitation can occur from each vibronic state of $^4A_{2g}$ to $^4T_{2g}$ and is represented by a dashed line.

gap higher.

From the data shown in Figs. 2.54 and 2.55, the linear relationship between excitation energy and fluorescent Stokes shift of Cr^{3+} in glass could be found. The radiative transition exhibits multiphonon-assisted behaviours [70].

Figure 2.57 shows the fluorescence spectra of Cr^{3+} in phosphate, silicate, germanate and borate glasses by using a He-Ne laser as the light souce at 632.8 nm. It can be seen from the figure that as for the shift to shorter wavelength the most noticeable is borate, the second is silicate or phosphate, and the final germanate. Similarly, as for the spectral narrowing the most noticeable one is also borate. These experimental results can be explained as follows: the ions selectively excited by an identical narrow line at 632.8 nm should possess a similar energy gap and D_q value; for D_q/B we have borate > silicate > phosphate (see Table 2.11); therefore, spectral shifting to the shorter wavelength is most pronounced for the borate glass.

Laser excited fluorescence spectra of the Mo^{3+} ion in phosphate glass have been studied by Weber *et al.* [71]. Only a small amount of line narrowing was observed and the Stokes-shifted emission peak changed monotonically with excitation wavelength. It is thus established that the emission from Mo^{3+} in glass is predominantly vibronic in nature.

The site effect can also be seen in Cu^+ doped glass [72]. Figure 2.58 shows fluo-

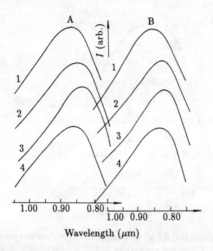

Fig. 2.57. 632.8 nm excited fluorescence spectra of Cr^{3+} in oxide glasses.

	A (77 K)		B (room temperature)	
1 Phosphate	$\lambda_p=0.862\ \mu m$	$\Delta\lambda=197$ nm	$\lambda_p=0.850\ \mu m$	$\Delta\lambda=200$ nm
2 Silicate	$\lambda_p=0.832\ \mu m$	$\Delta\lambda=206$ nm	$\lambda_p=0.820\ \mu m$	$\Delta\lambda=183$ nm
3 Germanate	$\lambda_p=0.825\ \mu m$	$\Delta\lambda=203$ nm	$\lambda_p=0.815\ \mu m$	$\Delta\lambda=176$ nm
4 Borate	$\lambda_p=0.850\ \mu m$	$\Delta\lambda=221$ nm	$\lambda_p=0.825\ \mu m$	$\Delta\lambda=188$ nm

rescence spectra of Cu^+ at room temperature by excitation at different wavelengths. The spectra are also recorded by using non-delay sampling technique, hence they represent the initial spectral distribution of the excited ions. Curves 1 to 3 in the figure represent the spectra by excitation at 250, 280 and 327 nm with a bandwidth of 10 nm, respectively. The peak position λ and bandwidth (FWHM) $\Delta\lambda$ are:

$$\lambda_{250} = 380 \text{ nm} \quad ; \quad \Delta\lambda = 90 \text{ nm},$$
$$\lambda_{280} = 385 \text{ nm} \quad ; \quad \Delta\lambda = 105 \text{ nm},$$
$$\lambda_{327} = 415 \text{ nm} \quad ; \quad \Delta\lambda = 132 \text{ nm}.$$

It shows that the site of excited ions possesses lower radiation energy and broader region of the spectral distribution with the excitation wavelength moving to the longer side. We tried to observe the spectra by excitation at still shorter wavelength than 250 nm which lies at the peak of the excitation spectrum [65]. However, as might have been expected, the very like result as that for 250 nm excitation was obtained because of the necessity of phonon emission accompanying the radiative decay of transition metal ions in glass host. This pattern of phonon-assisted radiative decay

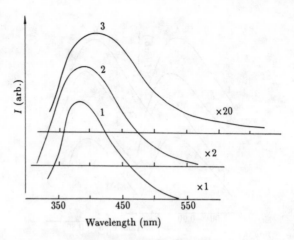

Fig. 2.58. Fluorescence spectra of Cu^+ in fluorophosphate glass at room temperature by excitation at different wavelengths. 1, 250 nm excitation; 2, 280 nm excitation; 3, 327 nm excitation.

Table 2.14. Data of fluorescence lifetime (μs) of Cu^+ in fluorophosphate glass at different temperatures for different excitations and emission wavelengths.

Temperature	Room temperature (°C)					77 K				
Emission wavelength (nm)	360	420	500	560	640	360	420	500	560	640
Excitation wavelength 220	(27)	(36)	(50)	(60)	(80)	(29)	(43)	(60)		
(nm) 250	33	37	42	48	—	(37)	52	72	88	115
280	35	43	62	84	90	40	55	85	125	130
330	—	44	62	106	114	41	55	85	135	145

is shown in Fig. 2.52.

Table 2.14 shows the values of fluorescence lifetime of Cu^+ in fluorophosphate glass at different temperatures as a function of excitation or emission wavelength. Note that the entire period for fluorescence decay lasts tens to hundreds of microseconds. As contrasted with non-delay sampling in fluorescence spectra, the fluorescence decay curves obtained include the contribution from energy transfer between sites.

Particularly for those ions having longer emission wavelength, which acts as an acceptor in the process of energy transfer, their lifetimes are prolonged by the contribution of non-radiative energy transfer.

As is shown in Table 2.14, for the constant emission wavelength (i.e. the detection wavelength is constant) the fluorescence decay varies with the excitation wavelength which corresponds to a subset of ion sites in an inhomogeneously broadened system as mentioned above. The value of fluorescence lifetime increases while the

Table 2.15. Spectral data on time-resolved fluorescence of Cr^{3+} in glasses for 510.6 nm and 578.2 nm excitation. Numbers in brackets refer to the delay time in μs.

Excitation wavelength	Temperature (K)	Fluorophosphate $\lambda_p(\mu m)$	$\Delta\lambda$(nm)	Phosphate $\lambda_p(\mu m)$	$\Delta\lambda$(nm)	Silicate $\lambda_p(\mu m)$	$\Delta\lambda$(nm)	Borate $\lambda_p(\mu m)$	$\Delta\lambda$(nm)
510.6 nm	77	0.886(10)	188	0.924(5)	172	0.890(4)	176	0.892(6)	158
		0.878(20)	181	0.920(8)	165	0.890(10)	168	0.886(12)	158
		0.868(30)	128	0.880(28)	214	0.868(24)	176	0.880(32)	202
	280	0.886(3)	145	0.874(2)	145	0.868(4)	157	0.844(6)	162
		0.868(7)	159	0.854(5)	159	0.832(13)	160		
		0.850(17)	161			0.838(23)	157		
		0.840(27)	176						
572.8 nm	77	0.850(10)	199	0.870(5)	143	0.844(4)	178	0.886(6)	194
		0.840(20)	187	0.860(10)	144	0.838(8)	170	0.886(12)	185
		0.836(30)	173			0.834(14)	174	0.862(32)	206
	280	0.850(3)	152	0.862(2)	140	0.8385(4)	164	0.848(8)	156
		0.846(7)	158	0.848(10)	166	0.822(13)	168		
		0.838(17)	170						
		0.830(27)	192						

excitation wavelength moves to the longer side. On the other hand, for the constant excitation wavelength, the fluorescence decay rate strongly depends on what wavelength the detection system is set. In parallel with the former case, the longer the emission wavelength, the larger the lifetime value obtained. In this case the values can greatly differ from each other.

In addition, a similar variation tendency could be found from Table 2.14 by comparing the data at room temperature with that at 77 K, which implies that the temperature variation has little effect on the spectral distribution provided that the initial site distribution of excited ions has not been changed.

Moreover, according to the description of the well-known Fuchtbane-Ladenburg formula [73], it is expected to have much longer lifetime for a spin-forbidden $^3E_g \rightarrow {}^1A_{1g}$ transition, and the value obtained is shorter by a factor of ten or more. The reason may be that in the absorption spectrum with a stronger oscillator, which arises from the configurational interaction that results in the increase of transition probability stolen from the spin-allowed transition.

2.2.4 *Time-Resolved Spectra and Transition of Multiple Sites*

Table 2.15 shows spectral data on the time-resolved fluorescence of Cr^{3+} in different glass hosts for 510.6 nm excitation. With the increase of delay time the spectral peak positions generally move toward shorter wavelength; the moving rate of the peak at room temperature is faster than that at low temperature. In addition, the variation in the bandwidth ($\Delta\lambda$) with delay time seems to be interesting. At low temperatures all the spectra become gradually narrowed; in contrast, at room temperature the bandwidth increases for the fluorophosphate glasses.

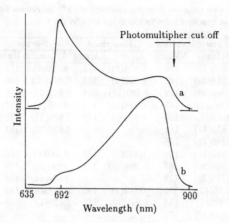

Fig. 2.59. Time-resolved emission spectra of cordierite glass under 580.0 nm excitation at 4.2 K. a, Delay=100 μs, gate=2 ms; b, delay=1 μs, gate=150 μs.

F. Durville *et al.* studied the time-resolved emission spectra of Cr^{3+} in cordierite glass under 580 nm excitation at 4.2 K. As shown in Fig. 2.59 clearly indicate that the two fluorescence bands have different decay times and that consequently the two emissions are coming from different sites [74].

In glasses there are differences from site to site, not only in the energy gap but also in the radiative and nonradiative transition probability. So with the increase of delay time after excitation the luminescence becomes more and more dominated by the contribution of the longer-lived ions. If the ions have different emission frequencies, this appears as a spectral shift in the time-resolved spectrum. The lifetime of sites (i) includes the total radiative (W_i^R) and nonradiative (W_i^{nR}) probabilities. In fluorophosphate glass the difference in nonradiative transition assisted by multiphonon processes is estimated to be several orders of magnitude. For different sites in glasses the ions having larger D_q values will have lower nonradiative transition probability and longer lifetimes, and this is contrary to the ions having smaller D_q value. Therefore, with the increase of delay time the luminescence comes mainly from contributions of the ions having larger D_q values and the spectral peak moves to the shorter wavelength.

In pure oxide glasses the shifting rate is strongly dependent on the temperature, which is rapidly enough at room temperature. The variation in the bandwidth differs from that for fluorophosphate; this shows that the energy transfer is rather a noticeable mechanism here in a sense.

Figure 2.60 shows time-resolved fluorescence spectra of Cu^+ in fluorophosphate glass at room temperature and 77 K. The peak position λ_p and line width $\Delta\lambda$ are listed in Table 2.16 [75].

It is shown in Fig. 2.60 that the fluorescent peak position shifts toward longer

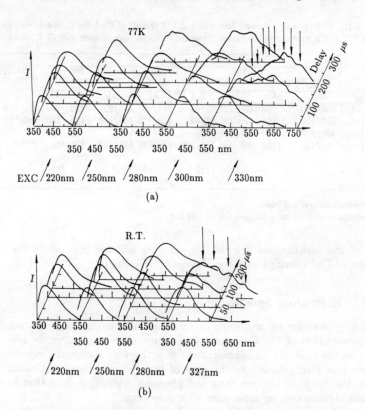

Fig. 2.60. a, Time-resolved fluorescence spectra of Cu^+ in fluorophosphate glass at 77 K (separation between the vertical down-arrow lines is 2000 cm^{-1}); b, time-resolved fluorescence spectra of Cu^+ in fluorophosphate glass at room temperature (separation between the vertical downward arrow lines is 1000 cm^{-1}).

wavelength with either increasing delay time or moving the excitation to a longer wavelength. The shifting rate rises with the delay time. In the case of excitation at a longer wavelength the low energy side band rises and the satellite line centred at 590 nm develops with delay time, besides the peak shifting to longer wavelength. The glass sample obtained contains both isolated Cu^+ ions and colloidal Cu_2O micro-crystallites in the case of dopant concentration: 0.5 wt%. The former has a d^{10}–d^9s absorption band at UV 250 nm and an emission at 400–420 nm. The latter has a band at 590 nm equally for either absorption or emission, which is attributed to the excitation on the Cu_2O band. Thus, there exists not only a transition of multiple sites of Cu^+ ions, but also the transfer from Cu^+ to colloidal Cu_2O, the former causes peak shifts to longer wavelength, whereas the latter behaves as the

Table 2.16. Peak position λ_p and line width $\Delta\lambda$ (FWHM) of Cu$^+$ for time-resolved fluorescence spectra R.T./77 K represents the values measured at room temperature and 77 K respectively.

Delay	λ_{ex} 220 R.T./77 K λ_p	$\Delta\lambda$	250 λ_p	$\Delta\lambda$	280 λ_p	$\Delta\lambda$	300 λ_p	$\Delta\lambda$	327/330nm λ_p	$\Delta\lambda$nm
0~2μs	380/390	110/95	380/390	90/85	385/400	105/92	/420	/100	415/430	132/a
60~100μs	390/400	105/103	390/395	98/92	400/420	105/103	/435	/140	440/b480	170/a
140μs	400/410	115/110	395/	103/	420/	127/				230
180~200μs	/420	/130	405/400	120/100	440/435	165/142	/510 590	/a	/b	/a
250~300μs	—	—	/420	/125	490 /590	/a	/520 590	/a		

[a] Represents spectral diffusion.
[b] Represents satellite with peak position at 590 nm.

rising of the satellite line around 590 nm with a delay time while the emission envelope of Cu$^+$ overlaps with the absorption of colloidal Cu$_2$O.

2.2.5 Relaxation Spectra

The energy transfer by multipole or exchange interactions may not occur when the concentration of the active ions and temperature are extremely low. Nevertheless, in the case of high concentration of active ions the short-range interaction between ions strengthens. In the case of high temperature, particularly in the glasses, the coupling between ions and phonons is stronger than that in crystals [76], and phonon-assisted cross relaxation may occur.

Figure 2.61 shows the fluorescence decay of Cr^{3+} in fluorophosphate glasses. Curve 1 represents the fluorescence decay at low temperature and excitation at 611.4 nm by a pulsed laser. At the beginning of the curve the fluorescence decays exponentially until about 40 μs; then $P_i(t)$ (at the time t, the probability of site i located at the excited state) changes remarkably when the decay time increases; hence the curve deviates from the initial exponential plot. Curve 2 represents the case at room temperature, which shows that the rate of $P_i(t)$ increases with temperature. Curve 3 is the case at room temperature, when excited at 510.6 nm by a pulsed laser at a repetition rate of 18 kHz, i.e. the interval between the successive pulses is shorter than the relaxation period of the activated ions; in that case the rate of $P_i(t)$ is high enough so that the curve deviates earlier than any of the former.

Figure 2.62 illustrates the time-resolved spectra of the Cr^{3+} ion measured at room temperature and excited at 578.2 nm. It can be seen that an additional peak at the low energy edge of the spectra appears after a delay time of 17 μs and rises with the delay time. When the delay time increases to 37 μs the spectra diffuse seriously due to the energy transfer, and the high energy edge of the spectrum

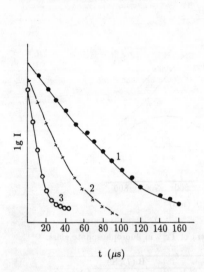

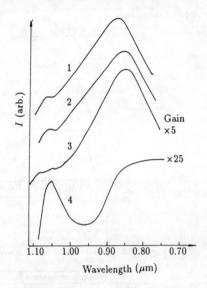

Fig. 2.61. Semilogarithmic plots of the fluorescence decay of Cr^{3+} in fluorophosphate glass (instrumental dispersion 1.2 nm/mm, slit 1 mm). Excitation and emission wavelengths were 611.4 nm and 862.0 nm respectively. 1, obtained at low temperature, 5 Hz repetition rate; 2, at room temperature, 5 Hz repetition rate; 3, the repetition rate of laser is 18 kHz at room temperature, excitation and emission wavelength are 510.6 nm and 951.8 nm, respectively.

Fig. 2.62. Fluorescence spectra of Cr^{3+} at room temperature in fluorophosphate glass. 1, Broadband excitation; 2, 632.9 nm CW excitation; 3, 578.2 nm excitation, delay 17 μs; 4, 578.2 nm excitation, delay 37 μs.

becomes structureless and divergent, whereas the additional peak at 1.042 μm becomes sharp. This shows that the interaction between the ions accompanied by phonon absorption and emission results in the energy transfer.

Similar phenomena can be observed in the time-resolved fluorescence spectra of Ti^{3+} ion in glasses (Fig. 2.63). At 77 K the line moves greatly from the peak position of 510 nm to 550 nm, and the band width from 225 μm to 220 nm within 20 μs. The high energy wing falls and the low energy wing rises. At 30 μs after the excitation the spectrum appears to be diffused due to self absorption [77].

The fluorescence lifetime of Cu^+ is also dependent upon the emission and excitation wavelengths as shown in Fig. 2.64. All the decays of luminescence by excitation at shorter wavelength are always non-exponential and faster than that by excitation at longer wavelength. Meanwhile, all the lifetime measured at longer wavelength are longer. Particularly, the lifetime measured at 560 nm is much longer, which is

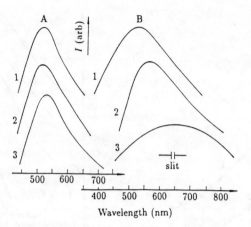

Fig. 2.63. Time-resolved fluorescence spectra of Ti^{3+} in fluorophosphate glass.

	A (room temperature)			B (77 K)		
	Delay (μs)	λ_p (nm)	$\Delta\lambda$ (nm)	Delay (μs)	λ_p (nm)	$\Delta\lambda$ (nm)
1	6	520	130	10	510	225
2	20	525	135	20	550	200
3	30	530	140	30	640	Diffused

originated from both contributions of isolated Cu^+ and colloidal copper. The energy transfer can be divided into two processes: D-D excitation migration and D-A transfer, where D is isolated Cu^+ ions and A is colloidal Cu_2O. D-A transfer is identified as a resonantly radiative one which finally results in the peak shift to 590 nm. The measured D-A transfer rate is 2.2–6.2×10^3 s^{-1}.

A more dramatic variation of multiphonon decay rates was observed for the ^2E lifetime of Mo^{3+} in phosphate glass [78]. The wavelength and temperature dependence of the decay time of ^2E–$^4A_2$ fluorescence of Mo^{3+} are shown in Table 2.17 [78]. The fluorescence decay is very non-exponential and the lifetime varies greatly with wavelength and temperature.

All these facts confirm that the coupling of the $3d(4d)$ electrons of TM ions via the time-dependent ligand field is strong. Both phonon-assisted radiative transition and multi-phonon emission contribute to the temperature dependence of the lifetime. The emission of TM ions in glasses is mainly vibronic in nature.

2.2.6 *Discussion*

From the above experimental results it can be seen that the difficulties for obtaining the stimulated emission of TM ions in glass hosts can be summarized as follows.

i) Due to the disorder of glassy state the site structure of TM ion in glass is

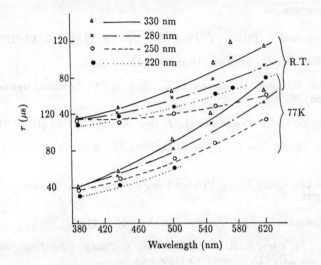

Fig. 2.64. Fluorescent lifetime of Cu^+ in fluorophosphate glass at R.T. and 77 K as functions of the emission and excitation wavelengths.

Table 2.17. Wavelength and temperature dependence of the decay time(μs) of Mo^{3+} in phosphate glass.

Decay	Fluorescence wavelength (nm)				at 1030 nm			
	990	1000	1020	1040	1060	1080	77 K	295 K
1st e-folding	980	835	495	285	200	135	385	70
2nd e-folding	2270	1830	1140	670	525	410	990	150
3rd e-folding	3450	2725	1720	1200	930	900	1585	305

quite different, the inhomogeneous line broadening is large and the interaction between the active ions at different sites is strong. The nonradiative energy transfer is intensive too.

ii) Due to disorder structure of glassy state the phonon spectra of glasses are more broad and diffuse than that of crystals, the nonradiative energy transfer from activated ion to host is large.

iii) In the sum the nonradiative energy transfer loss of excited state of TM ions is larger and the quantum efficiency of the emission in glass is lower than that in crystal. It is difficult to accomplish the population inversion at excited state.

It can be seen also from the above mentioned experimental results, when the chemical bond of glass host is more ionic in nature, it means that the glass structure in short range is more in order, the luminescent intensity of TM ion is more strong. Therefore, we have to concentrate our attention to TM ion doped fluoride, fluorophosphate and tellurate glasses, which possess lower phonon energy.

References

1. S. Sugano and Y. Tanabe, *J. Phys. Soc. Jpn.* **9** (1954) 766; **13** (1958) 880.

2. P. F. Moulton, *J. Opt. Soc. Amer.* **B3** (1986) 125.

3. A. Lupei, V. Lupei, C. Ionesco, H. G. Tang and M. L. Chen, *Optics Commun.* **59** (1986) 36.

4. A. Suginoto and Y. Segawa *et al.*, *J. Opt. Soc. Amer.* **B6** (1989) 2334

5. Xiurong Zhang, Yue Chai and Changhong Qi, *Acta Optica Sinica* **11** (1990) 270.

6. Jianhua Liu, Qiang Zhang, Peizhen Deng and Fuxi Gan, *Chinese J. Lasers* **A21** (1994) 576.

7. F. A. Zhdanov, *J. Appl. Spectro.* **XL11** (1985) 639.

8. C. Deka, M. Bass, B. H. T. Chai and X. X. Zhang, *OSA Proc. on Advanced Solid State Laser Conference* **13** (1992) 48.

9. C. Deka, B. H. T. Chai, Y. Shimony, X. X. Zhang, E. Munin and M. Bass, *Appl. Phys. Lett.* **61** (1992) 2141.

10. C. Deka, *Dissertation for Ph.D.* (University of Central Florida,1992).

11. Yuanqi Lin, Bacheng Yang, Peicong Pan and Peizhen Deng, *Chinese J. Lasers* **18** (1991) 599.

12. H. R. Verdum and L. M. Thomas, *et al.*, *OSA Proceedings on Tunable Solid State Lasers Conference* **5** (1989) 85.

13. R. Moncorge, G. Cormier, D. J. Simkin and J. A. Capobianco, *IEEE. J. Quant. Electron.*, **27** (1991) 114.

14. V. Petricevic, S. K. Gayen and R. R. Alfano, *OSA Proceedings on Tunable Solid State Lasers Conference* **5** (1989) 77.

15. T. H. Allik, B. H. T. Chai and L. D. Merkle, *OSA Proceedings on Advanced Solid State Lasers Conference* **10** (1991) 84.

16. B. M. Tissue, Weiyi Jia, Lizhu Lu and W. H. Yen, *J. Appl. Phys.* **70** (1991) 3775.

17. M. F. Weber and L. A. Riseberg, *J. Chem. Phys.* **55** (1971) 2032.

18. L. I. Krutova, N. A. Kulagin, V. A. Sandulenko and A. V. Sandulenko, *Proc. IX Symposium on Spectroscopy of Crystals Doped with Rare-earth and Transition Metals* (Leningrad, May 1990).

19. V. P. Mikkailov and N. V. Kuleshov, *et al.*, *Optical Materials* **2** (1993) 267.

20. J. A. Capobiano and G. Cormier, *et al.*, *OSA Proc. on Advanced Solid State Lasers Conference* **10** (1991).

21. J. A. Capobiano, G. Cormier, *et al.*, *Appl. Phys. Lett.* **60** (1992) 163.

22. J. A. Capobiano, G. Cormier, C. A. Morrison and R. Moncorge, *Optical Materials* **1** (1992) 209.

23. P. F. Moulton, in *Laser Handbook*, Vol. 5, ed. M. Bass and M.L. Stitch (North-Holland Publishers, Amsterdam, 1985) p. 203-289.

24. G. F. Imbusch and R. Kopelman, in *Laser Spectroscopy of Solids*, ed. W. M. Yen and P. M. Selze (Springer-Verlag, 1981) p. 12.

25. Yunkui Li, Shuchun Chen and Fuxi Gan, *Acta Optica Sinica* **11** (1991) 889.

26. Yunkui Li, Shuchun Chen and Fuxi Gan, *Chinese J. Luminescence* **12** (1991) 155.

27. Yunkui Li, Shuchun Chen and Fuxi Gan, *Chinese J. Luminescence* **12** (1991) 79.

28. R. C. Powell, *Optical Properties of Ions in Crystals* (Interscience Publishers, New York, 1967) p. 207.

29. G. G. P. Van Gorbon *et al.*, *Phys. Rev.* **B8** (1973) 955.

30. G. F. Imbusch and R. Kopelman, in *Laser Spectroscopy of Solids*, ed. W. M. Yen and P. M. Selze (Springer-Verlag, 1981) p. 26.

31. Zhiwei Hu, Guangzhao Wu, Xiaoshan Ma and Peicong Pan, *Acta Optica Sinica* **8** (1988) 601.

32. Zhiwei Hu, Guangzhao Wu, Xiaoshan Ma and Peicong Pan, *Chinese J. Lasers* **15** (1988) 250.

33. Zhiwei Hu, Guangzhao Wu, Xiaoshan Ma and Peicong Pan, *Acta Optica Sinica* **8** (1988) 601.

34. Honggao Tang, Yunkui Li, Minhua Miao and Yin Hang, *Acta Optica Sinica* **6** (1986) 155.

35. Genwang Wen and Luya Wang, *et al.*, *Chinese J. Lasers* **14** (1987) 476.

36. Genwang Wen, Songhao Liu and Luya Wang, *Chinese J. Lasers* **15** (1988) 412.

37. Guangzhao Wu, Xiurong Zhang, Shunxing Zhang and A. Lupei, *Acta Optica Sinica* **7** (1987) 563.

38. Yunkui Li and Honggao Tang, *Chinese J. Lasers* **17** (1990) 236.

39. Yunkui Li, Honggao Tang, Yin Hang and Shuchun Chen, *Chinese J. Lasers* **17** (1990) 750.

40. Jingcum Zang, Shaohua Wu and Yue Ma, *Chinese J. Lasers* **18** (1991) 446.

41. Suchao Chen and H. J. Weber, *Chinese J. Lasers* **15** (1988) 602.

42. U. Dürr, U. Brauch, W. Knierim and C. Schiller, in *Tunable Solid State Lasers*, ed. P. Hammerling, A. B. Budgor and A. Pinto (Springer Ser. Opt. Sci., Vol. 47, Springer, Berlin, 1985) p. 20.

43. S. T. Payne and L. L. Chase, *et al.*, *Proc. SPIE* **1223** (1990) 84

44. Xiaodong Liu, Peizhen Deng and Bing Hu, *Proc. SPIE* **1863** (1993) 90.

45. L. F. Johnson, H. J. Guggenheim, *J. Appl. Phys.* **38** (1967) 4837.

46. R. L. Greene, D. D. Sell *et al.*, *Phys. Rev. Lett.* **25** (1965) 656.

47. M. D. Sturge, *Phys. Rev.* **B8** (1973) 6.

48. R. M. Macfarlane, J. C. Vial, in *Mater. Sci. Forum*, ed. H. J. Von Bardeleban (Tran. Tech. Publications Ltd, Switzerland) **10-12** (1986) 845.

49. J. Ferguson, D. L. Wood and L. G. van Uitert, *J. Chem. Phys.* **51** (1969) 2904.

50. J. F. Donegan, F. J. Bergin, *et al.*, *J. Lumin.* **31/32** (1984) 278.

51. Waezhen Gin, Honggan Tang and Quesue Pai, private communication.

52. K. Knox, *et al.*, *Phys. Rev.* **130** (1963) 512.

53. J. Fergason, *Aust. J. Chem.* **21** (1968) 323.

54. Genwong Wen and Songhao Liu, *Chinese J. Lasers* **15** (1988) 510.

55. R. J. Tonucci, S. M. Jacobsen, W. M. Yen, *Chem. Phys. Lett.* **173** (1990) 456.

56. W. A. Weyl, *Coloured Glasses* (Society of Glass Technology, Sheffield, 1951).

57. A. N. Dauwalter, *Chrustalnie*, (Zvetnoe Onalovie Slekla, Moskva, 1959).

58. T. Bates, J. D. Mackenzie, *et al.*, *Modern Aspects of the Vitreous State*, Vol. 2, Chapt. 5 (Butterworths, London, 1962).

59. Fuxi Gan, *Acta Silicate Sinica* **6** (1978) 41.

60. A. Paul, *J. Mater. Sci.* **10** (1975) 693.

61. Fuxi Gan, He Deng, Huimin Liu, *J. Non-Cryst. Solids* **52** (1982) 135.

62. S. Sakka, K. Kamiya, H. Yoshikawa, *J. Non-Cryst. Solids* **27** (1978) 289.

63. Huimin Liu, Shuchun Chen, Fuxi Gan, *Acta Optica Sinica* **2** (1982) 393.

64. S. Parke, R. S. Webb, *Phys. Chem. Glass* **13** (1972) 157.

65. Huimin Liu, Fuxi Gan, *Lumin. Display Dev.* **5** (1984) 1.

66. Huimin Liu, Fuxi Gan, *Chinese J. Lasers* **11** (1984) 558.

67. L. J. Andrews, A. Lempicki, *et al.*, *J. Chem. Phys.* **74** (1981) 5526.

68. Fuxi Gan, Huimin Liu, *J. de Phys.* **43** (1982) c-9-903.

69. Fuxi Gan, Huimin Liu, *Acta Silicate Sinica* **11** (1983) 49.

70. Huimin Liu and Fuxi Gan, *Acta Optica Sinica* **7** (1987) 575.

71. M. J. Weber, S. A. Brawer, A. J. Degroot, *Phys. Rev.* **B23** (1981) 11.

72. Huimin Liu, Fuxi Gan, *Kexue Tonbao* **33** (1988) 1590.

73. S. Park, A. Watson, *Phys. Chem. Solids* **10** (1969) 37.

74. F. Durville, B. Champagnon, *et al*, *Phys. Chem. Glasses* **25** (1984) 126.

75. Huimin Liu, Fuxi Gan, *J. Luminescence* **40&41** (1988) 129; *Acta Optica Sinica* **8** (1988) 481.

76. L. J. Andrew, A. Lempick, B. C. McCollum, *J. Chem. Phys.* **74** (1981) 5326.

77. Huimin Liu, Fuxi Gan, *J. Non-Cryst. Solids* **80** (1986) 422.

78. M. J. Weber, S. A. Brawer, A. J. DeGroo, *Phys. Rev* **B23** (1981) 11.

3. Spectroscopy of Rare Earth Ions in Dielectric Solids

Since the middle of this century, study on spectroscopic properties of rare-earth (RE) metal ions in dielectric solids has been increasing. This attributed to the inorganic crystals and glasses doped with RE metal ions becoming important solid state laser materials. Now more experimental results have been accumulated. Because the electrons at inner shell $4f$ of the rare-earth ions are shielded by the electrons out of $5s$, p, d, so that the effect of ligand field becomes weak. The spectroscopic properties of rare earth ions in crystals and in glasses are rather similar, therefore, in this chapter we discuss them together.

3.1 Electronic Energy States

As we mentioned in Chapter 1, energy levels of rare-earth ions depend on, firstly, the static electric interaction between the electrons (H_e), and secondly on the interaction between the spin and the orbital (H_{so}). $4f^n$ electron configuration forms the energy level denoted with LSJ under the effects of the above two interactions. Figure 3.1 shows the energy level diagram of various rare-earth ions [1]. Their energy levels of RE ions in dielectric solids are similar to those of free ions.

The interaction of ligand field can be considered as the free ion energy being disturbed. The potential energy of ligand V_L may be expanded to be

$$V_L = \sum_{k,q,i} B_q^k (C_{-q}^k)_i, \tag{3.1}$$

where C_{-q}^k is the denotation of energy, B_q^k the interaction parameter of ligand field. The Stark splitting level number depends on the quantum number J at different symmetric lattices, to be determined by using the group theory (see Table 1.3). We may find from the Eq. 3.1 that the multistate splitting depends on the corresponding even part of parity. The types and properties of ligand around the rare-earth ions exert a dominant influence upon the Stark splitting.

Taking Nd^{3+} as an example, Nd^{3+} ions in the most of glasses and crystals are all located at low symmetric lattice ($\leq S_4$). According to the $J+1/2$ values, the numbers of spectral sub-terms of Stark splitting are found to be: 5 terms for ground

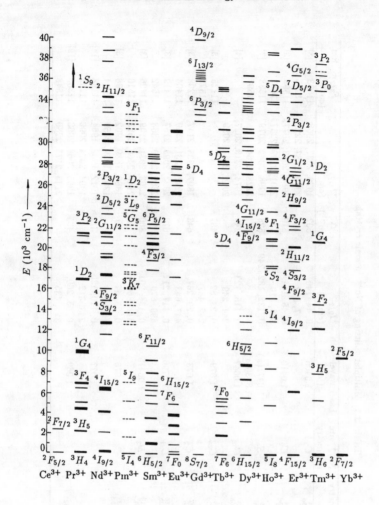

Fig. 3.1. Energy level diagram of rare-earth metal ions ($4f^n$).

state $^4I_{9/2}$, 6 terms for terminal state $^4I_{11/2}$, while 2 terms for metastable state $^4F_{3/2}$, as shown in Fig. 1.2. As an example, the Stark splitting terms and values Δ of $^4I_{9/2}$, $^4I_{11/2}$ and $^4F_{3/2}$ levels of Nd^{3+} ions, which can be obtained from the calculation of absorption and emission spectra in Nd doped glasses and crystals, will differ greatly, if the type of ligand changes. As shown in Table 3.1 the multiplet splitting values Δ of Nd^{3+} ions in various fluoride crystals are similar, but the values are much smaller in chlorides. When the Nd^{3+} ions locate in oxide ionic crystals (e.g. Y_2O_3, La_2O_3

Table 3.1. Multiplet splitting of Nd^{3+} in ionic crystals.

Host crystal	Symmetry	Nd^{3+} site symmetry	Laser wavelength (μm)	$^4I_{9/2}$ (cm^{-1})	Δ (cm^{-1})	$^4I_{11/2}$ (cm^{-1})	$^4F_{3/2}$ (cm^{-1})	Ref.
LaF_3	D_{3d}^4	C_2	1.0406	0, 44, 140, 297, 502	502	1978, 2037, 2068, 2090, 2187, 2222	11593, 11635	[3]
CeF_3	D_{3d}^4	C_2	1.0410	0, 45, 144, 303, 510	510	1986, 2043, 2077, 2101, 2199, 2237	11598, 11643	[3]
$LiYF_4$	C_{4h}^6	S_4	1.0471(π) 1.8530(σ)	0, 132, 182, 249, 528	528	1998, 2042, 2079, 2228, 2264	11538, 11597	[3]
$NdCl_3$	D_{6h}^3	C_{3h}	1.0647	0, 115, 123, 244, 249	249	1974, 2013, 2027, 2044, 2051, 2059		[4]
La_2O_3	D_{3d}^5	C_{3v}	1.079	0, 23, 82, 242, 487	487		11175, 11325	[3]
Y_2O_3	T_n^7	C_2	1.073	0, 29, 267, 447, 643	643	1897, 1935, 2147, 2271, 2331, 2359	11208, 11404	[3]
$YAlO_3$	D_{2h}^{16}	C_{1h}	1.0796	0, 117, 211, 499, 670	670	2023, 2098, 2157, 2269, 2320, 2368	11419, 11542	[3]
$Y_3Al_5O_{12}$	O_h^{10}	D_2	1.06415	0, 134, 201, 311, 860	860	2003, 2031, 2109, 2146, 2468, 2523	11426, 11510	[3]
La_2O_2S	D_{3d}^3	C_{3v}	1.075	0, 23, 47, 79, 90	90	1880, 1889, 1902, 1909, 1959, 2050	11198, 11214	[3]
La_2S_3			1.077	0, 110, 130, 245, 312	312	1912, 1932, 1964, 1994, 2074, 2090	11198, 11262	[18]
$Ca_3Ga_2Ge_2O_{14}$	P_{321}		1.069	0, 65, 260, 320, 435	435	1965, 2008, 2063, 2145, 2175, 2280	11363, 11548	[19]

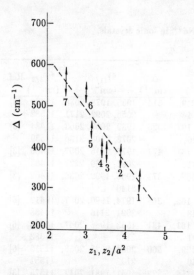

Fig. 3.2. Correlation of spectral splitting Δ and static interaction force F' in oxygen-polyhedra. 1, Phosphate; 2, tungstate, molybdate; 3, arsenate; 4, vanadate; 5, silicate; 6, borate; 7, niobate, tantalate.

Fig. 3.3. Absorption cross section of $^4I_{9/2}$-$^2P_{1/2}$ transition of Nd^{3+} in different glasses.

and $Y_3Al_5O_{12}$, etc.), the Δ values are greater than those in fluorides, but they are still small in sulfide-oxides (La_2O_2S). It can be seen that the energy levels splittings Δ of Nd^{3+} ions are relevant to the interaction of Nd^{3+} ions with the neighbouring ligands (negative ions). The electrostatic interaction is simply expressed as $F = \frac{z_1 z_2}{a^2}$, where z_1, z_2 are the valency of cation and anion, a is the distance between them. The value of Δ increases in proportion to F [2].

In the oxide hosts, the ligand having effect on rare-earth ions is the oxygen ion group (MO_x). The rule of this effect on Stark splitting value Δ of rare-earth ions is similar to that of transition metal ions mentioned before. With increasing interaction force of the central cation M and oxygen ion in group, the polarization of rare-earth ions by oxygen in group is decreased. The interaction force of M-O can be expressed by $F' = \frac{z_1' z_2'}{a^2}$, where z_1' and z_2' are the valences of anion and central cation respectively, a the spacing between them. We may find from Fig. 3.2 that the Δ value of ground state $^4I_{9/2}$ of Nd^{3+} is inversely proportional to F'. It can also be seen from Table 3.2 from phosphate, tungstate, vanadate to germanate, the values of Δ increases gradationally. Chemical bonding property of ligand field also exerts influences on the positions of absorption and luminescence lines of rare-earth ions, behaving Nephelauxetic Effect. Taking Nd^{3+} as example, the absorption peaks of $^4I_{9/2}$-$^2P_{1/2}$ transition of Nd^{3+} in different inorganic glasses at low temperature

Table 3.2. Multiplet splitting of Nd^{3+} in ionic crystals.

Host crystal	Symmetry	Nd^{3+} site symmetry	Laser wavelength (μm)	$^4I_{9/2}$ (cm^{-1})	Δ (cm^{-1})	$^4I_{11/2}$ (cm^{-1})	$^4F_{3/2}$ (cm^{-1})	Ref.
NdP_5O_{11}	C_{6h}^5	D_{4d}	1.051	0, 80, 219, 252, 314	314	1955, 1978, 2038, 2056, 2092, 2171	11470 11582	[3]
$LiNdP_4O_{12}$	C_{2h}^6		1.055	0, 106, 197, 268, 326	326	1939, 2003, 2053, 2078, 2108, 2136	11484 11539	[5]
$CaWO_4$	C_{4h}^6	$C_{4\nu}$	1.0582(π) 1.0652(σ)	0, 114, 161, 230, 471	471	1977, 2016, 2057, 2087, 2189, 2227	11406 11469	[3]
$SrMoO_4$	C_{4h}^6	S_4	1.061	0, 95, 158, 213, 377	377	1956, 1999, 2036, 2140	11402 11467	[3]
$YAsO_4$	D_{4h}^{19}		1.0597	0, 109, 165, 229, 368	368	1974, 1996, 2037, 2091, 2116	11412 11438	[6]
YVO_4	D_{4h}^{19}		1.06041	0, 107, 161, 212, 431	431	1963, 1986, 2047, 2150	11365 11383	[6]
$LaBO_3$			1.0545	0, 95, 129, 311, 560	560	2004, 2023, 2058, 2102	11515 11551	[6]
$Ca(NbO_3)_2$	D_{2h}^{14}	C_2	1.0615	0, 100, 158, 237, 522	522	1951, 1981, 2017, 2093, 2218, 2246	11375 11423	[3]
$Ba_2MgGe_2O_7$	D_{2d}^3		1.05436	0, 63, 105, 250, 700	700	1895, 2040, 2119	11379 11514	[3]

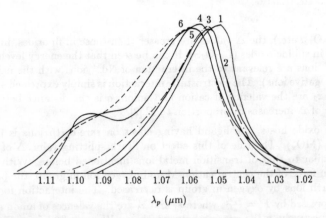

Fig. 3.4. Luminescence spectra of $^4F_{3/2}$–$^4I_{11/2}$ transition of Nd^{3+} in different glasses. 1, Fluoride; 2, fluorophosphate; 3, phosphate; 4, silicate; 5, germanate; 6, borate.

(5 K) are given in Fig. 3.3 [7]. Because there is no Stark splitting at the $^2P_{1/2}$ level, so it may represent the typical line position at low temperature. Figure 3.4 shows the luminescence peak positions of $^4F_{3/2}$–$^4I_{11/2}$ transition of Nd^{3+} in different inorganic glasses. The peak position shifts to a longer wavelength with the sequence

Table 3.3. Electro-negativity of Nd^{3+} ion and ligand and wavelength of luminescence.

Host	Difference of electro-negativity	Wavelength of luminescence λ_p (μm)
Fluoride	2.6	1.045~1.048
Oxide	2.2	1.060~1.062
Chloride	1.8	1.063~1.065
Chalcogenide	1.3	1.075

of fluoride, fluorophosphate, phosphate, silicate and borate. It is also indicated in Tables 3.1 and 3.2 the laser wavelength shifted to the longer wavelength from fluoride to oxide and from phosphate to niobate crystal hosts. As discussed previously, this indicates the increase of covalent bonding of rare-earth ions with the ligand (polarization of ligand to excited ions increases). We use the electric negativity difference Δx to denote the change of the chemical bonding property, as shown in Table 3.3. The peak wavelength of luminescence in Nd doped glasses and crystals shifts to a longer wavelength with the electric negativity difference decreasing.

Divalent rare-earth ions can be obtained at reduction atmosphere, the alkaline earth fluoride crystals have been the principal hosts for RE^{2+} ions, which occupy cubic Ca^{2+} (or Sr^{2+}, Ba^{2+}) sites. Figure 3.5 summarizes the energy levels and emission wavelengths of RE^{2+} ions. In divalent rare-earth ions the $5d$ energy levels are decreased, the $4f^n \rightarrow 4f^{n-1}5d$ transition can take place. It is electric-dipole allowed. The allowed radiative transition between $4f$ states are magnetic-dipole or vibronics.

In the lanthanide ions, Eu and Yb are more easy to reduce to the divalent state and remain stable, and this is less stable for Sm and Tm in crystals. The special treatments, such as irradiation with X-ray, β and γ rays, electrolysis and photochemical reaction, must be used to reduce trivalent to divalent state for the rest of lanthanide ions in crystals and all RE ions in glasses.

3.2 Radiative Transition

Light radiation of rare-earth ions mainly comes from the electric dipole transition. For transition from the $4f$ inner shell of free ions, it is forbidden because it does not correlate to the change of parity. In the ligand field, due to the odd part of parity in the potential expanded equation (Eq. 3.1) of ligand field and the vibration of structure network, a certain forbidden transition is removed by mixing $4f$ and $5d$ hence the radiative transition is allowed. Equation 3.2 gives the probability of spontaneous radiation transition W_r. For rare-earth ions, taking account of multiple terms splitting, its spontaneous radiative transition probability becomes

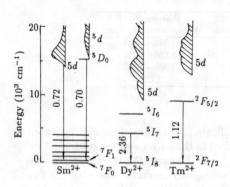

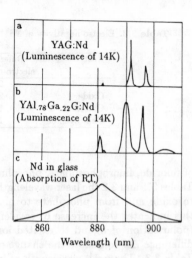

Fig. 3.5. Energy levels and laser transitions of divalent rare-earth ions. Wavelengths of transitions are in μm.

Fig. 3.6. $^4F_{3/2} \leftrightarrow {}^4I_{9/2}$ transition of Nd^{3+} in (a) $Y_3Al_5O_{12}$; (b) 78% $Y_3Al_5O_{12}$ and 22% $Y_3Ga_5O_{12}$ mixed crystal; (c) glass.

$$\overline{W}_r = \frac{64\pi^4 e^2}{3n(2J'+1)\lambda_p^3}\cdot[\frac{n(n^2+1)^2}{9}]\cdot S, \qquad (3.2)$$

where S is the spectral intensity. With use of the Judd and Ofelt theory [20, 21], it can be obtained approximately by

$$S = \sum_{\lambda=2,4,6} \Omega_\lambda |< S,L,J\|u_\lambda^2\|S',L',J' >|^2, \qquad (3.3)$$

where $\|u_\lambda^2\|$ is the operator of double simplified matrix, which is determined by $4f^n$ electronic configuration, but not by ligand interaction, Ω_λ the interaction parameter of ligand field for removing forbidden transition, which is determined by the property of ligand field and can be written as

$$\Omega_\lambda = (2\lambda+1)\sum_{p,t} A_{t,p}^2\cdot B^2(t,p)\cdot(2t+1)^{-1}, \qquad (3.4)$$

where $A_{t,p}^2$ is the odd part of parity of ligand field potential, B(t,p) the corresponding radial integrated part.

Generally, Ω_λ is obtained from the absorption spectra of rare-earth ions doped glasses and crystals, i.e. from the intensity of different absorption spectra S_i simulated with Eq. 3.3.

Table 3.4 gives the Ω_λ values of Nd^{3+} ion in different host glasses and crystals [8–11]. We find that Ω_2 is more sensitive to the symmetry and sequence of ligand

Table 3.4. Ω_λ values of Nd^{3+} in different glasses and crystals.

Host	$\Omega_2 \times 10^{-20}$ (cm^2)	$\Omega_4 \times 10^{-20}$ (cm^2)	$\Omega_6 \times 10^{-20}$ (cm^2)	References
LaF_3	0.35	2.57	2.50	[10, 11]
$NdW_{10}O_{35}$	0.99	7.19	9.13	[12]
$LiNdP_4O_{12}$	2.00	4.30	5.90	[13]
YAG	2.00	2.70	5.00	[10, 11]
$YAlO_3$	1.30	4.70	5.70	[10, 11]
γ-La_2S_3	2.85	5.08	4.72	[18]
$Ca_3Ga_2Ge_4O_{14}$	1.88	3.65	5.65	[19]
Fluoride glass	0.1~1.5	2.5~4.5	3.2~6.0	[9]
Chloride glass	5.0~6.0	6.0~7.5	4.5~5.0	[9]
Silicate glass	2.8~4.5	3.0~5.0	2.0~5.0	[8, 9]
Phosphate glass	3.0~6.5	3.5~6.0	3.5~6.7	[8, 9]
Borate glass	3.5~6.0	3.0~6.0	3.5~5.5	[8, 9]
Tellurate glass	3.0~5.6	2.8~5.0	3.4~5.0	[8, 9]

field, Ω_2 increases significantly in the sequence of fluoride, oxide and chloride, i.e. in the increment of covalent bond in nature. The probability of spontaneous radiative transition $\overline{W}_r$ is calculated from Eqs. 3.2, 3.3 with use of Ω_λ and $\|u_\lambda^2\|$ for transition from upper state to lower state (different $\|u_\lambda^2\|$ values correspond to different rare-earth ions and different transitions between energy levels, refer to the literature [14]). Fluorescence sub-terms ratio β can be derived from the ratio of Ω_λ.

The stimulated emission cross section σ_p of the rare-earth ion can be calculated from equation:

$$\sigma_p = \frac{\lambda_p^2}{8\pi c n^2} \cdot \frac{1}{\Delta\lambda_{eff}} \cdot \overline{W}_r. \tag{3.5}$$

Here n is the refractive index, $\Delta\lambda_{eff}$ half bandwidth of emission.

To accurately determine the real spontaneous radiative probability is a complicated problem. Because the Judd-Ofelt approach does not account J-J mixing and this does not give the required accuracy in estimation of intermanifold probabilities between high-located J-manifolds and its empirical character, which practically does not reveal relations between the observed spectroscopic (and lasing) characteristics and the microstructure of RE center. In the framework of point-exchange-charge-model [15] Kaminski proposed a method for calculating the spontaneous interstark transition probability and corresponding luminescence branching ratios of Nd^{3+} doped crystals. It seems that the theory and experiment are in good agreement with each other [16].

Table 3.5 lists the spectral parameters of Nd^{3+} ion in dielectric materials. In comparison to experimental data in Table 3.4, it can be shown that the radiative transition probability $\overline{W}_r$ (or the sum of spontaneous emission probability $\sum A_i$) is proportional to the sum of the ligand field interaction parameters $\sum \Omega_\lambda$. The change

Table 3.5. Fluorescent characteristics of Nd-doped glasses and crystals.

Host	λ_P (μ)	$\Delta\lambda_{eff}$ (cm^{-1})	$\sigma_p \times 10^{-20}$ (cm^2)	$\sum A_r$	$\beta_{1.06}$	n_d
YAG	1.0615	4	89	4380	0.52	1.836
YAlO$_3$	1.064, 1.073	10	44	6340	0.56	1.89, 1.905
LiNdP$_4$O$_{12}$	1.049, 1.0477	24	28	3152	0.48	1.585
LiYF$_6$	1.047, 1.053	12	~35	1650	0.53	1.445, 1.468
La$_2$S$_3$	1.077~1.079	140	11.5~16.0	17000	0.45	2.3
Ca$_3$Ga$_2$Ge$_4$O$_{14}$	1.069	120	10	4630	0.51	1.91
Fluoride glass (BeF$_2$-RF-RF$_2$)	1.046~1.050	190~280	2.0~3.5	1600~2500	0.5~0.57	1.28~1.38
Chloride glass (BiCl$_3$-RCl)	1.063~1.065	190~220	6.0~7.0	4500~5500	0.45~0.52	1.6~2.0
Silicate glass (SiO$_2$-R$_2$O-RO)	1.058~1.062	340~400	1.0~3.0	1100~3000	0.45~0.50	1.48~1.75
Phosphate glass (P$_2$O$_5$-R$_2$O-RO)	1.052~1.057	250~350	2.0~4.5	2000~3500	0.48~0.55	1.49~1.65
Borate glass (B$_2$O$_3$-R$_2$O$_3$-RO)	1.060~1.065	290~420	2.0~3.0	2200~3500	0.48~0.60	1.52~1.70
Tellurate glass (TeO$_2$-RO-R$_2$O)	1.057~1.063	260~310	3.0~5.0	5000~7000	0.46~0.55	1.8~2.2

of radiative transition probability $\sum A_r$ of Nd^{3+} ion in glasses and crystals is not so much, but the stimulated emission cross section σ_p is much larger in crystal than that in glass due to narrow fluorescence linewidth $\Delta\lambda_{eff}$ in crystal.

In general, the effective fluorescence bandwidth $\Delta\lambda_{eff}$ is given by

$$\Delta\lambda_{eff} = \frac{c}{\lambda_P} \int \frac{I\omega d\lambda}{I_{max}}, \tag{3.6}$$

where λ_P is the fluorescence peak wavelength, I_{max} is the maximum fluorescence intensity.

As we mentioned before, the fluorescence bandwidth consists of the homogeneous broadening part and inhomogeneous broadening part. The former includes the natural broadening due to spontaneous transition and nonradiative transition. In real laser materials, the homogeneous broadening mainly results from the crystalline lattice vibration, the electron-phonon interaction. In simple crystals with ordered structure the homogeneous broadening is the main aspect, but in glasses and mixed crystals with disordered structure the inhomogeneous broadening is the main aspect. It depends on the structure of the host materials rather than their composition. The adding or substituting Nd^{3+} ions into crystal lattices will result in distorting the lattice sites of Nd^{3+} ion and producing the defects in the crystals, thus the inhomogeneous broadening will be caused. Figure 3.6 shows the $^4F_{3/2}$-$^4I_{9/2}$ transition of Nd^{3+} in YAG single crystal, mixed crystal and glass. It can be seen

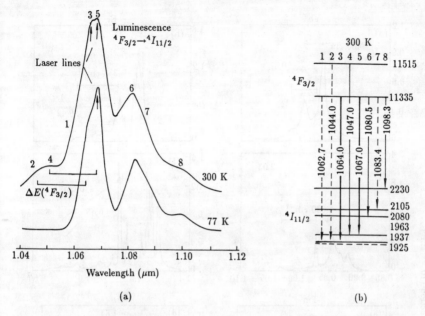

(a)

(b)

Fig. 3.7. Luminescence spectra and crystal-field splitting scheme of Nd^{3+} ions in $(La_{1-x}Nd_x)_3Ga_5SiO_{14}$ crystal. Energy of Stark levels are in cm^{-1} and transition between them are in nm.

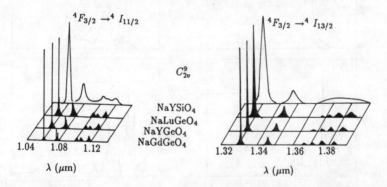

Fig. 3.8. Luminescence spectra of Nd^{3+} ions in olivin structure crystals at 77 K.

clearly that the line broadening and the splitting are quite different. The $\Delta\lambda_{eff}$ values are about 10–20 cm^{-1} for the simple crystals with ordered structure whether fluoride or oxide. In mixed crystals with disordered structure Nd^{3+} ions locate at

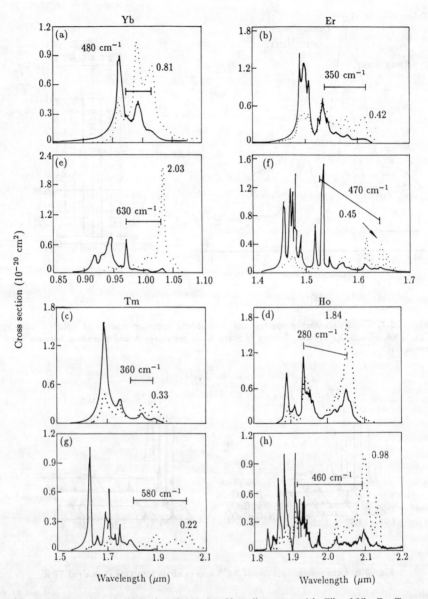

Fig. 3.9. Absorption (solid line) and emission (dotted) spectra with $E\|c$ of Yb, Er, Tm, and Ho, (a)–(d) in the crystal host LiYF$_4$, and (e)–(h) in Y$_3$Al$_5$O$_{12}$. All spectra appear on absolute cross-section scales, and the peak emission cross section is indicated. —: Absorption; ⋯⋯: emission.

Table 3.6. Absorption lines and corresponding transitions of Er^{3+}, Ho^{3+}, Tm^{3+} ions in ZBLAN fluoride glass.

Ion and ground state	Intensity parameters $(10^{-22}$ cm^2)	Absorption line wavelength (nm)	Transition
Er^{3+}, $^4I_{5/2}$	$\Omega_2 = 270$	361.6	$^2G_{9/2}$, $^2K_{15/2}$
	$\Omega_4 = 150$	374.4	$^2G_{11/2}$
	$\Omega_6 = 110$	403.2	$^2H_{9/2}$
		446.4	$^4F_{5/2}$, $^4F_{3/2}$
		483.2	$^4F_{7/2}$
		516.8	$^2H_{11/2}$
		537.6	$^4S_{3/2}$
		547.2	$^4F_{9/2}$
		794.4	$^4I_{9/2}$
		971.2	$^4I_{11/2}$
		1524.8	$^4I_{13/2}$
Ho^{3+}, $^5I_8$	$\Omega_2 = 210$	361	$^3G_5$, $^3H_{5,6}$
	$\Omega_4 = 200$	416	$^5G_5$
	$\Omega_6 = 140$	449	$^5G_6$
		477	$^3K_8$, $^5F_{2,3}$
		533	$^5S_2$, $^5F_4$
		640	$^5F_5$
		1171	$^5I_6$
		1920	$^5I_7$
Tm^{3+}, $^3H_6$	$\Omega_2 = 280$	352.8	$^1D_2$
	$\Omega_4 = 180$	462.4	$^1G_4$
	$\Omega_6 = 100$	680.0	$^3F_{2,3}$
		786.4	$^3F_4$
		1208.0	$^3H_5$
		1663.2	$^3H_4$

different sites, therefore the $\Delta\lambda_{\text{eff}}$ increase to 50–100 cm^{-1}. The character of noncrystalline materials is the long range disorder. Nd^{3+} ions may locate at many different sites. Because of the overlapping of numerous spectral lines, the $\Delta\lambda_{\text{eff}}$ values reach 250–400 cm^{-1}. According to [8], when the composition of host glasses changes in a wide range, the $\Delta\lambda_{\text{eff}}$ values will increase as the effects of host on Nd^{3+} ions strengthen (from fluoride to oxide glasses).

If there are two kinds of ligands in the host (e.g. F^- and $O^=$), then Nd^{3+} can locate at these different ligand sites, and their fluorescence band will be broadened. The $\Delta\lambda_{\text{eff}}$ values in fluorophosphate glasses are larger than those in fluoride glasses and phosphate glasses.

The crystal field Stark splitting can be obviously observed in the fluorescent spectra of stoichiometric single crystal. Figure 3.7 shows the Stark levels of $^4F_{3/2}$ and $^4I_{11/2}$, their luminescent transitions and spectra of Nd^{3+}-doped $(La_{1-x}Nd_x)_3Ga_5SiO_{14}$ crystals. The crystal lattice change can also influence the Stark splitting and luminescence spectra in a series of same structure crystals, it can be seen from Fig. 3.8 the luminescence spectra of Nd^{3+} ions in olivin structure

Table 3.7. Radiative transition probablity A_r, radiative lifetime τ_r, luminescence branching ratio $\beta_{JJ'}$, luminescence linewidth $\Delta\lambda$ and stimulated emission cross-section of Er^{3+}, Ho^{3+} and Tm^{3+} in fluoride glass.

Ion	Transition J–J'		λ (μm)	A_r (s^{-1})	τ_r (ms)	$\beta_{JJ'}$	$\Delta\lambda$ (nm)	σ_p (10^{-20}cm^2)
Er^{3+}	$^4I_{13/2}$	$^4I_{15/2}$	1.536	111.29	8.99	100	100	0.36
	$^4I_{11/2}$	$^4I_{13/2}$	2.695	120.53	8.30	18.4	30	2.27
		$^4I_{15/2}$	0.978			81.6	30	0.17
	$^4F_{9/2}$	$^4I_{9/2}$	3.256	977.00	1.02	0.2		
		$^4I_{11/2}$	1.930			4.9		
		$^4I_{13/2}$	1.125			4.5		
		$^4I_{15/2}$	0.649			90.4		
	$^4S_{3/2}$	$^4I_{9/2}$	1.627	1198.90	0.834	3.4	50	0.34
		$^4I_{11/2}$	1.211			9.8	25	0.59
		$^4I_{13/2}$	0.836			6.7	15	0.15
		$^4I_{15/2}$	0.541			80.1	15	0.32
	$^4H_{11/2}$	$^4I_{15/2}$	0.525	3224	0.310	100		
Ho^{3+}	$^5I_7$	$^5I_8$	2.030	58.12	17.21	100	80	0.72
	$^5I_6$	$^5I_7$	2.887	123.15	8.12	14.6	50	1.45
		$^5I_8$	1.190			85.4	50	0.24
	$^5F_5$	$^5F_4$	4.018	1640.71	0.61	0		
		$^5I_5$	2.221			0		
		$^5I_6$	1.411			4.4		
		$^5I_7$	0.948			18.6	15	0.95
		$^5I_8$	0.640			76.6	15	0.82
	$^5S_2+^5F_5$	$^5F_5$	3.176	4514.75	0.22	0		
		$^5I_4$	1.778			0.5		
		$^5I_5$	1.309			2.9		
		$^5I_6$	0.977			7.2		
		$^5I_7$	0.730			20.4	20	0.76
		$^5I_8$	0.542			69.0	20	0.78
	$^5K_8+^5F_{2,3}$	$^5I_8$	0.477	2916.71	0.34	100	10	0.88
Tm^{3+}	$^3H_4$	$^3H_6$	1.86	94.3	10.6	1.00	140	0.47
	$^3F_4$	$^3H_6$	0.802	622.3	1.45	0.90	40	0.37
		$^3H_4$	1.486	51.9		0.08	80	0.18
		$^3H_5$	2.44	13.2		0.02	100	2.68
	$^1D_2$	$^3H_6$	0.363	5815.0	0.07	0.40		
		$^3H_4$	0.453	6520.1		0.45		
		$^3H_5$	0.509	75.5		0.01		
		$^3F_4$	0.657	832.9		0.06		
		$^3F_3$	0.750	465.8		0.03		
		$^3F_2$	0.773	575.1		0.04		
		$^1G_4$	1.472	91.7		0.01		

crystals. The Stark splitting spectra are diffused in non-stoichiometric crystals and very difficult to identify in the glass hosts even at low temperature.

Er^{3+}, Ho^{3+}, Tm^{3+} and Yb^{3+} ions have energy levels, which could give rise to light emissions not only in the visible region, but also in the middle IR region (1.5–3

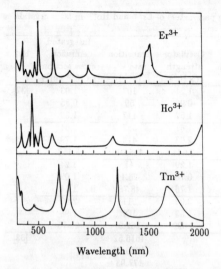

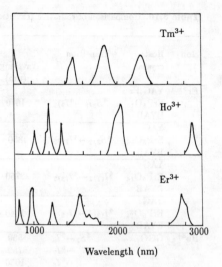

Fig. 3.10. Absorption spectra of Er^{3+}, Ho^{3+} and Tm^{3+} in fluoride glass.

Fig. 3.11. IR emission spectra of Tm^{3+}, Ho^{3+} and Er^{3+} in fluoride glass.

Table 3.8. Spectroscopic parameters (Ω_λ) of Er^{3+} and Ho^{3+} in laser crystals.

Ion	Host crystal	Ω_λ $(10^{-20}$ cm$^2)$			Ref.
		Ω_2	Ω_4	Ω_6	
	YAG	0.19	1.68	0.62	[31]
Er^{3+}	YAlO₃	1.06	2.63	0.78	[30]
	EYAB	2.37	1.08	1.97	[32]
Ho^{3+}	GGG	0.24	1.41	1.09	[33]

Table 3.9. Ω_λ values of Er^{3+} in inorganic glasses.

Host glass	Ω_λ $(\times 10^{-20}$ cm$^2)$			Ref.
	Ω_2	Ω_4	Ω_6	
P_2O_5–La_2O_3–Li_2O	4.91	0.78	0.89	[34]
AlF_3–RF_2–RF	2.99	0.92	0.87	[26]
ZrF_4–LaF_3–BaF_2	3.26	1.85	1.14	[26]
GeO_2–Na_2O–BaO	4.74	0.88	0.84	[34]
TeO_2–Li_2O	4.39	1.51	0.93	[14]

μm) by cascade energy transfer between intraionic or interionic energy levels, and emission can be obtained in the visible region (0.45–0.65 μm) by frequency up-conversion. Therefore, laser actions can be achieved the middle IR and the visible regions in Er^{3+}, Ho^{3+}, Tm^{3+} and Yb^{3+} ions doped dielectric solids.

Figure 3.9 shows the absorption and emission spectra of Yb^{3+}, Er^{3+}, Tm^{3+} and Ho^{3+} ions in $LiYF_4$ and YAG host crystals. It is apparent from Fig. 3.9 that the characteristic wavelength of the ions advances further into the infrared, in proceeding from Yb→Er→Tm→Ho (1.0→1.6→1.9→2.1 μm) [22].

Heavy metal fluoride glass are highly transparent in the middle IR region. Some composition can be easily prepared and doped with rare-earth (RE) ions. The optical spectra of Er^{3+}, Ho^{3+} and Tm^{3+} in fluoride glasses have been studied extensively

Table 3.10. Comparison of radiative transition parameters of Er^{3+} and Ho^{3+} in laser crystals.

Ion	Host crystal	Transition	Wavelength (nm)	Oscillator strength ($\times 10^6$)	Transition rate $\overline{W}_r$ (s^{-1})	Integrated emission cross section ($\times 10^{18}$ cm^2)	Ref.
Er^{3+}	YAG:Er			1.21	107	1.03	[32]
	ErP_5O_{14}	$^4S_{3/2} \to {}^4I_{9/2}$	1670	0.98	59	0.85	
	EYAB			1.57	117	1.36	
	YAG			1.61	557	2.61	
	ErP_5O_{14}	$^4S_{3/2} \to {}^4I_{13/2}$	850	2.03	479	1.79	
	EYAB			3.644	1056.1	3.153	
	YAG			1.29	41	1.24	
	ErP_5O_{14}	$^4I_{11/2} \to {}^4I_{13/2}$	2750	0.70	26.1	1.02	
	EYAB			1.76	48.78	1.54	
	YAG			2.04	211	2.00	
	ErP_5O_{14}	$^4I_{13/2} \to {}^4I_{15/2}$	1540	1.53	161	1.98	
	EYAB			3.099	262.9	2.575	
Ho^{3+}	GGG	$^5S_2 \to {}^5I_8$	550		1616.52		[33]
		$\to {}^5I_7$	750		1060.24		
		$\to {}^5I_6$	1030		177.93		
		$\to {}^5I_5$	1400	—	44.10	—	
		$^5I_6 \to {}^5I_8$	1260		229.4		
		$\to {}^5I_7$	2860		23.40		
		$^5I_7 \to {}^5I_8$	2080		94.80		

in recent years [23–28]. The recently measured absorption spectra of these ions in ZrF_4–BaF_2–LaF_3–AlF_3–NaF(ZBLAN) glass system are shown in Fig. 3.10. The absorption lines and the corresponding transitions are listed in Table 3.6 [29].

For RE^{3+} ions the probability of radiative transition $A_{jj'}$ from a given state J to the level of state J' is the sum of the induced electric dipole(ed) transition and magnetic dipole(md) transition. From the experimental values of the absorption intensity of spectral lines, according to the Judd-Ofelt equation a least-square-fitting approach is used to find out the intensity parameters Ω_t with the aid of a set of reduced matrix elements $(U^{(t)})$. The intensity parameters Ω_t values of Er^{3+}, Ho^{3+}, Tm^{3+} in fluoride glass ZBLAN are also shown in Table 3.6 [29].

The possible radiative emission transitions of Er^{3+}, Ho^{3+}, Tm^{3+} in fluoride glass can be seen from the energy level diagrams. The calculated radiative emission transition probability A_r, radiative lifetime τ_r, and luminescence band ratio $\beta_{JJ'}$, are shown in Table 3.7 for these ions. In these tables are also listed the measured emission linewidth $\Delta\lambda$ and stimulated emission cross-section σ_p [29].

The most attactive luminescence of Tm^{3+}, Ho^{3+} and Er^{3+} in fluoride glass are in the middle infrared region. The luminescence spectra of these ions are demonstrated in Fig. 3.11. Most of luminescence bands could not be observed in oxide glasses.

The spectroscopic intensity parameters (Ω_λ) of Er^{3+} in $YAlO_3$, YAG and

Table 3.11. Luminescence properties of Er^{3+} ion in glasses.

Type	Transition $J \to J'$	Average wavelength (μm)	Radiative transition probability (s^{-1}) (electric dipole +magnetic dipole)	Luminescence branching ratio $\beta(\%)$	FWHM (nm)	Stimulated cross section $\sigma_p(10^{-20} \text{ cm}^2)$
Fluoro-	$^4I_{13/2}$ $^4I_{15/2}$	1.536	111.29	100	100	0.3592
zirconate	$^4I_{11/2}$ $^4I_{13/2}$	2.695	22.22	18.4	30	2.2655
glass	$^4I_{15/2}$	0.978	98.31	81.6	30	0.1738
$n_D=1.5124$	$^4S_{3/2}$ $^4I_{9/2}$	1.627	41.31	3.4	50	0.3357
	$^4I_{11/2}$	1.211	117.57	9.8	25	0.5865
	$^4I_{13/2}$	0.836	80.17	6.7	15	0.1514
	$^4I_{15/2}$	0.541	959.85	80.1	15	0.3178
Fluoro-	$^4I_{13/2}$ $^4I_{15/2}$	1.566	87.14	100	100	0.3171
phosphate	$^4I_{11/2}$ $^4I_{13/2}$	2.510	23.49	22.11	30	1.8807
glass	$^4I_{15/2}$	0.964	82.75	77.89	30	0.1441
$n_D=1.4804$	$^4S_{3/2}$ $^4I_{9/2}$	1.669	29.92	3.27	50	0.2810
	$^4I_{11/2}$	1.234	85.30	9.32	25	0.4788
	$^4I_{13/2}$	0.827	63.52	6.94	15	0.1199
	$^4I_{15/2}$	0.547	736.22	80.46	15	0.2544
Phosphate	$^4I_{13/2}$ $^4I_{15/2}$	1.522	109.62	100	100	0.3427
glass	$^4I_{11/2}$ $^4I_{13/2}$	2.648	24.44	18.85	30	2.1585
$n_D=1.5688$	$^4I_{15/2}$	0.987	105.22	81.15	30	0.1794
	$^4S_{3/2}$ $^4I_{9/2}$	1.680	34.53	3.14	50	0.2695
	$^4I_{11/2}$	1.200	108.41	9.87	25	0.5177
	$^4I_{13/2}$	0.841	74.32	6.77	15	0.1336
	$^4I_{15/2}$	0.546	881.26	80.22	15	0.2814

$Er_xY_{1-x}Al_3(BO_3)_4$ crystals have been determined by using absorption spectra and Judd-Ofelt theory [30–32]. The calculation results are listed in Table 3.8. The Ω_λ values of Er^{3+} ions in inorganic glasses is given in Table 3.9. In comparison with the data shown in Table 3.8 it can be seen that the Ω_2 value in glassy state is much higher than that in crystalline state, this is due to low symmetry of Er^{3+} sites and high covalence of chemical bond in glassy state. According to Eq. 3.2, the radiative transition probability W_r can be calculated. Tables 3.10, 3.11 show the luminescence properties of Er^{3+} ions in different glass and crystal hosts. Similar calculation of radiative transition parameters of Ho^{3+} ions in $Gd_3Ga_5O_{12}$ has been performed [33], the results are summarized in Table 3.10.

3.3   Nonradiative Transition

For rare-earth ions in dielectric solids there exist two types of nonradiative transition processes: one is the interaction between the rare-earth ions, the other is that between the rare-earth ions and the hosts.

3.3.1   Interaction between Rare-earth Ions

The nonradiative transition caused by interaction among rare-earth ions is a process of resonant energy transfer or nonresonant energy transfer (phonon-assisted), which were investigated thoroughly both in theory and experiment [14, 36]. The interactions of same rare-earth ions bring about the intraionic energy transfer and concentration quenching effect (cross relaxation), while those of different rare-earth ions cause the dopant quenching and sensitizing.

1) *Intraionic energy transfer.* Cross relaxation and energy migration are an interesting case of energy transfer among ions in solids. Since the ions involved are of the same type, it presents features that are different than those of the energy transfer among different ions. The enhancement of IR emission of Er^{3+}, Ho^{3+} and Tm^{3+} by concentration effect is a pronounced evidence of intraionic energy transfer.

The emission in the middle IR region are accomplished by transition at lower energy levels. The photon energy is excited to the upper energy levels by absorption and then goes down by cascade nonradiative energy transfer between intraionic energy levels. It can be performed by ion-ion interaction. The energy levels and possible emission transitions of Er^{3+}, Ho^{3+} and Tm^{3+} are shown in Fig. 3.12. The emissions take place both in the visible and the IR regions. Figure 3.13 shows the effect of concentration of Er^{3+} ions in fluoride glasses on the luminescence intensity from the energy levels indicated in the figure to $^4I_{15/2}$. It can be seen from the figure, the lower the energy level, the higher the Er^{3+} ion concentration with strongest luminescence intensity. So the order of optimal concentration is $^4I_{13/2} \geq {}^4I_{11/2} > {}^4F_{9/2} > {}^4S_{3/2} > {}^2H_{11/2}$. It is shown that infrared luminescence of lower energy levels ($^4I_{11/2}$, $^4I_{13/2}$) can be enhanced by increasing Er^{3+} ion concentration. The similar results can also be observed in concentration effect of luminescence of Ho^{3+} ion doped fluoride glass. Figure 3.14 shows the effect of Ho^{3+} ion concentration on the luminescence intensity from excited states indicated in the figure to $^5I_8$. It can be seen that with increasing of Ho^{3+} ion concentration the luminescence intensity from $^5F_4 + {}^5S_2$ reduced monotonically, on the other hand, the intensities from $^5I_6$ and $^5I_7$ increase in the range of ion concentration lower than 8×10^{20} cm^{-3} [26, 27].

The RE ion-ion interaction in fluoride glass is characterized by electric dipole-dipole interaction. The experimental results indicate that nonradiative transition probabilities of $^2H_{11/2} + {}^4S_{3/2}$ states of Er^{3+} ions depend linearly on the square of ion concentration N_0^2 at different temperature [28].

G. Armagon *et al.* studied the excited state dynamics of Tm^{3+} ions in YAG crystal, and confirmed that the energy transfer from excited energy levels $^3H_4$,

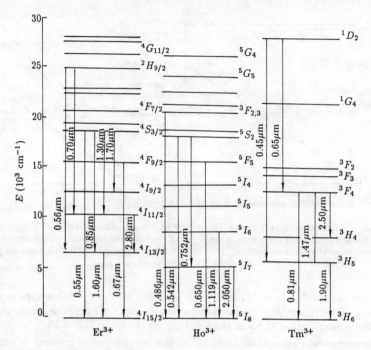

Fig. 3.12. Energy levels and emission transitions of Er^{3+}, Ho^{3+} and Tm^{3+} in fluoride glass.

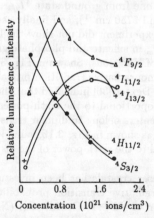

Fig. 3.13. Effect of Er^{3+} ion concentration on the luminescence intensty.

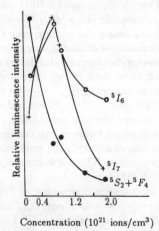

Fig. 3.14. Effect of Ho^{3+} ion concentration on the luminescence intensity.

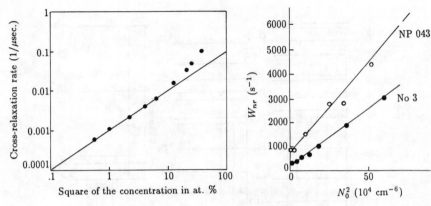

Fig. 3.15. Cross-relaxation rate among Tm ions in YAG as a function of the square of the Tm concentration at room temperature.

Fig. 3.16. Concentration dependence N_0^2 of nonradiative transition probability W_{nr} of Nd-doped glasses. NP 043—phosphate glass; No 3—silicate glass.

$^3H_6$ to the metastable level $^3F_4$ of 1.8 μm emission of Tm^{3+}:YAG is due to cross relaxation. The cross relaxation rate among Tm ions in YAG depends on the square of the Tm concentration at room temperature, as shown in Fig. 3.15.

2) *Concentration quenching.* Much effort was devoted to the research of concentration quenching of Nd doped glasses in the early 1970s [38–40]. With Nd^{3+} ion as an example, interactions of Nd^{3+} ions can be considered as that of an excited Nd^{3+} ion (1) with an unexcited Nd^{3+} ion (2) in the vicinity, resulting in the transition from metastable state $^4F_{3/2}-^4I_{15/2}$, and another one from ground state $^4I_{9/2}-^4I_{15/2}$ which has nearly the same energy (5730 cm^{-1} and 5750 cm^{-1}), and finally causing the nonradiative resonant energy transfer. Our experiment did not show the linear relationship between the luminescence lifetime τ_m in silicate and phosphate glasses and Nd_2O_3 content or volume concentration N_0 of Nd^{3+} ions. Shown in Fig. 3.16 is the linear change of $\overline{W}_{nr}$ vs N_0^2 by converting to nonradiative transition probability $\overline{W}_{nr}(Nd^{3+}-Nd^{3+})$. The theoretical calculation of Dexter [36] indicated that the resonant energy transfer probability is inversely proportional to the sixth power of d — the distance between two centers if the two centers belong to dipolar transition. It is in agreement with the experimental results, as shown in Fig. 3.16, value $\overline{W}_{nr}$ is proportional to N_0^2, so it is inversely proportional to the sixth power of the distance ($Nd^{3+}-Nd^{3+}$).

Concentration quenching of Nd^{3+} ions in laser crystals has been investigated extensively [41–43]. Figure 3.17 shows the nonradiative transition probability (in logarithm scale) with the Nd^{3+} ion concentration of Nd^{3+}:YAG, NdP_5O_{14} (NdPP) and $LiNdP_4O_{12}$ (LNP) crystals, the linear relation can also be observed.

Our recent work on high concentration doped crystals and glasses indicate the

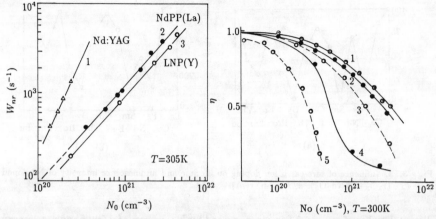

Fig. 3.17. Relationship between nonradiative transition probability W_{nr} and Nd^{3+} concentration N_0 of Nd-doped crystals.

Fig. 3.18. Concentration dependence of quantum efficiency of Nd-doped crystals and glasses. 1, $Nd_x La_{1-x} P_5 O_{14}$ crystal; 2, $LiLa_{1-x} Nd_x P_4 O_{12}$ crystal; 3, $La_{1-x} Nd_x P_3 O_9$ crystal; 4, $LiLa_{1-x} Nd_x P_4 O_{12}$ glass; 5, $Ba(PO_3)_2 \cdot xNd_2 O_3$ glass.

effect of hosts on concentration quenching of Nd^{3+} as shown in Fig. 3.18. In Nd doped phosphate glasses and crystals, with the content of $P_2 O_5$ in hosts decreases in the order of $(P_5 O_{14})^{-3}$, $(P_4 O_{12})^{-4}$, $(P_3 O_9)^{-3}$, $(PO_3)^{-}$, concentration quenching effect of Nd^{3+} increases greatly. This may account for gradual cracking of structural chains formed by P–O which has less and less isolation to Nd^{3+} ions [35].

Nd glasses are usually codoped with rare-earth ions. Shown in Fig. 3.19 are our experimental result, i.e. the influence of rare-earth ions doping on lifetime and luminescence intensity of Nd-doped glasses. Luminescence intensity changes in correspondence with lifetime, indicating the nonradiative transition of metastable state $^4 F_{3/2}$ of Nd^{3+}. The preliminary condition for resonant energy transfer is the energy gap ΔE between transition energy levels of two ions. For rare-earth ions La^{3+}, Ce^{3+}, Gd^{3+}, Lu^{3+} there are no energy levels of $\Delta E < 11000$ cm^{-1} ($^4 F_{3/2}$ of Nd^{3+} is at $\Delta E = 11400$ cm^{-1}), so the energy transfer is not possible from the above ions to Nd^{3+} ion, causing no effect on τ and I/I_0 of Nd^{3+}. For ions as Sm^{3+}, Dy^{3+}, Pr^{3+} and Yb^{3+} there are many corresponding energy level transitions, resonant energy transfer occurs more often. For example, energy of Nd^{3+}–$^4 F_{3/2}$ may transfer to enengy levels of $^1 G_4(Pr^{3+})$, $^4 F_{11/2,9/2}(Sm^{3+})$, $^4 F_{7/2}$, $^6 H_{5/2}(Dy^{3+})$ and $^2 F_{3/2}(Yb^{3+})$ thus resulting in a serious quenching effect on the luminescence lifetime and intensity of Nd-doped glass [40].

During resonant energy transfer, energy transfer by phonon assisted may also exist in the case of incomplete matching of gaps of energy levels. In this case, the

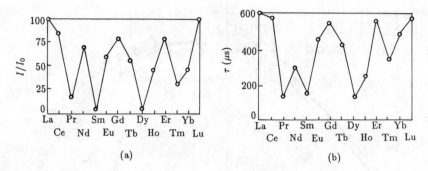

Fig. 3.19. Influence of rare-earth ion doping on lifetime and luminescence intensity of Nd-doped glass $(3\%Tr_2O_3+3\%Nd_2O_3)$. a, I/I_0; b, $\tau(\mu s)$.

dismatch value of energy $\Delta\delta$ is compensated by $\hbar\omega$, energy of a few phonons, energy transfer between Yb^{3+} and Nd^{3+} in phosphate glass is the exact example [44].

3) *Sensitized fluorescence*. Energy absorbed by sensitized ions can be transferred to fluorescence ions by means of resonant energy transfer. To illustrate sensitization process, ion pairs of sensitized fluorescence in solids are listed in Table 3.12. For example in phosphate glasses, Ce^{3+} has strong effects on the sensitized fluorescence of Tb^{3+}. As shown in Table 3.13, luminescence intensity of codoped $(Ce^{3+}+Tb^{3+})$ glass is tens of times that of single doped (Tb^{3+}) glass at the excitation wavelength of 312 nm. Measurement of the luminescence lifetime shows that in three glasses doped with different concentrations of Ce^{3+}, the luminescence lifetime of Ce^{3+} is 10 ns shorter than that of Ce^{3+} doped glass (the latter has the luminescence lifetime of 50 ns), the luminescence lifetime of Tb^{3+} $(^5D_4-^7F_5)$ is about 2.1 ms, (the same as that of Tb^{3+} doped glass). So, energy transfer from Ce^{3+} to Tb^{3+} is mainly a nonradiative resonant process [45, 46].

We have measured values of the luminescence intensity of donor ions (Ce^{3+}) with and without the existence of acceptor ions, I_d and I_d^0, and also calculated values of energy transfer efficiency η_t and probability P_{da} which are listed in Table 3.13. With the increment of Ce^{3+} concentration, ion gap of $Ce^{3+}-Tb^{3+}$ becomes shorter. Energy transfer probability P_{Tba} changes in proportion to the concentration of Ce^{3+}, $(N_{Ce}+N_{Tb})^2$ and I_{Tb}^{3+}, i.e. $P_{da}\propto\frac{1}{\gamma_{da}^6}$, it indicates that resonant energy transfer of $Ce^{3+}-Tb^{3+}$ is mainly a dipole interaction.

It has been noted that in codoped $(Tb^{3+}+Ce^{3+})$ glasses, although energy transfer propability P_{da} increases with the increment of Ce^{3+} concentration, luminescence intensity of Tb^{3+} does not increase with the increment of Ce^{3+} ion concentration, this is due to the strengthening of $Ce^{3+}-Ce^{3+}$ ion interaction when Ce^{3+} ion concentration increases, causing concentration quenching effect.

Another example for sensitized luminescence is the Nd–Yb couple [44]. The energy transfer for Nd–Yb is shown in the time-resolved fluorescence spectra and

Table 3.12. Ion pairs of sensitized fluorescence in glasses.

Fluorescence ions	Sensitized ions
Nd^{3+}	UO_2^{2+}, Mn^{2+}, Ce^{3+}, Tb^{3+}, Eu^{3+}, Cr^{3+}
Tb^{3+}	Dy^{3+}, Gd^{3+}, Ce^{3+}, Cu^+
Yb^{3+}	Nd^{3+}, Ce^{3+}, Cr^{3+}, $Ce^{3+}+Nd^{3+}$, $UO_2^{2+}+Nd^{3+}$
Ho^{3+}	Yb^{3+}, Er^{3+}, Tm^{3+}, Cr^{3+}
Er^{3+}	Yb^{3+}
Tm^{3+}	Yb^{3+}, Er^{3+}, Cr^{3+}

Table 3.13. Luminescence intensity, energy transfer efficiency and probability of Tb^{3+} doped glass at different Ce^{3+} concentration.

Concen-tration	wt%	$^5D_4\rightarrow{}^7F_6$	$^5D_4\rightarrow{}^7F_5$		$^5D_4\rightarrow{}^7F_4$						
Ce_2O_3	Tb_2O_3	I_{487}	I_{542}	I_{548}	I_{584}	I_{594}	γ_{da} (nm)	$P_{da}\cdot10^6$ (s^{-1})	η_t	I_d	I_d^0	τ_d (ns)
1	8	19.1	99.0	100	13.0	8.8	1.02	54.5	0.72	384	1378	47
3	8	15.6	80.8	80	11.1	7.4	0.95	61.8	0.73	316	1166	44
5	8	15.6	81.8	79.3	11.1	7.4	0.89	97.6	0.80	218	1101	41.5
0	8	0	1.7	1.7	0.3	0.4						

verified with fluorescence lifetime measurement. After the excitation within 50 μs, a majority of Nd^{3+} ions are at the $^4F_{3/2}(Nd^{3+})$ metastable state. Then, as a result of the gradual energy transfer into $Yb^{3+}(^2F_{5/2})$ state, the Yb^{3+} fluorescence (peak wavelength 974 nm) correspondingly increases. After 500 μs, the fluorescence is chiefly produced due to the $Yb^{3+}(^2F_{5/2})$ state. The gradual shortening of the lifetime of Nd^{3+} for codoped Nd^{3+}, Yb^{3+} glass magnifests the energy transfer for $Nd^{3+}\rightarrow Yb^{3+}$ (Fig. 3.20). As shown in Fig. 3.21, above 350 K, τ_{Yb} in codoped glass clearly decreases with the rise in temperature. This indicates the existance of a reversed transfer process for Yb^{3+} and Nd^{3+}. By calculation the transfer rate for Nd^{3+}, $^4F_{3/2}\rightarrow Yb^{3+}$, $^2F_{5/2}$ is 5×10^3 s^{-1} and the corresponding energy interval $\Delta\nu\sim1130$ cm^{-1}, while the rate for Yb^{3+}, $^2F_{5/2}\rightarrow Nd^{3+}$, $^4I_{15/2}$ is 75 s^{-1} and $\Delta\nu\sim4168$ cm^{-1}. The stretching vibration of O–P–O of phosphate glass is 1100 cm^{-1}, therefore, the energy transfer is characterized by phonon assistant one (single and four-phonon assisted transfer).

A. Kaminski summarized the more common types of sensitization in laser crystals illustrated in Fig. 3.22, and listed in Table 3.14 [3]. Sensitized luminescence in Nd-doped laser crystals is quite important for improving the pumping efficiency. The sensitizers must have broad absorption bands compatible with the emission spectra of the available pump sources, Cr^{3+} ions make it an ideal candidate for this purpose (Fig. 3.22d). However in Nd:YAG codoped with Cr^{3+}, the spectral overlap between the Cr^{3+} emission and Nd^{3+} absorption is poor, so sensitized pumping

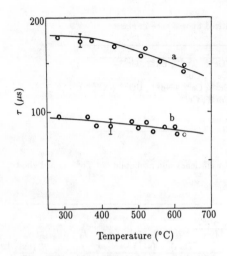

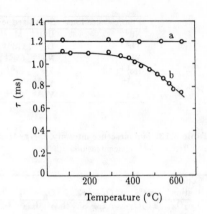

Fig. 3.20. Fluorescence lifetime τ_{Nd} for Nd^{3+} ($^4F_{3/2}$) state versus temperature. a, τ_{Nd} of the single-component doped with Nd^{3+}; b, τ_{Nd} of the two-component doped with Nd^{3+} and Yb^{3+}.

Fig. 3.21. Fluorescence lifetime τ_{Yb}. a, τ_{Yb} of the single-component doped with Yb^{3+}; b, τ_{Yb} of the two-component doped with Nd^{3+} and Yb^{3+}.

is not effective [47]. If the garnet host is modified to change the local crystal field strength at the site of Cr^{3+} ions, the effective spectral overlap and thus the efficiency of energy transfer can be greatly improved [48]. The Nd,Cr:GSGG laser crystal is a good example, and the attempt of exploring new crystal hosts has been going on [49].

3.3.2   Interaction between Rare-earth Ion and Host

At present, the nonradiative transition process is considered as a multiphonon relaxation process. Nonradiative transition probability $\overline{W}_{nr}$ of multiphonon can be described as in Eq. 1.15. The nonradiative transition probability $\overline{W}_{nr}$ and the quantum efficiency can be obtained by measuring fluorescence lifetime of $^4F_{3/2}$:

$$\overline{W}_{nr} = \frac{1}{\tau_m} - \frac{1}{\tau_r}, \tag{3.7}$$

where τ_r is the radiation lifetime, $\tau_r = 1/\overline{W}_{nr}$, which can be calculated or measured as described above.

A. Kaminski studied the multiphonon nonradiative transitions of rare-earth ions in YAG and LAG ($Lu_3Al_5O_{12}$) in detail [50]. The nonradiative transition probability $\overline{W}_{nr}$ is proportional to energy gap ΔE of transition between the two energy levels, as shown in Fig. 3.23, it is good in agreement with Eq. 1.15. The slope of the

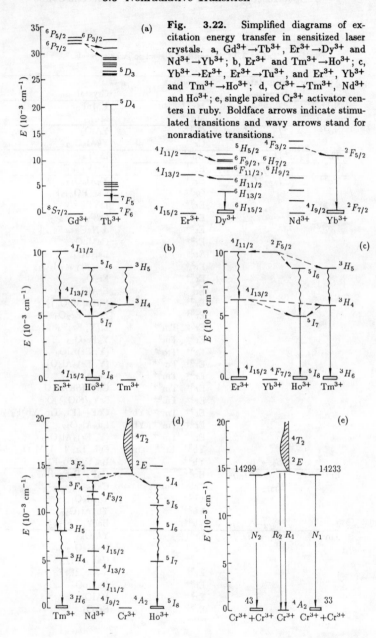

Fig. 3.22. Simplified diagrams of excitation energy transfer in sensitized laser crystals. a, $Gd^{3+} \rightarrow Tb^{3+}$, $Er^{3+} \rightarrow Dy^{3+}$ and $Nd^{3+} \rightarrow Yb^{3+}$; b, Er^{3+} and $Tm^{3+} \rightarrow Ho^{3+}$; c, $Yb^{3+} \rightarrow Er^{3+}$, $Er^{3+} \rightarrow Tu^{3+}$, and Er^{3+}, Yb^{3+} and $Tm^{3+} \rightarrow Ho^{3+}$; d, $Cr^{3+} \rightarrow Tm^{3+}$, Nd^{3+} and Ho^{3+}; e, single paired Cr^{3+} activator centers in ruby. Boldface arrows indicate stimulated transitions and wavy arrows stand for nonradiative transitions.

Table 3.14. Sensitized dielectric laser crystals.

Activator ion	Stimulated transition	Sensitizer ion	Crystal
Ni^{2+}	$^3T_2 \rightarrow ^3A_2$	Mn^{2+}	$KMnF_3$
		Mn^{2+}	MnF_2
Nd^{3+}	$^4F_{3/2} \rightarrow ^4I_{11/2}$	Cr^{3+}	$YAlO_3$
		Cr^{3+}	$Y_3Al_5O_{12}$
Tb^{3+}	$^5D_4 \rightarrow ^7F_6$	Gd^{3+}	$LiYF_4$
Dy^{3+}	$^6H_{11/2} \rightarrow ^6H_{15/2}$	Er^{3+}	$Ba(Y_{1.26}Er_{0.7})F_8$
Ho^{3+}	$^5I_7 \rightarrow ^5I_8$	Cr^{3+}	$Y_3Al_5O_{12}$
		Cr^{3+}	$Ca_5(PO_4)_3F$
		Er^{3+}	$LiYF_4$
		Er^{3+}	CaF_2–ErF_3
		Er^{3+}	α-$NaCaErF_6$
		Er^{3+}	$CaMoO_4$
		Er^{3+}	$LaNbO_4$
		Er^{3+}	$Y_{1.5}Er_{1.48}Al_5O_{12}$
		Er^{3+}	Er_2O_3
		Er^{3+}	$ErAlO_3$
		Er^{3+}	$NaLa(MoO_4)_2$
		Er^{3+}	ZrO_2–Er_2O_3
		Er^{3+}, Tm^{3+}	$Li(Y, Er)F_4$
		Er^{3+}, Tm^{3+}	$Y_3Fe_5O_{12}$
		Er^{3+}, Tm^{3+}	$(Y, Er)AlO_3$
		Er^{3+}, Tm^{3+}	$(Y, Er)_3Al_5O_{12}$
		Er^{3+}, Tm^{3+}	$Lu_3Al_5O_{12}$
		Er^{3+}, Tm^{3+}	$CaY_4(SiO_4)_3O$
		Er^{3+}, Tm^{3+}	$SrY_4(SiO_4)_3O$
		$Er^{3+}, Tm^{3+}, Yb^{3+}$	CaF_2–$(Er, Tm, Yb)F_3$
		$Er^{3+}, Tm^{3+}, Yb^{3+}$	$Lu_3Al_5O_{12}$
		$Er^{3+}, Tm^{3+}, Yb^{3+}$	$(Y, Er)_3Al_5O_{12}$
		$Yb^{3+}, Er^{3+}, Tm^{3+}$	$(Yb, Er, Tm)_3Al_5O_{12}$
	$^5S_2 \rightarrow ^5I_8$	Yb^{3+}	$Ba(Y, Yb)_2F_8^*$
Er^{3+}	$^4S_{3/2} \rightarrow ^4I_{13/2}$	Ho^{3+}	CaF_2–ErF_3–HoF_3
	$^4I_{13/2} \rightarrow I_{15/2}$	Color centers	CaF_2
		Yb^{3+}	$Y_3Al_5O_{12}$
		Yb^{3+}	$Yb_3Al_5O_{12}$
	$^4F_{9/2} \rightarrow ^4I_{15/2}$	Yb^{3+}	$Ba(Y, Yb)_2F_8^*$
Tm^{3+}	$^3H_4 \rightarrow ^3H_6$	Cr^{3+}	$YAlO_3$
		Cr^{3+}	$Y_3Al_5O_{12}$
		Er^{3+}	CaF_2–ErF_3
		Er^{3+}	α-$NaCaErF_6$
		Er^{3+}	$CaMoO_4$
		Er^{3+}	$YAlO_3$
		Er^{3+}	Er_2O_3

Table 3.14. (Continued)

Activator ion	Stimulated transition	Sensitizer ion	Crystal
		Er^{3+}	$ErAlO_3$
		Er^{3+}	$Er_3Al_5O_{12}$
		Er^{3+}	$Y_{1.5}Er_{1.48}Al_5O_{12}$
		Er^{3+}	$ZrO_2-Er_2O_3$
		Er^{3+}, Yb^{3+}	$(Yb, Er)_3Al_5O_{12}$
Yb^{3+}	$^2F_{5/2}\rightarrow{}^2F_{7/2}$	Nd^{3+}	CaF_2
		Nd^{3+}	$Y_3Al_5O_{12}$
		Nd^{3+}	$Lu_3Al_5O_{12}$
		Cr^{3+}, Nd^{3+}	$Y_3Al_5O_{12}$

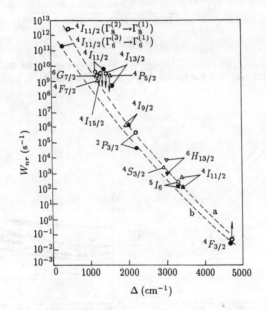

Fig. 3.23. Nonradiative transition probability of RE^{3+} ions in $Y_3Al_5O_{12}$ (a) and $YAlO_3$ (b) crystals as a function of energy gap ΔE.

straight lines depends on the phonon energy of hosts, Figs. 3.24 and 3.25 show the multiphonon emission rates from excited states of trivalent rare-earth ions as a function of energy gap to the next-lower level in different host crystals and glasses. The dominant phonon energies are shown in parentheses [14, 29].

We have systematically studied the effect of constitution and structure of inorganic glass hosts on the nonradiative transition process of Nd^{3+} ion, and measured

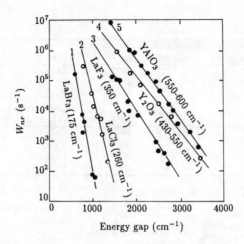

Fig. 3.24. Nonradiative transition probability W_{nr} vs energy gap ΔE for RE^{3+} ions in various crystals. 1, $LaBr_3$ (175 cm^{-1}); 2, $LaCl_3$ (260 cm^{-1}); 3, LaF_3 (350 cm^{-1}); 4, Y_2O_5 (430–550 cm^{-1}); 5, $YAlO_3$ (550–600 cm^{-1}).

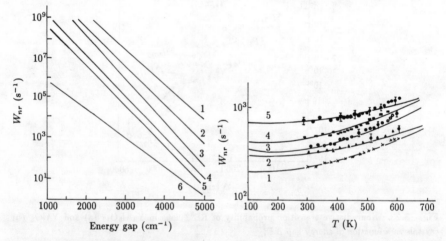

Fig. 3.25. Nonradiative transition probability W_{nr} vs energy gap ΔE for RE^{3+} ions in various glasses. 1, Borate glass (1340 cm^{-1}); 2, phosphate glasses (1290 cm^{-1}); 3, silicate glass (1080 cm^{-1}); 4, germanate glass (810 cm^{-1}); 5, tellurite glass (730 cm^{-1}); 6, fluoride glass (580 cm^{-1}).

Fig. 3.26. Temperature dependence of nonradiative decay process of Nd^{3+} ion in inorganic glasses. 1, Silicate glass; 2, fluoroberyllate glass; 3, fluorophosphate glass; 4, germanate glass; 5, phosphate glass.

Table 3.15. Luminescence properties of Nd-doped glasses.

No.	System	τ_m (μs)	$W_r 1.06$ (s^{-1}) m	c	$\sum W_r$ (s^{-1}) m	c	W_{nr} (s^{-1}) m	c	η m	c	$\sigma_p^{1.06}$ (10^{-20} cm^2) m	c
1	Borate	90	1220	1280	2420	2670	8700	8450	0.22	0.24	2.4	2.5
2	Silicate	510	860	910	1790	1850	170	110	0.91	0.94	1.9	2.0
3	Germanate	370	1150	1120	2430	2360	280	350	0.90	0.87	2.5	2.4
4	Tellurate	220	2910	2570	5950	5270	$\sim$0	$\sim$0	$\sim$1	$\sim$1	5.1	4.5
5	Phosphate	360	1410	1240	2830	2360	$\sim$0	420	$\sim$1	0.85	4.5	3.9
6	Aluminate	200	1730	2160	4030	4650	970	360	0.81	0.93	2.9	3.6
7	Alumino-silicate	200	1170	1500	2700	3120	2300	1880	0.54	0.62	1.9	2.4
8	Gallo-silicate	350	1150	1260	2300	2580	560	280	0.80	0.90	1.8	2.0
9	Fluorophosphate	475	1010	890	1890	1810	220	290	0.90	0.86	2.5	2.2
10	Fluorophosphate	540	860	820	1670	1620	230	290	0.90	0.86	2.3	2.4
11	Fluoroberyllate	650	700	700	1300	1290	230	250	0.85	0.84	2.2	2.2

m—measured, c—calculated.

Table 3.16. Numbers and energy of phonon in multi-phonon relaxation for Tm^{3+} and Nd^{3+} in inorganic glasses.

Glass system	Tm^{3+} ion $^3F_4$–$^3H_5$ Numbers of phonon	Energy of phonon (cm^{-1})	V_R (cm^{-1})	Nd^{3+} ion $^4F_{3/2}$–$^4I_{5/2}$ Numbers of phonon	Energy of phonon (cm^{-1})	V_{IR} (cm^{-1})
Phosphate	4	1100	1320	5	1060	1300
Silicate	4	1000	1100	4+1	1100+900	1100
Germanate	5	825	890	6	870	820
Tellurate	5	760	780	—	—	720
Fluorophosphate	—	—	—	2+4	1050+800	1100
Fluoride	—	—	—	6	900	800

the luminescence properties of Nd-doped glasses in 11 different systems, as shown in Table 3.15 [51]. It can be seen that the glass host has a very strong influence on the nonradiative transition probability $\overline{W}_{nr}$, which can be varied by 2–3 orders of magnitudes with the variation of the glass constitution. We have also measured the temperature dependence of nonradiative transition of $^4F_{3/2}$–$^4I_{15/2}$ of Nd^{3+}. As shown in Fig. 3.26, the variation curves of nonradiative transition probability depending on the temperature and the phonon order was simulated based on Eq. 1.15. Table 3.16 shows that the phonon energy is not in correspondence with the peak of IR spectrum although we employed two kinds of phonons to simulate the multi-phonon quenching process. Listed in the table are also the phonon energy, phonon order and Raman oscillation frequency in the multi-phonon relaxation process obtained by Layne from Tm^{3+} in four types of oxide glass [52]. The nonradiative energy transfer process in inorganic Nd-doped glasses can not be well explained using multi-phonon relaxation theory because the phonon mode is not easily defined

Table 3.17. Values of W_r, β, τ_r, τ_m and W_{nr} for Er^{3+}, Tm^{3+} and Tb^{3+} in various glasses.

	Transition levels	λ (μm)	W_r (s^{-1})	β	$\sum W_r$ (s^{-1})	τ_r (μs)	τ_m (μs)	W_{nr} (s^{-1})	$\Delta\lambda$ (nm) ($^5D_4 \to {}^7F_5$)	σ ($^5D_4 \to {}^7F_5$) ($\times 10^{-71}$) (cm^2)
Phosphate (Er^{3+}) $^4S_{3/2}$	$\to {}^4I_{15/2}$	0.545	751.5	0.664	1132.6	133	7.2	1.31 $\times$ 10^5		
	$\to {}^4I_{13/2}$	0.834	319.5	0.282						
	$\to {}^4I_{11/2}$	1.219	24.5	0.022						
	$\to {}^4I_{9/2}$	1.695	37.2	0.033						
Germanate (Er^{3+}) $^4S_{3/2}$	$\to {}^4I_{15/2}$	0.545	1026.3	0.647	1521.4	109	11	8.18 $\times$ 10^4		
	$\to {}^4I_{13/2}$	0.843	422.5	0.278						
	$\to {}^4I_{11/2}$	1.229	31.4	0.020						
	$\to {}^4I_{9/2}$	1.712	41.2	0.027						
Tellurate (Er^{3+}) $^4S_{3/2}$	$\to {}^4I_{15/2}$	0.543	1804.3	0.668	2699.3	62.9	15	5.10 $\times$ 10^4		
	$\to {}^4I_{13/2}$	0.841	739.8	0.274						
	$\to {}^4I_{11/2}$	1.224	58.5	0.022						
	$\to {}^4I_{9/2}$	1.698	96.7	0.036						
Phosphate (Tm^{3+}) $^1D_2$	$\to {}^3H_6$	0.358	4062.6	0.233	17468.4	57.2	17	4.14$\times$10^4		
	$\to {}^3H_4$	0.451	10850.4	0.621						
	$\to {}^3H_5$	0.508	80.0	0.005						
	$\to {}^3F_4$	0.655	1167.3	0.067						
	$\to {}^3F_3$	0.757	691.1	0.039						
	$\to {}^3F_2$	0.786	496.2	0.028						
	$\to {}^3G_4$	1.473	121.8	0.007						
Tellurate (Tm^{3+}) $^1D_2$	$\to {}^3H_6$	0.358	7324.4	0.215	34120	29.3	15.5	3.04$\times$10^4		
	$\to {}^3H_4$	0.453	21814.4	0.639						
	$\to {}^3H_5$	0.508	139.3	0.004						
	$\to {}^3F_4$	0.658	2238.0	0.066						
	$\to {}^3F_3$	0.756	1413.7	0.041						
	$\to {}^3F_2$	0.787	954.5	0.028						
	$\to {}^3G_4$	1.493	235.7	0.007						
Phosphate (Tb^{3+}) $^5D_4$	$\to {}^7F_6$	0.488	12.8	0.057	227.5	4400	2100	249	12.4	0.57
	$\to {}^7F_5$	0.542	144.9	0.635						
	$\to {}^7F_4$	0.583	19.9	0.089						
	$\to {}^7F_3$	0.621	19.7	0.087						
	$\to {}^7F_2$	0.647	7.2	0.032						
	$\to {}^7F_1$	0.664	8.6	0.038						
	$\to {}^7F_0$	0.676	13.5	0.060						

in the disorder glass structure (long range disorder and site disorder).

We have interpreted the influences of different host glasses on the probability of Nd^{3+} nonradiation transition by using interactions of the ligands (oxygen polyhedra) with the excited ions by the polarization effect. The stronger the polarization effect, the higher the nonradiation transition probability, which is discussed in the literature [40, 51] in detail.

Table 3.17 lists the probability of radiative and nonradiative transition of Er^{3+}, Tm^{3+} and Tb^{3+} in different host glasses [45].

3.4 Laser Selective Excited and Time Resolved Spectra

Absorption and emission of rare-earth ions mainly take place at the zero-phonon line because the effect of vibration state is less. So multi-state Stark splitting and different site condition could be clearly observed by using laser selective excitation with narrow linewidth. Most investigations are concentrated on the spectra of Eu^{3+}, Pr^{3+}, Sm^{3+} and Nd^{3+} ions in solids.

We have summarized the experimental results of the laser selective excitation spectra of Eu^{3+} and Nd^{3+} in inorganic glasses in reference [53]. Recently we studied the fluorescence line narrowing of Sm^{3+} in fluoride glass [54]. Figures 3.27 and 3.28 show the fluorescence spectra of transition $^4G_{5/2}-^6H_{5/2}$ of Sm^{3+} ions in ZBLAN glass under selective laser excitation and at different delay time after the exciting pulse respectively. The transition $^4G_{5/2}-^6H_{5/2}$ is a resonant one. The experimental results demonstrated the obvious narrowing effect in fluorescence transition $^4G_{5/2}-^6H_{5/2}$. A 2.5 cm^{-1} linewidth was achieved at 100 K. The gradually broadening of narrowed fluorescence line shows that energy transfer processes happen between Sm^{3+} ions at different sites.

From laser excited and time resolved spectra the energy transfer process can be analysed in detail. As an example, Figs. 3.29 and 3.30 show fluorescence spectra of transition $^4F_{3/2}-^4I_{9/2}$ of Nd^{3+} ions in $Nd_xY_{1-x}PO_4$ laser crystal with different delay times and different exciting laser wavelength respectively. It is shown that Nd^{3+} ions distributed at different sites in crystal lattice and energy transfer between ions at different sites take place by two-phonon assisted process. From experimental results, the energy transfer parameters, such as transfer rate, resonant energy, and diffusion coefficient, could be determined [55]. Similar experiments was carried out for neodymium phosphate glasses [56]. As shown in Figs. 3.31 and 3.32, the Nd ions located in various different local sites in NP-2 phosphate glass ($20Nd_2O_3 \cdot 80P_2O_5$) with N_D=4.2×10^{21} cm^{-1}. The energy transfer and diffusion between Nd^{3+} ions take place through forced electric dipole-dipole interaction. In the temperature range of 77~150 K, the transfer process involving a thermal activation energy of 330 cm^{-1} was found. By fitting experimental data with theory, several energy transfer parameters are also determined.

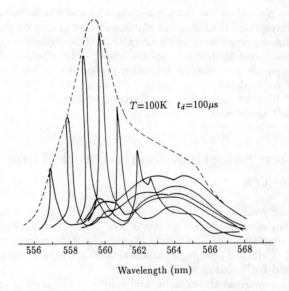

Fig. 3.27. Resonant fluorescence of transition $^4G_{5/2}-^6H_{5/2}$ of Sm^{3+} ions in ZBLAN glass at different exciting wavelength, the delay time was 100 μs (solid lines) and by wide band excitation (dashed line).

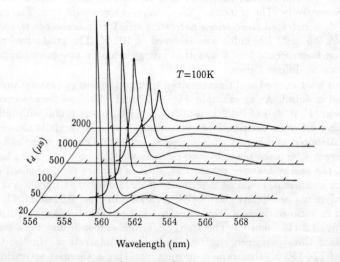

Fig. 3.28. Spectra of transition $^4G_{5/2}-^6H_{5/2}$ of Sm^{3+} ions in ZBLAN glass at different delay times after the exciting pulses.

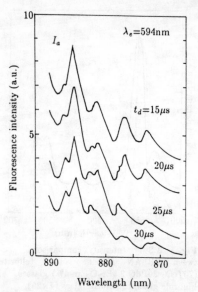

Fig. 3.29. Fluorescence spectra of $Nd_{0.03}Y_{0.97}PO_4$ crystal for transitions from $^4F_{3/2}$ to two lowest components of $^4I_{9/2}$ with different delay times t_d after laser pulse excitation for $\lambda_e = 594.0$ nm at 77 K.

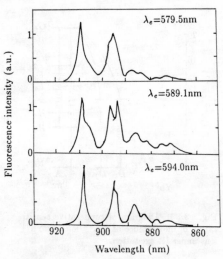

Fig. 3.30. $^4F_{3/2}-^4I_{9/2}$ transition fluorescence spectra of $Nd_{0.03}Y_{0.97}PO_4$ excited by three different laser narrow lines for $t_d = 10$ μs at 77 K.

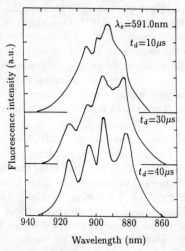

Fig. 3.31. Fluorescent spectra of NP-1 sample at different time t_D after laser pulse for $\lambda_e = 591$ nm at 77 K.

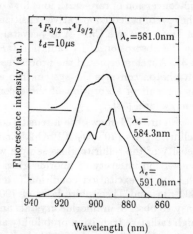

Fig. 3.32. Fluorescent spectra of NP-1 sample for three different laser narrow lines excitation at 77 K.

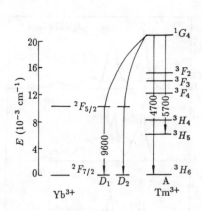

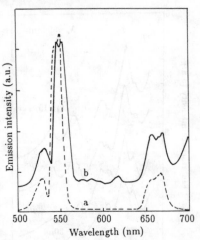

Fig. 3.33. Diagram of cooperative sensitization of fluorescence of Tm^{3+} ions by Yb^{3+} ions under IR excitation.

Fig. 3.34. Intensity comparison between ZBLAN fluoride and tellurite $(70TeO_2 \cdot 30PdO \cdot 2.5Er_2O_3$ mol%) glasses. a, Fluoride glass; b, tellurite glass ($I \times 25$);

3.5   Frequency Up-conversion

The frequency up-conversion is a multi-photon process, in which two or more IR photons are needed to generate one shorter wavelength photon. The phenomena of up-conversion of rare-earth ions first observed by P. Feofilov and V. Ovsyankin [57], they were called cooperative sensitization of fluorescence. The visible fluorescence of Tm^{3+} ions in a CaF_2–YbF_3 crystal in which the sensitizing Yb^{3+} ions are excited by the absorption of IR radiation ($^2F_{7/2} \rightarrow {}^2F_{5/2}$) as shown in energy diagram, Fig. 3.33. Auzel interpreted the process as a successive energy transfer whereby the first IR photon converts the system into an intermediate metastable state from which, as a result of absorption of the second photon, the system is excited to a higher energy level [58].

Figure 3.34 shows the intensity comparison between fluoride glass ZBLAN and tellurite glass ($70TeO_2 \cdot 30PbO \cdot 2.5Er_3O_3$) [29]. The up-conversion fluorescence intensity of the tellurite glass sample was multiplied by 25 times, therefore, the up-conversion intensity of fluoride glass is tens of times that of tellullite glass under the similar excitation condition (excited both by diode laser with wavelength of 804 nm amd Ti:sapphire laser of 790 nm). Therefore, the up-conversion effect is easy to observe in fluoride crystals and glasses. As we discussed above, it is due to high radiative transition probability and low multiphonon relaxation by the fluoride hosts rather than that of oxide hosts.

W. M. Yen demonstrated three photon absorption in the $^1S_0$ state of Pr^{3+} ions in LaF_3 [59]. The 3-photon sequence is illustrated in the inset in Fig. 3.35. Fluores-

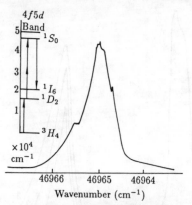

Fig. 3.35. Three photon excitation spectrum of the $^3H_4{\rightarrow}^1D_2{\rightarrow}^1S_0$ state of 1 at% Pr in LaF$_3$. Absorption sequence is shown in inset. Temperature is ~10 K.

cence can be observed from transition $^1S_0{\rightarrow}^1I_6$ at wavelength around 46965 cm^{-1}. The frequency up-conversion process takes place in one type of ions.

Recently, much attention have been paid in the heavy metal fluoride glasses (HMFG) which appear to be a favorable material for the frequency up-conversion. R. S. Quimby et al. [60] has studied the conversion of infrared (about 1 μm) to visible (0.55 μm) light in a series of fluoride glasses doped with YbF$_3$ and ErF$_3$, the frequency up-conversion efficiency of these bulk materials is four order of magnitude greater than that observed in doped oxide glasses, and compares very favorably with the results obtained for Yb^{3+}/Er^{3+}–containing crystals. In another work, D. C. Yeh et al. [61] have performed an efficient up-conversion emission at 805 nm in Tm^{3+}, Yb^{3+} codoped barium-thorium fluoride glass when excited at 973 nm. The efficiency for 805 nm up-conversion emission is 2×10^{-4} at RT for glass containing 0.05 mol% TmF$_3$ and 26 mol% YbF$_3$ when excited by 973 nm radiation with intensity of 16.5 mW/cm^2. For an absorbed power of 1 W/cm^2, a 1.2% up-conversion efficiency is predicted.

We have systematically studied the frequency up-conversion emission in Nd^{3+}:ZBLAN and Er^{3+}:ZBLAN glasses excited by Ti:sapphire laser at about 800 nm [62].

In Er:ZBLAN glass sample, the up-conversion emission at 550 nm is very strong when excited by Ti:sapphire laser at about 790 nm, especially in the sample with 5 mol% Er^{3+} doping. The measured spectrum of up-conversion emission has no obvious difference comparing with that in normal case but the mechanism is different. The fluorescence of up-conversion and the excitation spectra of the samples with the doping concentration 1 mol%, 2 mol% and 5 mol% Er^{3+} are measured and shown in Fig. 3.36. The peaks of all these excitation spectra are at 795 nm, but the widths have a little difference; the sample with the 1 mol% doping concentration

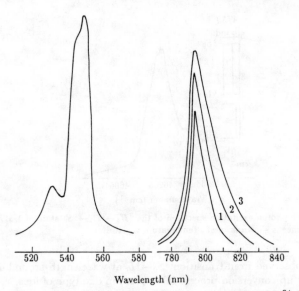

Fig. 3.36. Fluorescence and excitation spectra of up-conversion emission of Er^{3+}:ZBLAN glass. 1, 1 mol% Er:ZBLAN; 2, 1 mol% Er, 0.5 mol% Cr; 3, 5 mol% Er:ZBLAN.

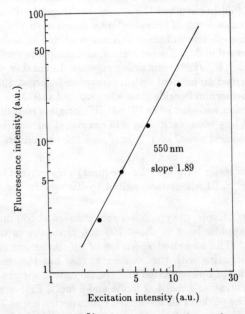

Fig. 3.37. Dependence of Nd^{3+} up-conversion emission on the excitation intensity.

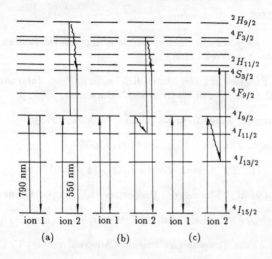

Fig. 3.38. Energy levels and excitation processes of Er^{3+} in ZBLAN glass.

has a full width at half maximum (FWHM) of 10 nm, while it is about 17 nm in 5 mol% doping sample. We regard this as the effect of inhomogeneous broadening of the absorption line and the higher efficiency of energy transfer in the high doping concentration sample.

It can be seen from Fig. 3.37 that the fluorescence intensity varied with the power of pumping laser source of Er^{3+}:ZBLAN glasses. The dependence of the fluorescence intensity on the power was approximately quadratic rather than linear, this quadratic dependence indicates that the two step excitation occurs. In Er^{3+} doped systems, the 550 nm up-conversion emission can be caused under 790 nm excitation, the possible mechanisms are shown in Fig. 3.38(a), (b) and (c). Through energy transfer between Er^{3+} ions, two Er^{3+} ions, ion 1 and ion 2 are initially be excited to $^4I_{9/2}$, ion 2 may decay to $^4I_{11/2}$ and $^4I_{13/2}$ through nonradiative transition processes, and then accept the energy transferred from ion 1, this will excite the ion 2 to upper states $^2H_{9/2}$, $^4F_{3/2}$ or $^2H_{11/2}$. The mismatches of energy of these energy transfer processes are 220 cm^{-1}, 40 cm^{-1} and 20 cm^{-1} respectively, and these parts of energy may be compensated by phonons. Considering from the energy level lifetimes and the energy mismatches, mechanisms shown in Fig. 3.38 (b) and (c) may be the domain mechanisms of the up-conversion emission when excited at 790 nm, especially in high doping concentration samples.

Recently S. Todoroki *et al.* observed strong up-conversion fluorescence in Er^{3+}-doped fluoride glasses containing a large amount of In and Pb. It was found that the lower phonon energy, generated the stronger up-conversion fluorescence. The phonon energy is associated with multiphonon relaxation of rare-earth ions [63].

References

1. G. H. Dieke, *Spectra and Energy Levels of Rare Earth Ions in Crystals* (Wiley-Interscience, New York, 1968).

2. Fuxi Gan, Progress in Laser Material Science, in *Proc. International Conference and School " Lasers and Applications"*, Bucharest, 1982, pp. 11–36.

3. A. A. Kaminski, *Laser Crystals* (Nauka, Moscow, 1975).

4. E. H. Carlson, *J. Chem. Phys.* **34** (1961) 1602.

5. K. Otsuka, *IEEE J. Quant. Electron.* **QE-11** (1975) 330.

6. Shunfu Liu *et al.*, "Fluorescent Properties of Nd-doped Oxide Crystals", 1974.

7. C. Brecher, L. A. Risebery and M. J. Weber, *Phys. Rev.* **B18** (1978) 5799.

8. Fuxi Gan, *Kexue Tongbao (Bulletin of Chinese Sciences)* **12** (1978) 723.

9. S. E. Stokowski and M. J. Weber, *Laser Glass Handbook*, Lawrence Livermore Lab., M-095 (1978, 1979).

10. F. Auzel, *IEEE J. Quant. Electron.* **QE-12** (1976) 256.

11. W. F. Krupke, *Phys. Rev.* **145** (1966) 325; *IEEE J. Quant. Electron.* **QE-7** (1973) 153.

12. R. D. Peacock, *Molecular Phys.* **25** (1973) 817.

13. Guanzhao Wu, Xiurong Zhang, *Chinese J. Lasers* **5** (1981) 12.

14. L. A. Risebery, M. J. Weber, *Progress in Optics* **14** (1976) 91.

15. B. Z. Malkin, Z. I. Ivanenko. I. B. Iazenberg, *Fiz-Tverdogo Tela* **12** (1970) 1873.

16. A. A. Kaminski, B. Z. Malkin, L. A. Bumagina, *Izv. Akad. Nauk. SSSR. Fizika* **46** (1982) 979.

17. A. A. Kaminski, V. A. Timofeeva, Agamalyan, A. B. Bykov, *Izv. Akad. Nauk. SSSR, Neorganicheskie Materialy* **17** (1981) 2278; *Kristallographiya* **27** (1982) 522.

18. A. A. Kaminski, *Izv. Akad. Nauk. SSSR, Neorganicheskie Materialy* **16** (1980) 1333.

19. A. A. Kaminski, *et al.*, *Phys. Stat. Sol.* **A86** (1984) 345.

20. B. R. Judd, *Phys. Rev.* **127** (1962) 750.

21. G. S. Ofelt, *J. Chem. Phys.* **37** (1962) 511.

22. 1992 ICF Annual Report of Lawrence Livermore National Laboratory (LLNL), UCRL-LR-105820-92, p. 55.

23. R. Reisfeld, G. Katz, C. Jacobini, M. G. Drexhage, R. N. Brown and C. K. Jorgensen, *J. Solid State Chem.* **48** (1983) 323; **41** (1982) 253.

24. J. Y. Allain, M. Moneric, H. Poignant, *Electron. Lett.* **26** (1990) 262.

25. R. Reisfeld, *J. Less-common, Met.* **93** (1983) 107.

26. Fuxi Gan and Haixing Zheng, *Chinese Physics Lett.* **2** (1985) 229.

27. Haixing Zheng, Guangzhao Wu and Fuxi Gan, *Acta Physica Sinica* **34** (1985) 1582.

28. Fuxi Gan and Haixing Zheng, *J. Non-Cryst. Solids* **95&96** (1987) 771.

29. Fuxi Gan, in *Proceedings "Ceramics: Toward the 21st Century"*, ed. N. Soga and A. Kato (Yokohama, Japan, 1992), p. 567.

30. M. J. Weber, *et al.*, *Phys. Rev.* **B8** (1973) 47.

31. Qingyuan Wang, *et al.*, *Acta Optica Sinica* **6** (1986) 307.

32. Mingguo Liu, *et al.*, *Acta Optica Sinica* **8** (1988) 1079.

33. Qingyuan Wang, *et al.*, *Chinese J. Lasers* **17** (1990) 31.

34. Changhong Qi and Fuxi Gan, *Chinese J. Lasers* **11** (1984) 648.

35. Changhong Qi and Fuxi Gan, *Chinese J. Lasers* **9** (1982) 692.

36. P. L. Dexter, *J. Chem. Phys.* **21** (1963) 876.

37. G. Armagan, A. M. Buoncristiani, B. Di Bartolo, *Opt. Mater.* **1** (1992) 11.

38. G. E. Peterson and P. M. Bidenbauch, *J. Opt. Soc. Amer.* **54** (1964) 644.

39. T. Komiyama, *J. Jpn. Ceram. Soc.* **82** (1974) 637.

40. Fuxi Gan, *Kexue Tongbao (Bulletin of Chinese Sciences)* **2** (1979) 59.

41. Xiurong Zhang, *Chinese J. Lasers* **17** (1990) 286.

42. Xiurong Zhang, Xiaoshan Ma, *et al.*, *Chinese J. Lasers* **17** (1990) 175.

43. M. M. Broer, D. L. Huber and W. M. Yen, *Phys. Rev. Lett.* **49** (1982) 394.

44. Shuchun Chen, Sen Mao and Fengmei Dai, *Acta Physica Sinica* **33** (1984) 515.

45. Changhong Qi and Fuxi Gan, *Lumin. Display Device* **5** (1985) 1.

46. Changhong Qi and Fuxi Gan, *J. Lumin.* **31&32** (1984) 339.

47. A. G. Avanesov, B. I. Denker, *et al.*, *Sov. J. Quant. Electron.* **12** (1982) 421.

48. V. G. Ostroumov, Yu. S. Privis, *et al.*. *J. Opt. Soc. Amer.* **B3** (1986) 81.

49. F. M. Hashmi, R. C. Powell and G. Boulon, *Opt. Mater.* **1** (1992) 281.

50. Yu. R. Perlin, A. A. Kaminskii, M. G. Blazha and V. N. Enakii, *Phys. Stat. Sol.* **B112** (1982) K125.

51. Fuxi Gan, Shuchun Chen, Hefang Hu, *Scientia Sinica (Science in China)* **3** (1981) 289.

52. C. B. Layne, M. J. Weber, *Phys. Rev.* **B16** (1977) 3259.

53. Fuxi Gan, *Optical and Spectroscopic Properties of Glasses* (Springer-Verlag, Berlin, 1992), pp. 194–200.

54. Fuxi Gan and Yihong Chen *J. Non-Cryst. Solids* **161** (1993) 282.

55. Shuchun Chen, *Acta Optica Sinica* **5** (1985) 785.

56. Shuchun Chen, Guanhong Yi, *et al.*, *Acta Optica Sinica* **4** (1984) 1074.

57. P. P. Feofilov and V. V. Ovsyankin, *Appl. Opt.* **6** (1967) 1828.

58. F. E. Auzel, *Proc. IEEE* **61** (1973) 758; *Phys. Rev.* **B13** (1976) 2809.

59. W. M. Yen, C. G. Levey, Shihua Huang and Shui T. Lai, *J. Lumin.* **24/25** (1981) 659.

60. R. S. Quimby, *Electron. Lett.* **23** (1987) 32.

61. P. C. Yeh, W. A. Sibly, M. J. Suscavage, *J. Appl. Phys.* **63** (1988) 4644.

62. Fuxi Gan and Yihong Chen, *Opt. Mater.* **2** (1993) 45.

63. S. Todoroki, K. Hirao and N. Soga, *J. Non-Cryst. Solids* **143** (1992) 46.

4. Laser Fundamentals

4.1 Spontaneous Emission and Stimulated Emission [1]

The interaction of radiation and matter can be considered as a perturbation problem in which the radiation field and the atomic system are assumed to be quantized. As we described in Section 1.2, when the perturbation theory is applied to the interaction of a quantized atomic system and the radiation field, the processes—absorption and spontaneous emission—can all be predicted, as expressed in Eqs. 1.11 and 1.13. The ratio of spontaneous transition probability to stimulated transition probability is shown in Eq. 1.14. Let us consider the stimulated transition between two levels 1, 2 in more detail.

The stimulated transition rate of $2 \rightarrow 1$ and $1 \rightarrow 2$ at radiation energy density $\rho(\nu)$ is as follows

$$(W'_{21})_i = B_{21}\rho(\nu), \tag{4.1}$$
$$(W'_{12})_i = B_{12}\rho(\nu). \tag{4.2}$$

The total transition rate from 2 to 1 involves stimulated and spontaneous transitions

$$W'_{21} = B_{21}\rho(\nu) + A. \tag{4.3}$$

Here
$$\rho(\nu) = \frac{8\pi n^3 h\nu^3}{c^3}\left(\frac{1}{e^{h\nu/kT} - 1}\right). \tag{4.4}$$

Considering the thermal radiation at temperature T in thermal equilibrium, Boltzmann relation between the two levels, we have

$$\frac{n_2}{n_1} = \frac{g_2}{g_1}e^{-\frac{h\nu_{12}}{kT}}, \tag{4.5}$$

here n_2 and n_1 is the population of the upper level and the lower level respectively. At given time interval, that is,

$$n_2 W'_{21} = n_1 W'_{12}. \tag{4.6}$$

109

From the above mentioned equations, we obtain

$$\frac{8n^3\pi h\nu^3}{c^3(e^{h\nu/kT-1})} = \frac{A(g_2/g_1)}{B_{12}e^{h\nu/kT} - B_{21}(g_2/g_1)}, \tag{4.7}$$

when
$$B_{12} = B_{21}\frac{g_2}{g_1}, \tag{4.8}$$

and
$$\frac{A}{B_{21}} = \frac{8\pi n^3 h\nu^3}{c^3}. \tag{4.9}$$

The stimulated transition rate can be written as

$$(W'_{21})_i = \frac{Ac^3}{8\pi n^3 h\nu^3}\rho(\nu) = \frac{c^3}{8\pi n^3 h\nu^3 \tau_R}\rho(\nu). \tag{4.10}$$

From Eqs. 4.4 and 4.9, the ratio of the rate of the spontaneous emission to that of the stimulated emission is given by

$$R = \frac{A}{B\rho(\nu)} = \frac{4\pi\nu^3 n^3}{C^3\rho(\nu)} = \exp(h\nu/kT - 1). \tag{4.11}$$

For the microwave region, $h\nu \ll kT$, the effect of spontaneous emission is negligible. But under the condition of thermal equilibrium, the proportion of stimulated emission to spontaneous emission is entirely negligible for transition in the optical region. It is clear that if we want to use stimulated emission in the optical region, the radiation density should be built up to such a value enormously exceeding the thermal equilibrium value. If the rate of stimulated emission exceeds that of the absorption, it should be $n_2 > n_1$. This condition is described as population inversion, which is sometimes referred to as a negative temperature, $T < 0$.

4.2   Oscillation Condition

For increasing the radiation energy density, a resonator cavity should be used. One of the simplest is the Fabry-Perot interferometer, with the active material lying between the parallel plates. The wave will gain in energy during each traverse between the two reflecting mirrors. The threshold condition for laser oscillation is given by

$$R_1 R_2 \exp(2Gl) = 1, \tag{4.12}$$

where R_1 and R_2 is the reflectivity of the mirrors in optical resonator cavity, respectively, G is the gain per unit length, and l is the length of active material. The gain is determined by

$$G = \beta - \alpha, \tag{4.13}$$

where β is the gain coefficient of the active medium, α is the loss coefficient due to scattering and absorption of the medium.

To obtain net gain from the active medium, a population inversion between two levels is necessary. This is usually achieved by exciting into a third level (4), which rapidly and efficiently transfers its energy to a metastable upper laser level (3). Figure 4.1 shows a generalized energy level scheme for the laser action. If the terminal level (2) is the ground state (1), this is a three level laser system represented by Cr^{3+} in Al_2O_3(ruby) laser, wherein more than one half of the ions must be excited to obtain an inverted population. If the terminal level 2 is above the ground state, then only the population in an excited state sufficient to overcome the Boltzmann thermal distribution in the terminal level is needed. This reduces the pumping requirements. This four level system is represented by most of the rare-earth ion-doped materials, such as Nd^{3+} in glass and crystal.

If N_3 and N_2 is the population in the upper and lower laser levels respectively, shown in Fig. 4.1, the net gain of the laser medium is

$$\beta = N_3\sigma_{32} - N_2\sigma_{23} - N_3\sigma_{ex}, \tag{4.14}$$

where σ_{32} and σ_{23} is the cross section for stimulated emission and absorption, respectively. σ_{ex} is the excited state absorption from an upper laser level to a higher excited state. For narrow-line absorption and emission spectra, σ_{32} and σ_{23} are equal. For broadband spectra, such as in glass medium, with an emission bandwidth greater than kT, σ_{32} and σ_{23} are connected by a generalized Einstein relation. The σ_{ex} is a possible excited state absorption cross section from the upper level (3) to higher excited states. If $\sigma_{ex} > \sigma_{32}$, the absorption from level 3 dominates the stimulated emission and the laser action is not possible.

Combining Eqs. 4.12 and 4.13, the oscillation condition is

$$\beta \geq \alpha + \frac{1}{2l}\ln\frac{1}{R}. \tag{4.15}$$

From the absorption equation

$$\int \sigma(\nu)d\nu = \frac{n}{c}h\nu_{23}(\beta_{23}N_2 - \beta_{32}N_3) = \sigma_{23}N_2 - \sigma_{32}N_3. \tag{4.16}$$

Neglecting the excited absorption loss and combining Eqs. 4.14 and 4.16, we have

$$\beta = \frac{h}{C\Delta\nu_{32}}h\nu_{32}\beta_{32}(N_3 - \frac{g_3}{g_2}N_2) = \frac{\sigma_{32}}{\Delta\nu_{32}}, \tag{4.17}$$

where $\Delta \equiv (N_3 - g_3/g_2 N_2)$ population inversion, g_3 and g_2 represents degeneracy of the metastable and terminal states. Substituting Eq. 4.17 into Eq. 4.15, we have

$$\Delta \geq \frac{\Delta\nu_{32}}{\sigma_{32}}(\alpha + \frac{1}{2l}\ln\frac{1}{R}). \tag{4.18}$$

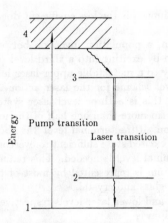

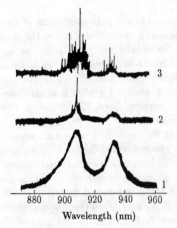

Fig. 4.1. Energy level diagram for a four-level laser scheme.

Fig. 4.2. Change process of the output of a Nd^{3+}-doped silica glass fiber from the superfluorescence to laser action.

4.3   Laser Oscillation Frequency and Linewidth

The laser oscillation frequency ν depends on the atomic resonant frequency (the central frequency of spontaneous emission) ν_0 and the resonant frequency of optical cavity of m mode ν_m

$$\nu \approx \nu_m - (\nu_m - \nu_0)(\frac{\Delta\nu_{1/2}}{\Delta\nu}), \qquad (4.19)$$

where $\Delta\nu$ is the emission linewidth, $\Delta\nu_{1/2}$ full linewidth of the optical resonator

$$\Delta\nu_{1/2} = \frac{c(\alpha - \frac{1}{l}\ln\sqrt{R_1 R_2})}{2\pi n}. \qquad (4.20)$$

The atomic resonant frequency ν_0 does not coincide with the passive resonance frequency ν_m, therefore the laser oscillation frequency will be shifted from ν_m toward ν_0, which is called "frequency pulling".

For the multi-mode oscillation, there are many oscillation peaks taking place. Figure 4.2 shows the change process of the output of a Nd^{3+}-doped silica glass fiber from the superfluorescence to laser action [2]. Below the laser threshold only the superfluorescence could be observed (Fig. 4.2-1). At the beginning of the oscillation, the single mode of LP_{01} oscillated (Fig. 4.2-2) and the multi-mode appeared at higher pumping level (Fig. 4.2-3).

Generally, if a system oscillates in single mode, the linewidth of the oscillation is mainly governed by the spontaneous emission in that mode. When the system oscillates at a power level P, then the linewidth of the oscillation signal is related to the normal linewidth $\Delta\nu_L$ of the transition by the expression

$$\Delta\nu = \frac{2h\nu}{P}(\Delta\nu_L)^2. \tag{4.21}$$

4.4 Laser Amplification and Gain Coefficient [3]

From the stimulated transition rate equation Eq. 4.10 the stimulated emission power P (per unit volume) can be expressed as follows

$$\frac{P}{V} = (n_2 - \frac{g_1}{g_2}n_1)\frac{\lambda^2 g(\nu)I_\nu}{8\pi n^2 \tau_R}, \tag{4.22}$$

where I_ν is the input laser signal intensity, which increases exponentially with the amplification length z at unsaturated condition

$$I_\nu(z) = I_\nu(0)e^{r_0(\nu)z} = \frac{P}{V}, \tag{4.23}$$

$r_0(\nu)$ is the amplification gain coefficient

$$r_0(\nu) = \frac{(N_2 - N_1\frac{g_2}{g_1})\lambda^2}{8\pi n^2 \tau_R}g(\nu). \tag{4.24}$$

At saturated condition and in the case of homogeneous broadening the amplification gain coefficient $r(\nu)$ is

$$r(\nu) = \frac{r_0(\nu)}{1 + \frac{I_\nu}{I_s(\nu)}}, \tag{4.25}$$

where $I_s(\nu)$ is called as saturation intensity, which is the intensity (FWHM) of r_0.

$$I_s(\nu) = \frac{4\pi n^2 h\nu}{(\tau/\tau_R)\lambda^2 g(\nu)}, \tag{4.26}$$

τ is the lifetime of the upper level (level 3 in Fig. 4.1). From Eq. 4.26 it can be seen that the saturation intensity is inversely proportional to $g(\nu)$. It means that the saturation is not easy at the side of spectral line profile.

At saturated condition and in the case of inhomogeneous broadening the amplification gain coefficient $r(\nu)$ is

$$r(\nu) = r_0(\nu)\frac{1}{\sqrt{1 + \frac{I_0}{I_s}}}, \tag{4.27}$$

and I_s is given by

$$I_s = \frac{2\pi^2 n^2 h\nu\Delta\nu}{\lambda^2(\frac{\tau}{\tau_R})}. \tag{4.28}$$

4.5 Threshold Energy and Oscillation Energy

According to the rate equation for the four level system, the pumping threshold energy E_p^0 and oscillation energy E at steady-state can be expressed as follows:

$$E_p^0 = 4\pi n^2(\sigma + \ln\frac{1}{R})h\nu_p \cdot \frac{\Delta\nu_L \cdot \tau}{\eta\lambda^2 K_p \Delta\nu_p}, \tag{4.29}$$

$$E = \frac{1-R}{(2l+\ln 1/R)}\frac{\lambda_p}{\lambda}\eta_1 K_p \Delta\nu_p(E_p - E_p^0), \tag{4.30}$$

where E_p is the pumping energy, λ_p, λ is the central wavelength of the excitation band and the laser oscillation respectively, $\Delta\nu_p$, $\Delta\nu_L$ is the bandwidth of the excitation and the luminescence respectively, K_p is the absorption coefficient of active medium at wavelength λ_p, τ is the life time of luminescence, η is the luminescence quantum efficiency, η_1 is the energy conversion efficiency from the excited state to metastable state, α is the optical loss of active medium, R is the reflectivity of output mirror.

4.6 Q-switching and Mode-locking [3]

By Q-switching and mode-locking techniques laser beam can be compressed to very short pulses (ns, ps and fs), therefore, high laser pulse power (MW, GW and TW) can be obtained. It is rather important for solid state lasers.

4.6.1 Q-switching

The loss of the optical cavity is caused by absorption and scattering of active medium, transmittance of reflectors and diffraction, which is always expressed by Q-value (the figure of merit).

$$Q = \frac{\omega E}{P} = -\frac{\omega E}{dE/dt} = \omega t_c, \tag{4.31}$$

where E is the storage energy, $P = -\frac{dE}{dt}$ loss power, t_c the photon lifetime in the optical cavity (the mode lifetime of optical cavity) given in Eq. 4.32 as

$$t_c \simeq \frac{n_0 l}{C[\alpha l - \ln\sqrt{R_1 R_2}]}. \tag{4.32}$$

Q-switching technique is accomplished by suddenly changing the Q-value or t_c. At low Q-value the gain can not exceed the oscillation threshold, the energy is stored in the cavity. As the population inversion reaches its peak, the Q-value raises abruptly to its high value, at that time the gain greatly exceeds the oscillation threshold, the energy stored at the excited state exhausts at the moment, thus high laser peak

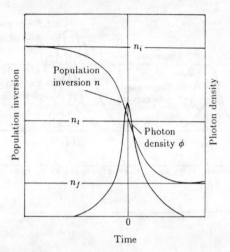

Fig. 4.3. Change of population inversion and photon density with time during a giant pulse formation.

power can be obtained. Figure 4.3 shows the change of population inversion and photon density with time during the formation of giant pulse. The peak power can be expressed as follows:

$$P_{\max} \simeq \frac{n_i h\nu}{2t_c},$$ (4.33)

where n_i is the maximum stored photons.

There are several methods to realize the Q-switching, such as using a rotating prism for reflector at one side, inserting a saturable absorber in the cavity, and applying an electro-optic gate (liquid Kerr cell or opto-electric crystal).

4.6.2 Mode-locking

In the case of inhomogeneous broadening when the pumping power exceeds the threshold several oscillation modes occur. Figure 4.4 illustrates the difference of the single pass gain curves, mode spectra and oscillation lines in the case of homogeneous and inhomogeneous broadenings. Therefore, the mode-locking takes place easily in the case of inhomogeneous broadening. The interval of mode-locked pulse chain is

$$\omega_{q+1} - \omega_q = \frac{\pi c}{l} \equiv \omega.$$ (4.35)

The length of the mode-locked pulses (τ) is inversely proportional to the gain linewidth ($\Delta\nu$)

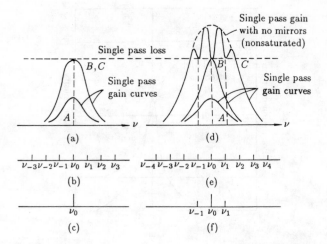

(a)   (d)

(b)   (e)

(c)   (f)

Fig. 4.4. (a) Single-pass gain curves for a homogeneous atomic system (A—below threshold; B—at threshold; C—well above threshold). (b) Mode spectrum of optical resonator. (c) Oscillation spectrum (only one mode oscillates). (d) Single-pass gain curves for an inhomogeneous atomic system (A—below threshold; B—at threshold; C—well above threshold). (e) Mode spectrum of optical resonator. (f) Oscillation spectrum for pumping level C, showing three oscillating modes.

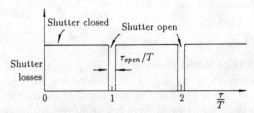

Fig. 4.5. Periodic losses introduced by a shutter to induce mode-locking. The presence of these losses favors the choice of mode phases that results in a pulse passing through the shutter during the open intervals—that is, mode-locking.

$$\tau \sim \frac{2\pi}{\Delta\omega} = \frac{1}{\Delta\nu}. \tag{4.34}$$

Thus mode-locking can be accomplished by modulating the loss (or gain) of the laser with angular frequency $\omega = \frac{\pi c}{L}$. Any optical shutter can realize the mode-locking, as shown in Fig. 4.5. The optical shutter is controlled by external signal, so it is active mode-locking. So called passive mode-locking is performed by a saturable absorber.

References

1. Fuxi Gan, *Optical and Spectroscopy Properties of Glass* (Springer-Verlag, Berlin, 1992) pp. 204–207.

2. Fuxi Gan and Yihong Chen, *Pure Appl. Opt.* **2** (1993) 359.

3. A. Yariv, *Quantum Electronics* (Second Edition, John Wiley & Sons Inc., New York, 1975).

5. Physical Properties of Laser Materials

The solid state laser materials are composed of doping ions (RE and TM ions) and host dielectrics. As described above, the spectroscopic and laser properties of laser materials are mainly dependent on doping ions although the hosts have a great influence on them. The physical properties of laser materials are dominated by host materials. During laser operation the laser induced damage, nonlinear optical effects and laser beam quality are much related with the physical properties of host materials, among them the optical, thermal and mechanical properties are most important ones.

5.1 Thermal and Mechanical Properties

To evaluate the thermal stability and mechanical strength the following thermal and mechanical properties are important.

Coefficient of thermal expansion	α	$(^\circ C^{-1})$
Coefficient of specific heat	C_p	$(cal/g \cdot ^\circ C)$
Coefficient of thermal conductivity	k	$(cal/cm \cdot s \cdot ^\circ C)$
Melting temperature	T_m	$(^\circ C)$
Transition temperature	T_g	$(^\circ C)$
Young's modulus	E	(Pa)
Shearing modulus	G	(Pa)
Poisson's ratio	μ	
Elastic coefficient	C_{ij}	(Pa)
Tensile strength	P_t	(kg/cm^2)
Density	ρ	(g/cm^3)

The coefficient of thermal stability k can be expressed as

$$k = \frac{P_t}{\alpha E}\left(\frac{\lambda}{c_p \cdot \rho}\right)^{1/2}.$$

For isotropic media the elastic properties are related by the following equations.

$$E = C_{11}, \quad G = \frac{E}{2(1+\mu)}.$$

Tables 5.1 and 5.2 list the values of the above mentioned physical properties of typical host crystals and glasses respectively [1–3].

5.2 Optical Properties [1–4]

The optical properties of transparent dielectric are determined by the optical dispersion characteristics. The solid state laser host materials can be regarded as transparent dielectrics in visible region. On the basis of dispersion theory, the polarizability of dielectric can be described as

$$\gamma = \frac{c^2}{m} \sum_k \frac{f_k}{\omega_k^2 - \omega^2}, \tag{5.1}$$

where m and e are the mass and the charge of the electron, respectively. f_k is the oscillator strength corresponding to the inherent frequency ω_k, and ω is the frequency of the incident beam. For the representation of the dispersion curve of dielectric in the visible region it is sufficient to take account of only two inherent frequencies, i.e. the electron transition frequency ω_1 in the ultra-violet region and the vibration frequency of the structural lattice or network ω_2 in the infrared region. The former plays a more important role than the latter. Therefore,

$$\gamma = \frac{c^2}{m} \left[\frac{f_1}{\omega_1^2 - \omega^2} + \frac{f_2}{\omega^2 - \omega_2^2} \right]. \tag{5.2}$$

The change in the polarizability of dielectric under the actions of an external field is due to the change of ω_1, ω_2, f_1 and f_2. The change of ω_2 results in a shift of the dispersion curve only, which could either increase or decrease the refractive index. In the case where both f_1 and ω_1 values change, the shape of the dispersion curve will also vary. The change in the energy of inherent electron transition of ions, induced by the actions of stress, electric, thermal, and acoustic fields, as well as powerful light, gives rise to changes in the polarizability of dielectric. This change in the macroscopically observed refractive index of inorganic dielectric as a transparent nonferromagnetic substance, can be deduced from the material equation:

$$H_1 = \frac{\varepsilon - 1}{4\pi + \Gamma(\varepsilon + 1)}, \tag{5.3}$$

$$\varepsilon_\infty = n_0^2. \tag{5.4}$$

Here H_1, ε and n_0 are the linear susceptibility tensor, the dielectric constant and the linear refractive index, respectively. Γ is the partial effective field coefficient,

Table 5.1. Thermal and mechanical properties of host crystals.

Crystal	Space group	α $\times10^{-6}$ °C^{-1}	C_{11} Pa$\times10^5$	C_{12} Pa$\times10^5$	C_{44}	C_p cal/g·°C	T_m °C	ρ g/cm^3	H kg/mm^2	μ	$k\times10^{-4}$ cal/cm·s·°C
$Y_3Al_5O_{12}$	$Ia3d$	6.9	27.8				1950	4.55	100	0.25	150
$YAlO_3$	$P6mn$	10.8, 9.5, 4.3					1875	5.18	100		120
Al_2O_3	D_{3d}^6-R3c	4.78, 5.31	49.68	16.36	14.74	0.18	2040	3.98	1360		350
KDP	D_{2d}^{12}	24.9, 44.0	7.14	−0.49	1.27			2.34	20	0.23	29
YVO_3	$D_{4h}-I4/2md$	7.3					1730	4.23	460		60
KCl	$Fm3m-O_h^5$	37.4	38.25		5.88	0.168	776	1.98			156
$CaWO_4$	$I4_1/a$	11.2, 18.7					1580	6.06	420		48
LiF	O_h^5	36	9.74	4.04	15.86	0.374	1130	2.64	110		
$Ca_5(PO_4)_3F$	$P6_3/m$	9.4, 10					170	3.2			40
CaF_2	O_h^5	18.4	16.4	5.3	3.37	0.21	1360	3.18	160		69

Table 5.2. Thermal and mechanical properties of glasses.

Glass type	Glass system	α 10^{-6} °C^{-1}	$k\times10^{-4}$ cal/cm·s·°C	C_p cal/g·°C	$E\times10^3$ kg/cm^2	T_g °C	ρ g/cm^3	H kg/mm^2	μ	P_t kg/mm^2
N_{10}	$K_2O(Na_2O)-CaO-Al_2O_3-SiO_2$	8.9	19	0.26	750	525	2.52	585	0.23	8.9
N_{03}	$K_2O(Na_2O)-CaO-SiO_2$	8.8	17	0.27	759	590	2.51	606	0.22	11.8
N_{01}	$K_2O-BaO-SiO_2$	104	16	0.28	679	500	2.90	560	0.22	9.7
N_{21}	$BaO-P_2O_5$	117	18	0.17	555	510	3.38	420	0.26	4.5
LFP	$RF_2-AlF_3-Al(PO_3)_3$	157	17	0.18	830	420	3.52	470	0.28	4.0

its value is equal to zero for a covalent substance and $4\pi/3$ for an isotropic ionic substance. For inorganic dielectric, a substance with polar-covalent bonds, the value of Γ is between zero and $4\pi/3$. Using $\Gamma=0$ and $4\pi/3$, we obtained the following two equations for the refractive indices at the two extremes:

$$\Gamma = 0, \quad n^2 - 1 = 4\pi N\gamma(\mathscr{H}_1 - N\gamma), \tag{5.5}$$

$$\Gamma = \frac{4}{3}\pi, \quad \frac{n^2-1}{n^2+2} = \frac{4}{3}\pi N\gamma[\mathscr{H}_1 = \frac{N\gamma}{1 - \frac{4}{3}\pi N\gamma}], \tag{5.6}$$

where γ is the summation of the polarizability of ions or atoms in dielectric, N is the number of ions or atoms, $N = N_0/V = \rho N_0/M$, here V, M and ρ are the gram-molar volume, the molecular weight and the density of dielectric, respectively. N_0 is the Avogadro number. The change in refractive index of dielectric can be expressed as:

$$\Gamma = 0, \quad \delta n = \frac{n^2-1}{2n}[\frac{\delta\rho}{\rho} + \frac{\delta\gamma}{\gamma}], \tag{5.7}$$

$$\Gamma = \frac{4}{3}\pi, \quad \delta n = \frac{(n^2-1)(n^2+2)}{6n}[\frac{\delta\rho}{\rho} + \frac{\delta\gamma}{\gamma}]. \tag{5.8}$$

Therefore, the change in the refractive index of transparent dielectric can be attributed to the variation in either density or polarizability.

The variation in density is caused by the expansion under the action of a thermal field, and the strain under the stress field, as well as the stretch under the actions of electric, acoustic and powerful optical fields.

5.2.1 Refractive Index n_d and Mean Dispersion n_F-n_C

The $n_d{\sim}v_d$ $(= \frac{n_d-1}{n_F-n_c})$ diagram is always used for classifying the refractive index-dispersion characteristic of optical materials. Figures 5.1 and 5.2 show the $n_d{\sim}v_d$ diagram of laser host crystals, as well as laser and optical glasses respectively. It can be seen that the fluoride crystals and glasses possess low refractive index and mean dispersion, which are located at the lower-left corner, and the oxide crystals and glasses with high atomic weight elements are distributed at the upper-right corner of the $n_d{\sim}v_d$ diagrams.

5.2.2 Temperature Coefficient of Refractive Index and Thermo-optical Coefficients

The change in optical constants of optical materials under the actions of external temperature field results from effects on both density and polarizability.

By differentiating Eq. 5.6 with respect to the temperature, one can obtain

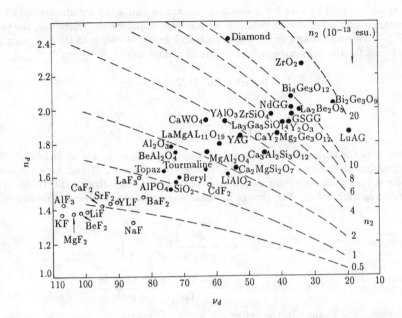

Fig. 5.1. $n_d \sim \nu_d$ diagram of laser host crystals and optical crystals (constant $n_2 \times 10^{-13}$ esu lines).

$$\beta = \frac{\partial n}{\partial T} = \frac{(n^2 - 1)(n^2 - 2)}{6n} \left(\frac{1}{\gamma} \frac{d\gamma}{dT} - 3\alpha \right), \qquad (5.9)$$

where α is the linear coefficient of thermal expansion.

From Eq. 5.9, we can see when the temperature rises, the density of material will decrease as a result of thermal expansion, but on the other hand the decrease in the inherent frequency of electron transition, i.e. the absorption edge in the ultra-violet region shifts toward the longer wavelength range, gives rise to an increment in the polarizability. Therefore, whether the sign of β is positive or negative is determined by the resultant of the two effects mentioned above. It is shown that most oxide crystals and glasses have a positive temperature coefficient of the refractive index and the fluoride crystals and glasses possess negative values (See Tables 5.3 and 5.4).

When a beam of light propagates through a glass which is subjected to a thermal field or powerful light, the optical distortions of the wave front will appear. These optical distortions are described by the following thermo-optical coefficients.

In the case of no thermal stress, the thermo-optical constant W is defined as:

$$W = \beta + (n - 1)\alpha. \qquad (5.10)$$

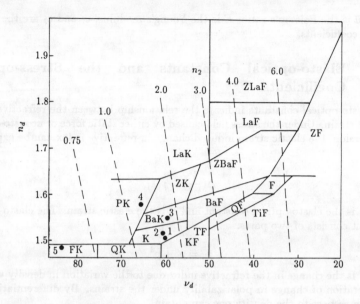

Fig. 5.2. $n_d \sim \nu_d$ diagram of laser host glasses and optical glasses (constant $n_2 \times 10^{-13}$ esu lines). 1, N_{10}; 2, N_{03}; 3, N_{01}; 4, N_{21}; 5, LFP.

By comparing it with Eq. 5.9, the thermo-optical constant W indicates physically the temperature dependence of polarizability. Its value is mainly determined by the temperature coefficient of refractive index. In case there is thermal stress rather than free expansion, the optical distortions for a glass rod ($L \gg D$) are conveniently described by the stress thermo-optical coefficient P and the stress birefringence coefficient Q:

$$P = \beta - \frac{\alpha E}{2(1-\mu)}(c_1 + 3c_2), \tag{5.11}$$

$$Q = \frac{\alpha F}{2(1-\mu)}(c_1 - c_2). \tag{5.12}$$

While for a glass disk ($D \gg L$), it is described by the stress thermo-optical coefficient R and stress birefringence coefficient Q':

$$R = \alpha(n-1)\mu - \alpha E(\frac{c_1 - c_2}{2}), \tag{5.13}$$

$$Q' = \alpha E(\frac{c_1 - c_2}{2}), \tag{5.14}$$

where μ is the Poisson's ratio, E is the Young's modulus, c_1 and c_2 are the stress-optical coefficients.

5.2.3 Elasto-optical Constants and the Stress-optical Coefficients

The elasto-optical constants indicate the relationship between the refractive index and the strain. Under the stress field caused by either static force or acousto-optical compression, the elastic strain induced change in refractive index can be expressed as

$$\Delta(\frac{1}{n_1^2}) = p_{ij}\varepsilon_j, \tag{5.15}$$

here p_{ij} is the elasto-optical constant and ε_j is the elastic strain. The elasto-optical constant consists of two parts:

$$p_{ij} = p^c + p^d, \tag{5.16}$$

here p^d is the change in the refractive index due to the variation in density without consideration of change in polarizability under the strains. By differentiating Eq. 5.6 with respect to the density one can obtain

$$\rho\frac{\partial n}{\partial \rho} = \frac{(n^2 + 2)(n^2 - 1)}{6n}. \tag{5.17}$$

For three-dimensional homogeneous stress and isotropic medium the relationship between the elasto-optical constants and the change of refractive index with the density is

$$\rho\frac{dn}{d\rho} = \frac{n^3}{6}(p_{11} + 2p_{12}). \tag{5.18}$$

According to Eq. 5.17, $\rho\frac{dn}{d\rho}$ has a relation only with the refractive index. Let Λ_0 represent the polarizability variation with the density, then Eq. 5.17 can be rewritten as

$$\rho\frac{dn}{d\rho} = \frac{(n^2 + 2)(n^2 - 1)}{6n}(1 - \Lambda_0). \tag{5.19}$$

The stress-optical coefficients c_1 and c_2 are related to the elasto-optical constants as

$$c_1 = -\frac{n^3}{2E}[p_{11} - 2\mu p_{12}], \tag{5.20}$$

$$c_2 = -\frac{n^3}{2E}[(1 - \mu)p_{12} - \mu p_{11}]. \tag{5.21}$$

Table 5.3. Optical properties of host crystals.

Crystal	n_d	$(n_F - n_C)$ $\times 10^{-5}$	$\frac{dn}{dT} \times 10^{-5}$ $^\circ C^{-1}$	p_{11}	p_{12}	p_{44}	υ	$n_2 \times 10^{-14}$ esu
$Y_3Al_5O_{12}$	1.83		0.73	−0.029	0.009	−0.061	7	34
$YAlO_3$	1.97, 1.96, 1.94							
Al_2O_3	1.760, 1.7684	1020	1.3	0.20	0.078	0.25 (p_{33})	75.3	15
KDP	1.5092, 1.4681	893, 609		0.25	0.249	0.22 (p_{33})	57, 70	10
YVO_3	1.86, 1.88							
KCl	1.4904	1114	−3.8	0.22	0.171	−0.026	44	20
$CaWO_4$	1.92							
LiF	1.4012	396	−1.4	−0.43	0.19	−0.164	99	4
$Ca_5(PO_4)_3F$	1.63							
CaF_2	1.4335	457	−1.1				94.8	5

Table 5.4. Optical properties of host glasses.

Glass type	Glass system	n_d	$(n_F - n_C)$ $\times 10^{-5}$	υ	$\frac{dn}{dT} \times 10^{-5}$ $^\circ C^{-1}$	$-C_1$ $(10^{-6}$ $cm^2/kg)$	$-C_2$	$n_2 \times 10^{-14}$ esu
N_{10}	$K_2O(Na_2O)$–CaO –Al_2O_3–SiO_2	1.5171	880	58.8	0.08	0.09	0.35	18
N_{03}	$K_2O(Na_2O)$–CaO –Al_2O_3-SiO_2	1.5224	874	59.8	0.16	0.11	0.36	18
N_{01}	K_2O–BaO–SiO_2	1.5424	933	58.1	−0.09	0.15	0.38	20
N_{21}	BaO–P_2O_5	1.574	889	64.5	−0.48	0.26	0.40	13
LFP	RF_2–AlF_3 –$Al(PO_3)_3$	1.481	573	83.9	−0.8	0.09	0.32	6.8

Tables 5.3 and 5.4 list the linear optical properties of typical host crystals and glasses [1, 4].

5.2.4 Nonlinear Refractive Index

Under the actions of intense electric field and powerful light (electro-magnetic field) the refractive index of dielectric will change. The refractive index which depends on the intensity of electric field is defined as the nonlinear refractive index n_2.

$$\delta n = n_2 |E|^2 = n_2 I, \tag{5.22}$$

where E is the intensity of electric field (V/m), I is the intensity of light beam (W/cm^2). The units of n_2 and n_2' are in esu and cm^2/W, respectively.

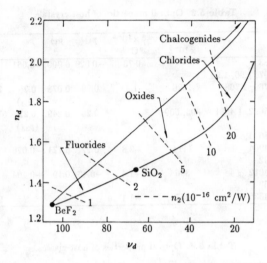

Fig. 5.3. Constant n_2' value lines on $n_d \sim \upsilon_d$ diagram for different laser host materials.

The change in the polarizability of dielectric under the action of an intense electric field is due to the nonlinear polarizations of ions, i.e. the nonlinear distortions of electron orbits surrounding the average position of the nucleus. The nonlinear refractive index $n_2(E)$ is expressed as

$$n_2(E) = \frac{2\pi}{n_0}\mathscr{H}_3 = \left(\frac{2\pi N}{n_0}\right)\theta, \tag{5.23}$$

where $\mathscr{H}_3$ is the third-order susceptibility tensor, θ is the nonlinear polarizability, i.e. the electric field induced change of polarizability $\frac{\partial \gamma}{\partial E}$. In the next chapter we will discuss the measurement, mechanism and calculation of nonlinear refractive index in detail.

On the $n_d \sim \upsilon_d$ diagrams (Figs. 5.1 and 5.2) we also expressed the lines of constant n_2 value. For different laser host materials the n_2 value can be predicted from Fig. 5.3. A large reduction in n_2 is achieved by using low index, low dispersion materials in the lower left-hand corner of Fig. 5.3.

The laser glasses are of multicomponent in chemical composition, and the main physical properties of different kinds of multicomponent glasses can be calculated from the chemical composition by the method proposed by us [1].

References

1. Fuxi Gan, *The Calculation of Physical Properties and Design of Chemical Composition of Inorganic Glasses* (Shanghai Press for Science and Technology, Shanghai, 1981).

2. Fuxi Gan ed., *Optical Glasses*, Vol. 1, (Science Press, Beijing, 1982), pp. 203–257.

3. Fuxi Gan ed., *Science and Technology of Modern Glass*, Vol. 1, (Shanghai Press for Science and Technology, Shanghai, 1988), pp. 310–380.

4. Fuxi Gan, *Optical and Spectroscopic Properties of Glass* (Springer-Verlag, Berlin, 1992), pp. 78–125.

6. Nonlinear Optical Properties of Laser Materials

Nonlinear optical effects in laser materials can be easily observed when the laser intensity is increased. The even-order nonlinear optical susceptibility of glassy materials should be zero, there are no second nonlinear optical effects in them due to isotropy of glassy materials. By the second order optical nonlinearity the frequency doubling, electro-optic modulation and switching, as well as scanning and mode-locking can be performed in crystalline materials. So-called optical nonlinear crystals have been specially introduced and reviewed in some books [1, 2].

As the laser intensity increases, the third-order nonlinear optical susceptibility will play an important role and a series of nonlinear optical effects can be found in both crystalline and glassy materials. The nonlinear refractive index is a very important property, which determines all nonlinear optical effects in materials. The mechanism, measurement and calculation of the nonlinear refractive index of laser materials will be presented in this chapter, and the frequency dependence of nonlinear refractive index will also be discussed.

6.1   Laser Interaction with Matter

The interaction of a powerful laser beam with laser media results in three effects: the thermal effect and the electrostriction, as well as the nonlinear polarization. Because the response times are different from each other, the main reason for a nonlinear change in the refractive index also differs with differing pulse duration of the laser beam [3].

6.1.1   *Thermal Effect*

As the optical medium absorbs laser energy and causes a temperature field, both the density and the polarizability will change, and these changes depend on the laser beam intensity. Because most of laser materials are thermal insulators, the heating process is treated as adiabatic for the laser pulses with millisecond and microsecond duration or less. In the simple case, the nonlinear refractive index $n_2(T)$ caused by a thermal effect is expressed as

$$n_2(T) = \frac{dn}{dT}\cdot\alpha\cdot\frac{t}{\rho}\cdot C_{\mathrm{p}}, \tag{6.1}$$

where, t is the laser pulse duration, α, C_{p} and ρ is the optical absorption coefficient, specific heat and density of glass, respectively, and $(\frac{dn}{dt})$ is the temperature coefficient of the refractive index.

6.1.2 Electrostriction Effect

The change in refractive index under the action of intense electric field is also attributed to the effect of electric field on the density and polarizability of the material. Brillouin scattering, which appears under the action of strong high frequency electric field or powerful laser light, generates an electrostriction that gives rise to the variation in density. The wave equation of the steady state electrostriction is:

$$(\Delta^2 + \frac{2\Gamma}{\nu^2}\frac{\partial}{\partial t} - \frac{1}{\nu^2}\frac{\partial^2}{\partial t^2})\delta\rho = \frac{1}{8\pi}\frac{K}{\nu^2}(\Delta^2|E|^2), \tag{6.2}$$

where ν is the velocity of sound, $2\Gamma/\nu^2$ is the extinction coefficient of sound damping, and $K = 2n_0\rho_0\cdot dn/d\rho$. If we neglect the sound damping and express Eq. 6.2 by the change of refractive index, then

$$(\Delta^2 - \frac{1}{\nu^2}\frac{\partial^2}{\partial t^2})\delta n = \frac{1}{8\pi}\frac{K}{\nu^2}\frac{dn}{d\rho}(\Delta^2|E|^2). \tag{6.3}$$

Therefore, the electrostriction induced nonlinear refractive index $n_2(s)$ can be expressed as:

$$n_2(s) = \frac{1}{8\pi}\cdot\frac{K}{\nu^2}(\frac{dn}{d\rho}), \tag{6.4}$$

for isotropic medium, $\nu = \sqrt{E/\rho}$. Taking the edge effect into consideration, then

$$n_2(s) = \frac{1}{4\pi}\cdot\frac{n_0}{E}[\frac{(1+\mu)(1-2\mu)}{(1-\mu)}](\rho\frac{dn}{d\rho})^2 \text{ (esu)}. \tag{6.5}$$

It is shown that the electrostriction induced nonlinear refractive index is strongly dependent on $\rho\cdot dn/d\rho$, which can be calculated from the elasto-optical constants or the stress-optical coefficients of materials.

6.1.3 Nonlinear Polarization

The change in the polarizability of material under the action of an intense electric field is due to the nonlinear polarization of ions or atoms, i.e. the nonlinear distortions of electron orbits surrounding the average position of the nucleus. The nonlinear refractive index $n_2(E)$ is expressed as

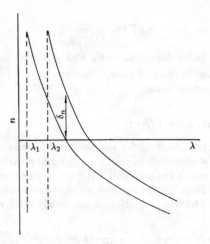

Fig. 6.1. Stokes shift of dispersion curve.

$$n_2(E) = \frac{2\pi}{n_0}\mathscr{H}_3 = (\frac{2\pi N}{n_0})\theta, \qquad (6.6)$$

where $\mathscr{H}_3$ is the third-order susceptibility tensor, θ is the nonlinear polarizability, i.e. the electric field induced change of polarizability $\frac{d\gamma}{dE}$. As the linear refractive index in the visible region is mainly dependent on the inherent frequency ω_1 in the ultraviolet region, we can consider the nonlinear refractive index as an increment in the refractive index (δ_n) due to the inherent frequency in the ultra-violet region, which dominates the dispersion curve, generates a Stokes displacement and shifts toward the longer wavelength as the intensity of electric field increases, as shown in Fig. 6.1.

$$\delta n = n_2|E|^2. \qquad (6.7)$$

Thus, the nonlinear refractive index $n_2(E)$ can be directly derived from the dispersion data.

Tables 6.1 and 6.2 presents the nonlinear refractive index of laser glass N03 and crystal KCl and YAG under the action of different pulse duration laser beams respectively. It is obvious from the tables that the thermal effect is the main reason for the change in refractive index induced by a laser beam with a pulse duration of a millisecond or more, such as a CW laser beam. While for the laser beam with a pulse duration of the order of a few microseconds, all three effects will play a role in the nonlinear refractive index. Among these, the electrostriction is less

Table 6.1. Nonlinear refractive index of N_{03} Nd-doped glass under the action of a laser beam with different pulse duration (absorption coefficient: 0.0015 cm^{-1}).

Nonlinear effect	Response time (s)	Nonlinear refractive index (cm^2/W)		
		ms $(10^{-3}$ s$)$	μs $(10^{-6}$ s$)$	ns $(10^{-9}$ s$)$
Thermal effect	$10^{-6} \sim 10^{-7}$	6×10^{-12}	6×10^{-15}	—
Electrostriction	$10^{-7} \sim 10^{-8}$	1.06×10^{-16}	1.06×10^{-16}	1.06×10^{-16}
Nonlinear polarization	$10^{-15} \sim 10^{-16}$	0.92×10^{-15}	0.92×10^{-15}	0.92×10^{-15}

Table 6.2. Nonlinear optical constants of KCl and YAG in the laser radiation with different pulse widths (absorption coefficient : KCl, 0.001 cm^{-1}; YAG, 0.01 cm^{-1}).

Effect	Response time (second)	n_2 (cm^2/W)					
		KCl			YAG		
		ms$(10^{-3}$s$)$	μs$(10^{-6}$s$)$	ns$(10^{-9}$s$)$	ms$(10^{-3}$s$)$	μs$(10^{-6}$s$)$	ns$(10^{-9}$s$)$
Thermal	$10^{-6} \sim 10^{-7}$	2×10^{-11}	2×10^{-14}	2×10^{-17}	5.6×10^{-11}	5.6×10^{-14}	5.6×10^{-17}
Electrostriction	$10^{-7} \sim 10^{-8}$	2.1×10^{-16}	2.1×10^{-16}	2.1×10^{-16}	3×10^{-18}	3×10^{-18}	3×10^{-18}
Nonlinear polarization	$10^{-15} \sim 10^{-16}$	9.2×10^{-15}	9.2×10^{-15}	9.2×10^{-15}	4.6×10^{-14}	4.6×10^{-14}	4.6×10^{-14}

important than the other two, however, some authors considered the electrostriction as a major effect [4, 5]. In the case of a laser beam with a pulse duration of the order of a nanosecond or sub-nanosecond, the change of refractive index is mainly caused by nonlinear polarization, and both thermal and electrostriction effects can be neglected because their response time is too long.

Therefore, the phenomena caused by the interaction of a powerful laser beam with the laser materials, such as thermal blooming, self-focusing and laser induced damage etc., are related to the nonlinear change in the refractive index.

6.2 Measurement of Nonlinear Refractive Index

In recent years we have developed several methods for measuring the nonlinear refractive index of laser and optical materials [6].

6.2.1 *Self-focusing Damage*

Using several lenses with different focal lengths to focus a TEM$_{00}$ laser beam, the self-focusing threshold I_{max} and filament lengths L_{sf} of laser materials can be measured. By the following equation the self-focusing critical power P_{cr} was determined. The nonlinear refractive index $n_2(E)$ was calculated by Eq. 6.10.

$$I_{max} = \frac{P}{\pi r_0^2}(1 - \frac{P}{P_{cr}})^{-1}, \tag{6.8}$$

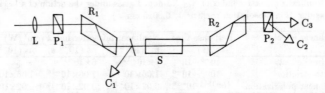

Fig. 6.2. Optical arrangement for SIPC measurement. L, lens; P_1, P_2, polarimeters; S, sample; R_1, R_2, Fresnel rhomb; C_1, C_2, C_3, detectors.

$$L_{sf} = \frac{R^2}{R_d}(\frac{P}{P_{cr}} - 1)^{1/2}, \tag{6.9}$$

$$n_2(E) = c\lambda_0^2/(64\pi^2 P_{cr})^{-1}, \tag{6.10}$$

where P is the power of the laser beam, r_0, the radius of focal spot without self-focusing, $R_d = k\omega_0^2$, k, the wave vector, ω_0 the radius of the laser beam at the entrance to glass, and R, the curvature radius of laser beam at the same point.

6.2.2   *Self-induced Polarization Change*

While a high power laser beam with elliptical polarization transmits through an isotropic medium, the main axis of the polarization ellipse will rotate. The rotation angle depends on the nonlinear property of the material and the laser beam power:

$$\theta = \frac{48\pi^2\omega}{n_0^2 c^2}C_{1221}PL\cos2\phi, \tag{6.11}$$

where ω is the frequency of the laser beam, c is the velocity of light in vacuum, L, n_0 and C_{1221} is the length, refractive index and cubic nonlinear polarizability of laser materials, respectively. Figure 6.2 shows the optical arrangement of the experiment. The lens in Fig. 6.2 is used to converge the laser beam in order to increase the power density of the beam.

6.2.3   *Laser-induced Birefringence*

High-power laser beams can also make another weak probe beam to demonstrate birefringence in material. In general, the effect in insulator materials is quite small, but the laser beam power is so high that the laser-induced birefringence of laser materials in a high power laser system is not negligible. Using a weak light at 1.06 μm as probe beam, the laser-induced birefringence in laser materials was investigated, and the optical Kerr-coefficients n_k were measured. Figure 6.3 is the experimental apparatus for laser-induced birefringence measurement.

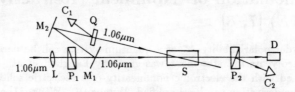

Fig. 6.3. Experimental arrangement for LIB measurement at 1.06 μm. L, lens; P_1, P_2, polarimeters; M_1, beam splitter; S, sample; M_2, directing mirror; C_1, C_2, D, detectors; Q, rotation plate.

Table 6.3. Nonlinear refractive coefficients of glasses measured by various methods.

Glass	n_2 (10^{-13} esu)		
	SF	SIPC	LIB
ZF-7	6.4	7.5	8.3
ZF-2		5.0	
BaF-2	4.1	3.0	3.0
QK-3	1.3		
N03	1.9	2.0	2.0
N08	1.6	1.8	2.0
N10	1.6	1.8	2.0
N21	1.3	2.0	
N24	1.2	1.6	1.2

Three nonlinear parameters of P_{cr}, C_{1221} and n_k were measured using the above methods. Therefore, the nonlinear refractive coefficients of isotropic media were determined from these parameters by the following equations:
For the SF method,

$$n_2 = C\lambda_0^2/(64\pi^2 P_{cr}). \qquad (6.12)$$

For the SIPC method,

$$n_2 = \frac{360\pi}{n_0}C_{1221}. \qquad (6.13)$$

For the LIB method,

$$n_2 = \frac{3}{2}nk. \qquad (6.14)$$

As the glassy materials are typical isotropic ones, we measured the nonlinear refractive index of several laser and optical glasses by the above mentioned three methods. All the results are listed in Table 6.3, and they are comparable [6].

6.3 Calculation of Nonlinear Refractive Index $n_2(E)$ [7, 8]

The third order polarizability $\mathscr{H}_3$ of a quantum system can be derived from perturbation theory, its full statement was given by Langhoff *et al.* [9]. If we are only concerned with the electronic nonlinearity of a dielectric solid far from any resonance, the derivation can be simplified. Wynne [10], Wang [11] and Boling [12] proposed the calculation methods for the third order polarizability or nonlinear refractive index of transparent dielectric from the first order polarizability or linear refractive index and dispersion equation.

As mentioned above, we consider the nonlinear refractive index $n_2(E)$ induced by nonlinear electronic polarization effect is resulted by Stokes shift of dispersion curve, therefore, it is mainly determined by dispersion equation:

$$\gamma = \frac{e^2}{m} \sum_{K} \frac{f_K}{\omega_K^2 - \omega^2}, \tag{6.15}$$

where, m and e is the mass and the charge of the electron, respectively, f_K is the oscillator strength corresponding to the inherent frequency ω_K, and ω is the frequency of the incident beam. For representation of the dispersion curve of insulating solids in the visible region it is sufficient to take account of only two inherent frequencies, i.e. the electron transition frequency ω_1 in the ultra-violet region and the vibration frequency of the lattice ω_2 in the infra-red region. Therefore,

$$\gamma = \frac{e^2}{m}[\frac{f_1}{\omega_1^2 - \omega^2} + \frac{f_2}{\omega^2 - \omega_2^2}]. \tag{6.16}$$

The former plays a more important role than the latter. The simplified dispersion equation can be expressed as follows:

$$n_\omega^2 - 1 = f[\frac{\omega_0^2}{\omega_0^2 - \omega^2}]^{1/2}, \tag{6.17}$$

where $f = (n_\infty^2 - 1)$, when $\omega = 0(\lambda = \infty)$. Equation 6.17 is rewritten as follows:

$$\frac{1}{(n_\lambda^2 - 1)^2} = \frac{1}{(n_\infty^2 - 1)}[1 - \frac{\lambda_s^2}{\lambda^2}]^{1/2}. \tag{6.18}$$

Therefore, the ultra-violet inherent wavelength λ_s can be found by plotting $1/(n_\lambda^2 - 1)^2$ vs $1/\lambda^2$. Figures 6.4 and 6.5 show this relationship for inorganic glasses and halide crystals. The deviation from straight line in infra-red region is due to the resonant absorption of structural lattice vibration. The ensuring straight line can be extrapolated to zero to give λ_s. On the basis of some simplification, the λ_s value can also be obtained from the next equation by two linear refractive index values:

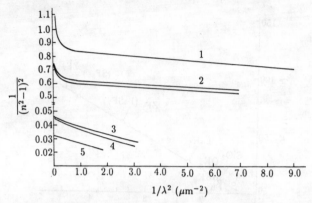

Fig. 6.4. Optical dispersion curves of inorganic glasses. 1, Fused silica; 2, ZrF_4-based fluoride glass; 3, As_2S_3; 4, $GeSe_4$; 5, $AsSe_4$.

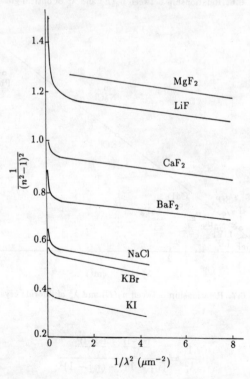

Fig. 6.5. Optical dispersion curves of halide crystals.

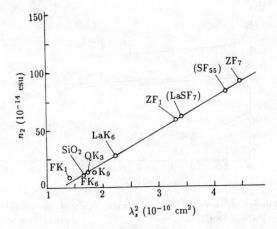

Fig. 6.6. Relationship between $n_2(E)$ and λ_s^2 of optical glasses.

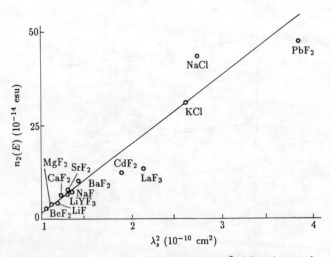

Fig. 6.7. Relationship between $n_2(E)$ and λ_s^2 of fluoride crystals.

$$\frac{1}{\lambda_s^2} = \frac{\frac{(n_1^2-1)}{\lambda_1^2} - \frac{(n_2^2-1)^2}{\lambda_2^2}}{(n_1^2-1)^2 - (n_2^2-1)^2}. \tag{6.19}$$

Figures 6.6–6.8 show the relation between $n_2(E)$ and λ_s^2 of optical glasses as

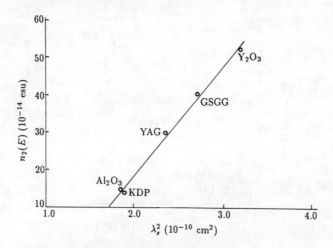

Fig. 6.8. Relationship between $n_2(E)$ and λ_s^2 of oxide crystals.

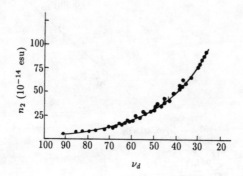

Fig. 6.9. Relationship between $n_2(E)$ and ν value of optical and laser glasses.

well as oxide and halide crystals. Thus we can empirically derive several formulas for calculating nonlinear refractive index $n_2(E)$ of optical and laser materials. There also exist some simple relationship between nonlinear refractive index $n_2(E)$ and dispersion coefficient ν (Abbe value). Figures 6.9–6.11 show the relationship between n_2 and ν value of optical glasses, oxide crystals and halide crystals. Therefore, we can also estimate $n_2(E)$ value from ν value.

We proposed the following equations for calculating nonlinear refractive index of laser and optical materials from λ_s or ν_d values.

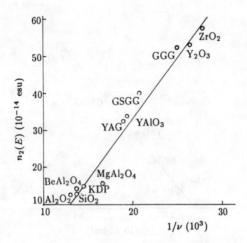

Fig. 6.10. Relationship between $n_2(E)$ and ν value of oxide crystals.

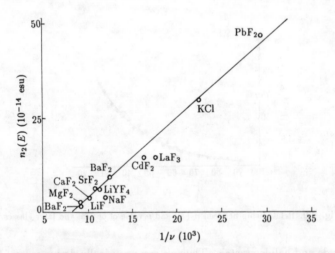

Fig. 6.11. Relationship between $n_2(E)$ and ν value of fluoride crystals.

$$n_2 = -39 \times 10^{-14} + 29 \times 10^{-10} \times \lambda_s^2 \text{ (esu)}, \qquad \text{(oxide glasses)} \tag{6.20}$$

$$n_2 = -11 \times 10^{-14} + 12 \times 10^{-10} \times \lambda_s^2 \text{ (esu)}, \qquad \text{(fluoride glasses)} \tag{6.21}$$

$$\nu_d \text{ at } 68 - 90, \quad n_2 = 25.24 - 0.23\nu_d \; (10^{-14} \text{ esu}), \qquad \text{(glasses)} \tag{6.22}$$

$$\nu_d \text{ at } 25 - 68, \quad n_2 = 3210/\nu_d - 34.67 \; (10^{-14} \text{ esu}), \qquad \text{(glasses)} \tag{6.23}$$

Table 6.4. Comparison between calculated and measured values of $n_2(E)$ of optical and laser glasses.

Glass	Cal. values (10^{-13} esu)			Exp. val. of n_2 (10^{-13} esu)
	$n_2(T)$	$n_2(S)$	$n_2(E)$	
ZF-7	1.05	1.09	9.0	6.4±1.1
BaF-2	0.37	0.22	3.6	4.1±0.4
QK-3	0.17	0.21	1.2	1.3±0.4
N0312	0.32	0.19	1.7	1.9±0.4
N0812	0.12	0.17	1.9	1.6±0.5
N1012	0.29	0.17	1.7	1.6±0.5
N2120	0.04	0.28	1.5	1.3±0.3
N2420			1.4	1.2

Table 6.5. Comparison between calculated and measured values of $n_2(E)$ of halide and oxide crystals.

Crystal	n_2 (10^{-14} esu)			Crystal	n_2 (10^{-14} esu)		
	Calculated value		Measured value		Calculated value		Measured value
	Eqs. 6.24 and 6.25	Eqs. 6.26 and 6.27			Eqs. 6.24 and 6.25	Eqs. 6.26 and 6.27	
LiF	4.0	4.3	4	LaF_3	25.0	20.0	14.9
NaF	7.5	7.8	7	$LiYF_3$	6.5	5.7	5.9
BeF_2	2.0	2.7	2.3	NaCl	40	32.3	45
MgF_2	4.0	3.5	3.0	KCl	36	30.3	31
CaF_2	6.0	5.2	6.5	KBr	53	45.9	71
SrF_2	6.5	5.9	6.0	KI	90	74.9	102
BaF_2	10.0	8.9	10	Al_2O_3	19	15	15
CdF_2	21.5	17.6	14.4	$Y_3Al_5O_{12}$	31	23	35
PbF_2	63.0	44.9	48	KH_2PO_4	14	15	10

Table 6.6. Values $n_K, n_2, \lambda_s, \nu_d$ of several glasses.

Glass	n_K (10^{-13} esu)		n_2 (10^{-13} esu)		λ_s (μm)	ν_d
	1.06 μm	0.53 μm	1.06 μm	0.53 μm		
ZF-7	5.7	14.4	8.3	21.4	0.216	25.4
BaF-2	2.0	5.3	3.0	8.0	0.172	49.4
N03	1.4	2.5	2.0	3.8	0.142	59.8
N24	0.77		1.2		0.130	66.6
SiO_2			0.97	1.16	0.130	67.5

$$n_2 = -15.6 \times 10^{-14} + 18.3 \times 10^{-10} \times \lambda_s^2 \text{ (esu)}, \quad \text{(fluoride crystals)} \quad (6.24)$$

$$n_2 = -36.1 \times 10^{-14} + 28.8 \times 10^{-10} \times \lambda_s^2 \text{ (esu)}, \quad \text{(oxide crystals)} \quad (6.25)$$

$$n_2 = 2206/\nu_d - 18.6 \ (10^{-14} \text{ esu}), \quad \text{(fluoride crystals)} \quad (6.26)$$

$$n_2 = 2875/\nu_d - 24.6 \ (10^{-14} \text{ esu}), \quad \text{(oxide crystals)} \quad (6.27)$$

Tables 6.4 and 6.5 demonstrate the comparison between the calculated and the

Table 6.7. Measured results of n_2 value at 1ω and 3ω.

Glass	$n_2 \times 10^{-14}$ esu at 3ω $(0.34 \ \mu m)$	$n_2 \times 10^{-14}$ esu at 1ω $(1.06 \ \mu m)$
Fused silica	$0.9 \pm 85\%$	0.95
BK-10	$0.6 \pm 100\%$	0.49

measured values of $n_2(E)$ of glasses and crystals respectively. The calculation error is about 15%.

6.4 Frequency Dependence of Nonlinear Refractive Index $n_2(E)$

The dispersion of nonlinear refractive index of laser materials is an important property in both practical and theoretical meaning. Hellwarth [13] remarked that on the basis of the classical anharmonic oscillator, one would expect the dispersion of n_2 to vary as $(\omega_0^2 - \omega^2)^{-4}$, where ω_0 is the zero-amplitude, resonant frequency of the oscillator. Glass proposed a useful functional form derived from the third-order perturbation theory [14].

On the basis of above proposed concept, the nonlinear refractive index is, as an increment in the refractive index change, due to Stokes displacement of the dispersion curve toward longer wavelength as the electric field intensity increases. Empirically, n_2 is proportional to λ_s^2 or inversely proportional to Abbe value ν_d. Therefore, the dispersion of nonlinear refractive index $dn_2/d\lambda$ will be related to λ_s or ν_d.

The dispersion of laser induced birefringence of laser glasses and optical glasses has been measured by the interferometric method [15]. The probe beam at 0.53 μm was the second harmonic of an intense laser beam at 1.06 μm and was applied to measure the laser induced birefringence. The experimental results are shown in Table 6.6. It is found that the optical Kerr coefficients n_K at 0.53 μm are larger than those at 1.06 μm. The values of λ_s and ν_d of these glasses are also listed in the table. It can be found that the increment of nonlinear refractive index of glasses $(n_2(0.53)/n_2(1.06))$ is proportional to λ_s, as shown in Fig. 6.12.

A. J. Glass predicated that the value of n_2 at 3ω (0.34 μm) of Nd^{3+}:YAG laser would be 1.5 times that of n_2 at 1ω (1.06 μm) in fused silica [16]. But W. T. White and P. Milam measured n_2 value at 3ω in fused silica and borosilicate optical glass BK-10 by time-resolved interferometer [17], the results are shown in Table 6.7. The n_2 was no greater at 3ω rather than at 1ω. It is obvious due to the beam quality problem at 3ω generation.

Figure 6.13 shows the wavelength dependence of nonlinear refractive index of LiF crystal. The n_2 value increased by more than 1.5 times when the laser wavelength was changed from 1ω to 3ω.

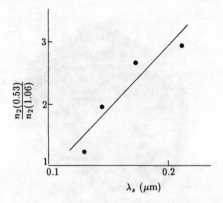

Fig. 6.12. Relation of increment of nonlinear refractive index with λ_s.

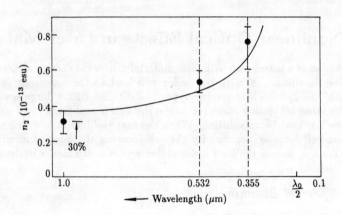

Fig. 6.13. Nonlinear refractive index of LiF at three Nd laser wavelengths.

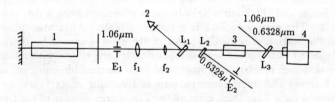

Fig. 6.14. Apparatus for studying thermal blooming.

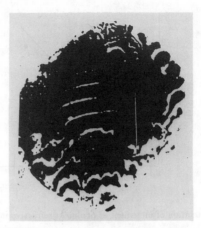

Fig. 6.15. Interference pattern of thermal blooming.

6.5   Nonlinear Optical Effects in Laser Materials

The interaction of a laser beam with laser material causes the change of its refractive index. The fluctuation of refractive index will lead to the wavefront (or phase) distortion of the light. There are two types of light distortions: large scale (at cm level or the integral) lensing effect (of course the lens is aspherical) and small scale (at mm level or less) beam splitting. When the material interacts with high power laser beam, self-focusing occurs. By the self-focusing the laser induced damage appears. We will discuss the laser induced damage in next chapter in detail.

6.5.1   *Thermal Blooming*

The thermal blooming effect is a small scale beam splitting due to absorption of longer pulse laser beam.

The interaction of glass with multimode ms laser pulse is now under study. The experimental set-up is illustrated in Fig. 6.14. A He-Ne laser is used as diagnostic beam and the intense light beam is generated from a free running Nd-glass laser 1. The glass sample 3 is a rod or a plate. After absorption of laser energy, the fluctuation of refractive index of the glass is caused by thermal effect. Multimode laser beams induce several aspherical lensing effects with a linearity less than cm level, which overlap each other, and form thermal blooming of laser beam. Figure 6.15 shows the interference pattern of blooming effect taken by a Jamin interferometer 4 for a 1 cm thick glass plate. The energy density of the free running laser is about 1.65 J/mm^2 and pulse duration is 10 ms. The thermal blooming

Fig. 6.16. The thermal diffusion process of focusing spot in BaK$_7$ glass when a laser pulse is passing through (average energy density 0.422 J/cm^2, pulse duration 5 ms).

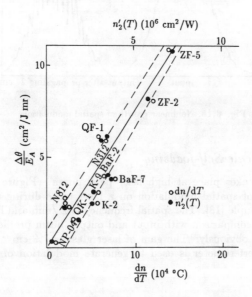

Fig. 6.17. Thermal blooming in glasses as a function of $n_2(T)$.

effect is time integrated, its response time is equal to that within ns range for the establishment of the temperature gradient. Figure 6.16 illustrates the thermal blooming process during a 10 ms laser pulse passing through a BK-7 optical glass ($\alpha_{1.06}=0.01$ cm^{-1}). After terminating the laser pulse, the disappearance process of the blooming effect is just that of thermal relaxation within the order of seconds range.

The relationship between the distortion (or divergence) of the laser beam and the physical properties of glass was also established. Figure 6.17 shows that in ms range of pulse duration, the divergence of the output laser beam is proportional to the thermal nonlinear refractive index $n_2(T)$ for different kinds of optical and laser glasses.

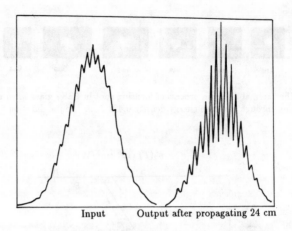

Input   Output after propagating 24 cm

Fig. 6.18. Nonlinear growth of spatial modulation.

6.5.2   *Small-scale Self-focusing*

The self-focusing takes place at high laser peak power. Figure 6.18 shows the nonlinear growth of spatial modulation on a laser pulse during the propagation through a glass sample [18]. The spatial frequency is 2 /mm and laser peak power is 5 GW/cm². In comparison with input and output beam profile the small-scale self-focusing arose obviously. The gain of laser glass is 7% cm^{-1} at 1.06 μm. A shearing plate interferometer is used to generate modulation of a single spatial frequency.

Reference

1. F. F. Arecchi, E. O. Schulz-Dubois, *Laser Handbook* (North-Holland, 1972).

2. H. J. Juretschke, *Crystal Physics: Macroscopic Physics of Anisotropic Solids* (Benjamin, Reading, Mass., 1974).

3. Fuxi Gan and Fengying Lin, *Chinese Phys.* **2** (1982) 462.

4. R. M. Waxler, *IEEE J. Quant. Electron.* **QE-7** (1971) 166.

5. V. Shatiliv *et al.*, *Soviet J. Opt. Tech.* **39** (1972) 203.

6. He Deng and Fuxi Gan, *Proceedings of 1981 Symposium on Laser Induced Damage in Opical Materials*, NBS Special Publication **638** (1983) 568.

7. Fuxi Gan and Fengying Lin, *Chinese J. Lasers* **6** (1979) 12.

8. Fuxi Gan and Fengying Lin, *Chinese J. Physics* **8** (1979) 385.

9. P. W. Langhoff, S. T. Epstein and M. Karplus, *Mod. Phys.* **44** (1972) 602.

10. J. J. Wynne, *Phys. Rev.* **178** (1969) 1295; *Appl. Phys. Lett.* **12** (1968) 191.

11. C. C. Wang, *Phys. Rev.* **B2** (1970) 2045.

12. N. L. Boling, A. J. Glass and A. Owyoung, *IEEE J. Quant. Electron.* **QE-14** (1978) 601.

13. R. W. Hellwarth, *Prog. Quant. Electron.* **5** (1977) 2.

14. A. J. Glass, *Digest of 1983 International Conf. on Lasers (Guangzhou)*.

15. Chengfu Li and He Deng, *Chinese J. Lasers* **10** (1983) 110.

16. A. J. Glass, *LLNL. Intern. Memorandum*, AJG-80-377A (Oct. 8, 1980).

17. D. Milam and M. J. Weber, *J. Appl. Phys.* **47** (1976) 2497.

18. E. Bliss, D. R. Speck, I. F. Holzrichter, J. H. Erkkila and A. J. Glass, *Appl. Phys. Lett.* **25** (1974) 448.

7. Laser Induced Damage in Laser Materials

The development of high power laser systems depends greatly on progress of laser and optical materials. With shortening the laser pulse duration the laser peak power increases, that can induce damage in solid state materials. The continuous wave laser output power can also reach tens of thousands of watts, therefore, the thermal break of laser medium and window materials is an important problem for high power laser technology. The study of the laser induced damage in optical and laser crystals are insufficient in comparison with optical and laser glasses and thin films, and the application of optical and laser crystals in high power laser systems is more widespread recently, so that it is worth investigating the laser induced damage problem in laser materials in detail.

7.1   Laser Induced Damage in Glasses

When an intense laser beam passes through a glass medium, the refractive index of glass is changed owing to the inhomogeneous distribution of the laser intensity, which leads to self-focusing in glass. The local intensity of laser beam will be increased by several orders of magnitude due to the self-focusing, so laser induced damage occurs. Filamentary damage in the glass prepared in a ceramic crucible and spot damage in the glass melted in a platinum crucible are shown in photograph (Fig. 7.1).

7.1.1   *Inclusion Induced Damage*

Considering the small size of the inclusions, and the approximate energy required to fracture glass, one can conclude that only metallic inclusions can provide sufficient absorption over small distances to damage glass. Up to now high quality phosphate and fluorophosphate laser glasses must be melted in platinum crucibles, metallic platinum is the likely culprit. A very small amount of Pt in the glass can result in high density of inclusions, e.g. 1 ppm of Pt can produce 4×10^7 particles/cm^3 with an average particle diameter of 0.1 μm.

146

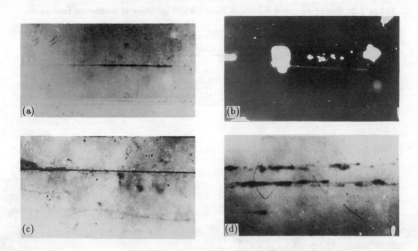

Fig. 7.1. Filamentary damage in the glass melted in a ceramic crucible and spot damage in the glass melted in a platinum crucible. a, Filamentary damage in the glass melted in platinum crucible by mono-mode laser light at 1–2 J/cm^2, 30 ns; b, spot damage in the glass melted in a platinum crucible; c, filamentary damage in the glass melted in a ceramic crucible by mono-mode laser light at 30–50 J/cm^2, 30 ns; d, filamentary damage in the glass melted in platinum crucible by multi-mode laser light at 1–2 J/cm^2, 30 ns.

Table 7.1. Laser induced damage threshold of glasses made in ceramic and platinum crucibles.

Type of laser glass	Crucible	Absorption coeff. (cm^{-1})	Laser pulse width	Damage thershold J/mm^2
N312	Pt	0.001	3 ms	3.6
N320	Pt	0.002	3 ms	2
N312	Ceramic	0.003	3 ms	37
N320	Ceramic	0.004	3 ms	24
N312	Pt	0.001	4 ns	0.065
N324	Pt	0.0015	4 ns	0.049
N312	Ceramic	0.004	4 ns	0.082
N320	Ceramic	0.006	4 ns	0.068

From the technological point of view the main goal is to eliminate the metallic inclusions in the laser glasses for improving the laser induced damage threshold, as the high pure laser glasses are always melted in platinum crucible. As shown in Table 7.1 [1], the damage threshold E_d of silicate laser glasses made in ceramic crucible was much higher than that of glasses melted in platinum crucible at long laser pulse duration (ms range), but the damage threshold difference was not so

Table 7.2. Size of damaged volume in fluorophosphate glass vs number of laser shots.

Energy fluence (J/cm²)	No. of shots	Diameters of damaged volumes (μm)
10 to 12	2	10, 10, 15, 25
10 to 12	4	10, 10, 30, 50
10 to 12	8	30, 30, 40, 40, 120, 140
6 to 8		No damage observed after 8 shots

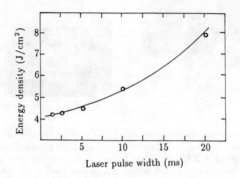

Fig. 7.2. Relationship between the damage threshold and laser pulse width.

large at short laser pulse duration.

Table 7.2 shows the size of damage volumes in fluorophosphate glass vs number of laser shots [2]. Damage measurements by Gonzales and Milam on Pt inclusions in phosphate laser glass showed that the damage threshold E_d scales with pulse duration τ as $E_d = 2.2\tau^{0.3}$, where E_d is in J/cm² and τ is in ns [3]. Model calculations of inclusion damage agree with the experimental results [4].

7.1.2   Damage at Long Laser Pulse Duration (ms)

As shown in Figs. 7.2 and 7.3, at long laser pulse duration (ms) the damage threshold E_d increases with prolonging the laser pulse width, and damage thershold P_d decreases with increasing the absorption coefficient. Therefore, the origin of damage at long laser pulse (ms) is a thermal one.

The laser damage thershold at long laser pulse width (ms) is dependent on the laser beam quality, as shown in Table 7.3 the laser damage threshold decreased pronouncedly with minimizing the divergence angle of laser beam.

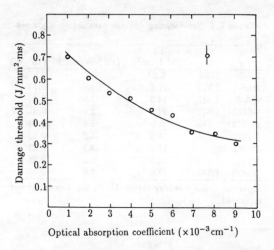

Fig. 7.3. Influence of optical absorption coefficient on damage threshold (laser pulse duration∼5 ms).

Table 7.3. Influence of the divergence angle of laser beam on laser damage threshold (laser pulse duration∼5 ms).

No.	Divergence angle (mrad)	Cavity length (m)	Damage threshold (J/mm²·ms)
1	1.5	14	0.2∼0.3
2	2.5	14	3.6
3	6.6	2.6	8

7.1.3 Damage at Short Laser Pulse Duration (ns)

The laser induced damage threshold at short laser pulse width (ns) of a series of optical and laser glasses has been measured by us, the laser pulse duration is 5 ns at 1.06 μm wavelength. The results are listed in Table 7.4.

The damage under the action of nanosecond laser pulse mainly depends on $n_2(E)$ caused by nonlinear polarization effect. This dependence is shown in Fig. 7.4. Usually, the materials with smaller nonlinear refractive index possess higher damage thresholds, but they are not completely consistent because the laser induced damage of materials also depends on other physical properties besides the nonlinear refractive index n_2.

The laser induced damage threshold at short laser pulse duration (∼ns) is also dependent on laser pulse width. The experimental results are shown in Table 7.5 [5], the damage threshold (laser power density P_{th} or electric field strength E_{th}) decreases with prolonging the pulse width τ. The relation can be expressed as $E_{th} \sim \tau^{-0.256}$. It means that the damage threshold E_{th} is inversely proportional to the fourth order of magnitude of pulse duration τ.

Table 7.4. Self-focusing damage parameters of glasses.

Glass	n_0	E_d (10^2 J/cm^2)	P_d (10^{10} W/cm^2)	P_{cr} (10^6 W)
ZF-7	1.806	6.6	1.3	0.082
BaF-2	1.570	19.5	4.9	0.13
QK-3	1.487	14.3	2.9	0.4
N0312	1.522	14.5	2.9	0.3
N0812	1.535	9.5	1.9	0.33
N1012	1.517	16.2	3.2	0.33
P7701		17.2	3.7	0.33
N2120	1.578	26.5	5.3	0.38
N2420	1.543	18.0	3.8	0.40

E_d is the damage energy threshold; P_d the damage power
threshold.

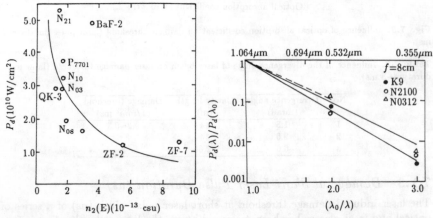

Fig. 7.4. Laser induced damage threshold
as a function of $n_2(E)$ of glasses.

Fig. 7.5. The comparison between experimental
and theoretical results of laser damage.

Table 7.5. Threshold of damage in silicate laser glass.

Pulse width τ (ns)	13	27	40	48	70
Power density $P_{th} \times 10^9$ (W/cm^2)	577	508	305	193	90.6
Electric field strength E_{th} (MV/cm)	16.9	15.8	12.3	9.8	6.7

7.1.4 Wavelength Dependence of Laser Induced Damage

As interest grows in shifting the operation wavelength of laser application toward
shorter wavelength, the question of frequency dependence of laser induced damage
becomes more important.

Table 7.6. Experimental results of laser damage in several glasses at different wavelengths.

Glass	Focal spot area ($cm^2 \times 10^{-6}$)	13.3			5.9		
	Laser wavelength (μm)	1.064	0.532	0.355	1.064	0.532	0.355
	Damage threshold ($10^9 \times W/cm^2$)						
K9		24	2.9	0.15	32.0	1.6	0.1
N2100		23.0	1.5	0.20	29.0	2.9	0.35
N0312		25.0	2.2	0.1	45.0	2.8	0.10

Laser induced damage thresholds in some silicate and phosphate glasses at the fundamental, the second and the third harmonic generation of Nd laser light have also been measured. Table 7.6 shows the results. It can be seen that when the frequency of laser beam is doubled, the laser damage threshold is decreased by about one order of magnitude [6].

The present theoretical models for laser induced damage in transparent dielectrics include electronic avalanche, multiphoton ionization and composite models. According to the electron avalanche model, the damage threshold increases with increasing frequency, but is independent of frequency at longer wavelengths [7]. Multiphoton ionization theory asserts that the damage threshold decreases with increasing frequency [8]. Our experimental results can be summarized by empirical formula shown below :

$$\ln[\frac{P_d(\lambda)}{P_d(\lambda_0)}] = 2.55(1 - \frac{\lambda_0}{\lambda}).\qquad (7.1)$$

Here, $P_d(\lambda_0)$ is the damage threshold at $\lambda_0 = 1.06$ μm. Figure 7.5 shows the comparison of the theoretical calculations and the experimental results of optical glass K9 and laser glass N2100 and N0312 at different wavelengths. It can be seen that the experimental results agree with the multiphoton ionization theory. Therefore, we conclude that in laser glasses, the laser induced damage is mainly due to multiphoton ionization.

7.2   Laser Induced Damage in Crystals

Laser induced damage in crystals can be classified as surface damage and bulk damage. The bulk damage in crystals can be raised by defects and intrinsic causes, the latter is caused by nonlinear optical effects, such as self-focusing, which increases the laser power density.

Figure 7.6 shows the morphology of surface damage of Ti:sapphire Al_2O_3 crystals. The surface damage is induced by optical absorption of defects and impurities

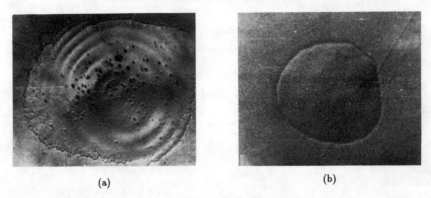

(a)   (b)

Fig. 7.6. Surface damage morphology of Ti-doped sapphire (0.04 wt% Ti) at laser wavelength 0.53 μm. (a) Around damage threshold power, without surface treatment; (b) around damage threshold power, after Ar$^+$ laser polishing.

on the surface, the high temperature plasma always occurs during laser damage process. There exists local damage or damage points.

Figure 7.7 demonstrates the bulk damage induced by defects and impurities in Ti:sapphire crystal. It is also caused by absorption of impurities, specially, metallic particles. The size of point damage is around tens of micrometers.

Figure 7.8 shows the filamentary damage in sapphire, which is characterized by self-focusing at high laser power density.

As the large size harmonic crystals, KDP and KD*P have been concentrated for laser damage study [9, 10]. A little work on laser crystals was carried out [11]. The thermal induced damage of optical window crystals, such as LiF, CaF$_2$, KCl, have been studied in recent years [12, 13]. Recently we studied the laser induced damage in sapphire, Ti:Al$_2$O$_3$, LiF and KDP [5, 14, 15].

7.2.1 *Laser Induced Surface Damage*

The laser induced damage appears more easily in the surface than in the bulk. It is obvious that more impurities, such as polishing powders, moisture, air particles adhered to the surface. Therefore, the surface treatment can improve the laser damage threshold greatly [16, 17]. We treated the Ti:sapphire crystal surface by molten borex etching and Ar$^+$ laser polishing, and measured the damage threshold before and after treatment [18]. Figure 7.9 shows the distribution curves of surface damage probability with laser energy density. The doping concentration of samples A and B are 0.062% and 0.04% wt respectively. The laser damage threshold can be increased about 1.5 times after molten borex etching, and about 1 time after Ar$^+$ laser polishing. The effective thickness for surface treatment is around 3–5 μm. It can be seen from Fig. 7.6 that the surface morphology of the crystal before and

Fig. 7.7. Bulk damage in Ti-doped sapphire crystal (0.04 wt% Ti) caused by metallic particle at laser wavelength 0.53 μm. Incident surface (0001), damage face (1100), breaking along the direction (1120).

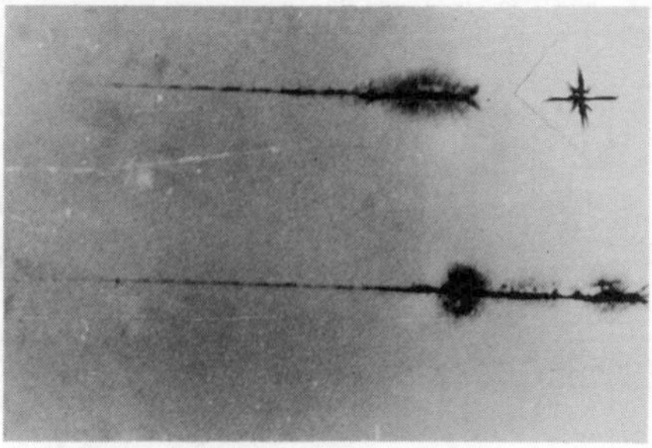

Fig. 7.8. Filamentary damage in sapphire crystal at laser wavelength 0.53 μm, pulse width 10 ns.

Table 7.7. Surface damage thresholds measured for sapphire (laser: 20 ns, 1.06 μm).

Radiation type	Surface damage threshold (j/cm^2)	
	E_{ent}	E_{exit}
Normal incident	58	31
Brewster incident	72	60

after the treatment is different, the surface is smooth and no impurity exists after treatment.

Besides the surface polishing condition, the laser damage threshold depends on doping Ti concentration or absorption coefficient. In Fig. 7.10 it shows that large influence happens at low doping concentration.

According to Fresnel's reflection equation, at normal incidence the intensity of exit light from back surface (I_{ex}) is stronger than that of incident light at front surface (I_{in}). The intensity ratio is [19],

$$R = \frac{I_{\text{ex}}}{I_{\text{in}}} = \left(\frac{4n}{(n+1)^2}\right)^2. \tag{7.2}$$

Here n is refractive index of crystal, for sapphire $n = 1.76$, $R = 1.63$. Table 7.7 lists the surface damage threshold of sapphire at the front and the back surfaces at normal incidence and Brewster angle. The experimental results are fitted with the above equation Eq. 7.2. The threshold ratio is about 1.8 at normal incidence.

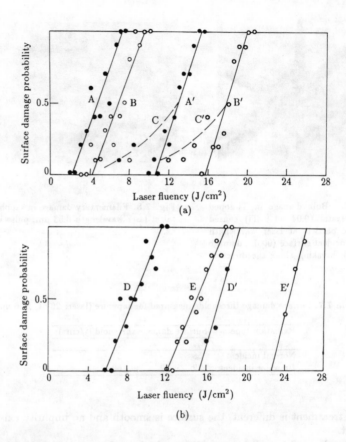

Fig. 7.9. Probability distribution curves of laser damage threshold with pulsed laser energy density of Ti-doped sapphire ($\lambda=0.53\ \mu m$, $t=10$ ns). (a) Molten borex etching; (b) Ar$^+$ laser polishing. A, B, C, D, E: before treatment; A', B', C', D', E': after treatment.

The surface damage is also dependent on laser pulse duration and wavelength. As shown in Table 7.8, the laser induced surface damage in different crystals at sub-nanosecond pulse duration and UV wavelength is very low [20]. The first six fluoride samples had an average damage threshold of 1 J/cm^2 at 0.1 ns and 2 to 3 J/cm^2 at 0.7 ns. We hypothesize that the general clustering of these thresholds may be due to the surface-finishing techniques rather than to any intrinsic material behavior.

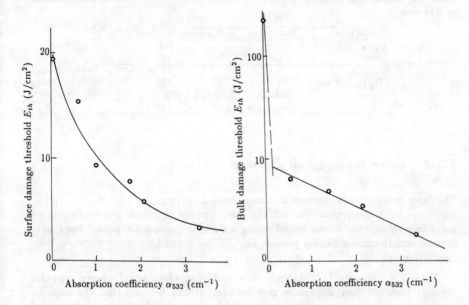

Fig. 7.10. Relationship between surface damage threshold and absorption coefficient of Ti-doped sapphire (λ=0.53 μm, t=10 ns).

Fig. 7.11. Relationship between bulk damage threshold and absorption coefficient of Ti-doped sapphire (λ=0.53 μm, t=10 ns).

Table 7.8. Damage thresholds of bare surfaces at two laser pulse lengths and wavelength of 0.266 μm.

Material	Fluence (J/cm^2)	
	0.1 ns	0.7 ns
LiF	0.60±0.39	—
BaF$_2$	0.50±0.13	2.06±0.72
SrF$_2$	0.47±0.12	1.54±0.51
CaF$_2$	0.65±0.15	2.37±0.83
MgF$_2$	1.32±0.32	2.69±0.41
LaF$_3$	1.80±0.80	3.10±0.70
Al$_2$O$_3$	0.75±0.26	—
CD*A	1.10±0.27	—
KDP	>(1.70±0.25)	6.48±1.40
LiF	—	14.30±2.50
LiF (hot forged)	—	14.40±2.50

Table 7.9. Bulk damage thresholds measured for sapphire and titanium doped sapphire (laser: 10 ns, 0.53 μm).

Crystal type	Defect type	Bulk damage threshold (J/cm^2)
Al_2O_3		210
(pure)	Ir particle	11
$Ti:Al_2O_3$		31
(Ti_2O_3; 0.04 wt%)	Bubble	73
	Inclusion	21
	Mo particle	14

7.2.2 *Laser Induced Bulk Damage*

The laser induced bulk damage is always caused by impurity particles in crystal, which are light absorption centers. The laser damage threshold can be calculated according to thermal stress model, using adiabatic process for pulsed laser irradiation and thermal diffusion process for CW laser and long pulse duration laser irradiation.

Table 7.9 shows the bulk damage threshold of sapphire and Ti-doped sapphire crystals with different impurities and defects. It can be seen that the metallic particles, such as iridium, are more harmful because the absorption coefficient of metals are so high. Impurity inclusions, such as TiO_2, TiAlO, with higher absorption coefficient decrease the laser damage threshold, dislocations and bubbles have less influence on damage threshold. Figure 7.11 demonstrates the dependence of damage threshold on the Ti-doping concentration, which is obvious due to light absorption of Ti ions.

For the most of nonlinear optical crystals their mechanical and thermal strength are too low, and the stress characterizes anisotropic, the laser induced damage causes the break of crystals [21, 22].

Figure 7.12 shows the histogram of one on one KDP bulk-damage threshold measured with 1.06 μm and 1 ns pulse. The average damage threshold is around 8 j/cm^2, and it strongly depends on contamination of the crystal. As shown in Fig. 7.13, the damage threshold of KDP crystals with contamination<10 ppb is 3 times that of crystals with 1 ppm contamination [23]. Table 7.10 listed the typical impurity concentration ranges in KDP crystals, but the organic impurities are most harmful. The growth of crystals with UV irradiation, laser irradiation and thermal treatment can improve the resistance to laser induced damage. Table 7.11 shows some experimental results [24].

The intrinsic damage can be raised by nonlinear optical effects at short pulse laser. The filamentary damage always occurs. We have systematically studied the intrinsic damage in KDP crystals [5, 25].

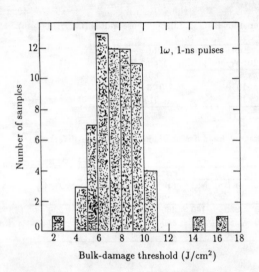

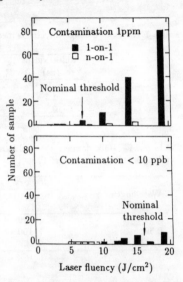

Fig. 7.12. Histogram of one-on-one KDP bulk-damage thresholds measured with 1 ω, 1-ns pulses.

Fig. 7.13. Histogram of damage threshold of KDP crystals with different contamination.

Table 7.10. Typical impurity concentration ranges in KDP crystals.

Element	Concentration	Element	Concentration
Na	1 to 3 ppm	Mg	1 to 2 ppm
Al	2 to 30	Si	10 to 20
Ca	4 to 15	Cr	1
Fe	1 to 7	Rb	1 to 3
Cs	3 to 10	Mo	1 to 10 ppm

Table 7.11. Relationship between thermal treatment of laser irradiation and the bulk damage threshold measured for KDP (laser: 1 ns, 1.06 μm).

Treatment	Laser irradiation 20		Thermal treatment 110–160°C	
Threshold (J/cm^2)	Single	Multiple	Single	Multiple
Before treatment	2.9	9.1	7	11
After treatment	7.7	11.8	10	15

Table 7.12. Bulk damage thresholds measured for KDP in different pulse widths (laser: 1.06 μm).

Pulse width (ns)	Spot area (10^{-6} cm^2)	Bulk damage threshold	
		P_{th} ($10^9 \cdot$W/cm^2)	E_{th} (J/cm$^2 \cdot 10^2$)
13	7.33	281	3.7
27	7.33	27.8	7.5
40	7.33	21.8	8.4
48	7.33	10.7	5.1
70	7.33	7.46	5.2

Table 7.13. Bulk damage thresholds measured for BBO in different pulse widths (laser: 1.06 μm).

Pulse width (ns)	Bulk damage threshold P_{th} ($10^9 \cdot$W/cm^2)	E_{th} (J/cm$^2 \cdot 10^2$)
10	5	0.5
1.3	10	0.13
0.1	15	0.015

Table 7.14. Bulk damage thresholds measured for different kinds of crystals in different laser wavelength (GW/cm^2) (laser: 10 ns).

Wavelength (μm)	Crystal							
	NaCl	KCl	KBr	NaF	LiF	CaF$_2$	Al$_2$O$_3$	SiO$_2$ (C)
1.06	120	70	50	140	360	200	400	230
0.69	150	80	58	140	360	—	400	230
0.266	45	50	50	18	240	380	18	70

7.2.3 *Relationship between Damage Threshold and Laser Pulse Duration*

Tables 7.12 and 7.13 show the experimental results of damage threshold of KDP and BBO nonlinear optical crystals respectively. The damage power density decreases and energy density increases with shortening the laser pulse duration in the range of <40 ns [5].

The relationship between damage threshold and laser pulse duration depends on laser induced damage mechanism. In references [26, 27] different kinds of damage mechanism have been discussed. According to theoretical calculation the relation between laser pulse width t and damage threshold (in electric field intensity) P_{th} can be expressed as $P_{th} \propto t^{-0.035}$, $t^{-0.155}$, $t^{-0.25}$, for ionization avalanche, multiphoton absorption and ionization, as well as plasma heating respectively. According to our experimental results the multiphoton absorption and ionization is the main source for laser induced damage at short pulse (<30 ns). At longer pulse duration the laser damage is accompanied with plasma heating. The thermal effect is pronounced for long laser pulse.

7.2.4 *Relationship between Damage Threshold and Laser Wavelength*

As shown in Table 7.14, the experimental data about light wavelength dependence of damage threshold of different optical and laser crystals were rather dispersive in previous references [12]. We measured the damage threshold of KDP crystals at three different wavelengths at the same experimental apparatus. The measured

Table 7.15. Relationship between bulk damage threshold and laser wavelength on KDP.

Wavelength (μm)	Spot area $(cm^2 \cdot 10^{-6})$	Bulk damage threshold	
		$(W/cm^2 \cdot 10^{12})$	$(Volt/cm \cdot 10^6)$
1.0642	10	8.4	2.0
	5.3	10.3	2.2
0.532	5	1.93	0.98
	2.6	2.22	1.0
0.355	3.1	0.54	0.52
	1.71	0.69	0.58

Table 7.16. Relationship between bulk damage threshold and laser wavelength measured for urea and BBO.

Wavelength (μm)	Bulk damage threshold (GW/cm^2)	
	BBO	Urea
1.064	5	5
0.532	1	3

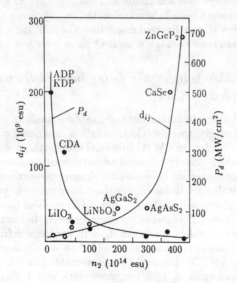

Fig. 7.14. Relationship between laser damage threshold P_d and frequency doubling coefficient d_{ij} with nonlinear refractive index $n_2(E)$ of nonlinear optical crystals.

results are shown in Table 7.15. It can be seen that the damage threshold decreases with the shortening laser light wavelength [25]. Similar results have been obtained for BBO crystal and urea organic crystal (Table 7.16).

It has been pointed out that laser damage due to plasma heating is wavelength

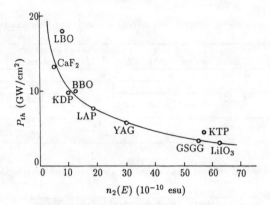

Fig. 7.15. Laser damage threshold dependence P_{th} with nonlinear refractive index $n_2(E)$ of optical and laser crystals (at 1.06 μm and 1 ns).

independent, but multiphoton absorption and ionization causes lowering the damage threshold with decreasing laser light wavelength [28, 29]. From the above two laser damage relationships it can be confirmed that the multiphoton absorption and ionization is the main laser damage process at short laser pulse width ($<$20 ns).

7.2.5 *Relationship between the Laser Induced Damage and Nonlinear Refractive Index*

We have developed the calculation method for three kinds of nonlinear refractive index $n_2(T)$, $n_2(S)$, $n_2(E)$: thermal, electrostriction, and nonlinear polarization [30–32]. For long laser pulse (ms or CW) the thermal effect is the dominative one in laser induced damage, which mainly concerns with thermal nonlinear refractive index $n_2(T)$. Good relationship exists between laser damage power threshold and $n_2(T)$ of optical window crystals for CW lasers. For short laser pulse (ns or μs) the refractive index caused by nonlinear polarization change $n_2(E)$ plays an important role in laser induced damage. Figure 7.14 shows relationship between the laser damage threshold P_d and the frequency doubling coefficient d_{ij} with nonlinear refractive index $n_2(E)$ calculated by us. The laser damage threshold data were collected from reference published at early stage [33]. Figure 7.15 demonstrates this relationship $P_d \sim n_2(E)$ from recent measured data at 1.06 μm wavelength and 1 ns pulse width. It is obvious that with increasing nonlinear refractive index $n_2(E)$ the damage threshold drops rapidly.

References

1. Research Report of Shanghai Institute of Optics and Fine Mechanics, Vol. 2, 1974, p. 142–154.

2. S. E. Stokowski, Laser Program, Annual Report 1979, LLNL, UCRL-50021-79, p. 2–154.

3. R. P. Gonzales and D. Milam, in *Laser Induced Damage in Optical Materials*, NBS Special Pub. **746** (1985) 128.

4. J. H. Pitts, in *Laser Induced Damage in Optical Materials*, NBS Special Pub. **746** (1985) 537.

5. Meizhen Zhang and Chengfu Li, *Acta Optica Sinica* **5** (1985) 667.

6. He Deng and Chengfu Li, *Acta Optica Sinica* **3** (1983) 766.

7. A. Vaidyanathan, *IEEE J. Quant. Electron.* **QE-16** (1980) 89.

8. A. Schmid, P. Kelly, *et al.*, *Phys. Rev.* **B16** (1977) 4569.

9. L. E. Swain, *et al.*, Report of National Bureau Standard (NBS) **638** (1981) 119.

10. D. Eimerl, S. P. Velsko, Annual Report of Lawrence Livermore National Laboratory (LLNL), UCRL 50021-87 (1987), p. 5–27.

11. D. Milam, I. M. Thomas, Annual Report LLNL, UCRL 50021-84 (1984), p. 6–43.

12. B. G. Gorshkov, *et al.*, Report of NBS **638** (1981) 76.

13. D. Milam *et al.*, Annual Report of LLNL, UCRL-85 (1985), p. 7–56.

14. Jiang Zhou, Jingwen Qiao and Peizhen Deng, *Chinese J. Lasers* **16** (1989) 364, 432.

15. Qiang Zhang, Ph.D. Thesis of Shanghai Institute of Optics and Fine Mechanics (SIOFM), 1990.

16. C. C. Wang *et al.*, *Phys. Rev.* **185** (1969) 1079.

17. J. B. Frank and M. J. Soileau, Report of NBS **356** (1981) 114.

18. Qiang Zhang, Peizhen Deng and Fuxi Gan, *Materials Lett.* **9** (1990) 128.

19. M. D. Crisp *et al.*, *Appl. Phys.* **21** (1972) 364.

20. F. Rainer, T. F. Deaton, D. Wirtenson, Annual Report LLNL, UCRL-50621-80, **1** (1980), p. 2–231.

21. M. F. Singleton, Annual Report of LLNL, UCRL 50021-87 (1987), p. 5–23.

22. A. Yokotaini *et al.*, Report of NBS **746** (1985) 101.

23. T. Sasaki, A. Yokotani *et al.*, *Jap. J. Appl. Phys.* **26** (1987) L1767.

24. Y. Nishida, A. Yokotani *et al.*, *Appl. Phys. Lett.* **52** (1988) 420.

25. Chenfu Li and Meizhen Zhang, *Chinese J. Lasers* **12** (1985) 54.

26. I. Guenther, A. Vaidyanahan *et al.*, *IEEE J. Quant. Electron.* **QE-16** (1980) 89.

27. I. R. Betlis *et al.*, Report of NBS **464** (1976) 338.

28. A. Vaidyn *et al.*, *IEEE J. Quant. Electron.* **QE-16** (1980) 89.

29. A. Schid *et al.*, *Phys. Rev.* **B16** (1977) 4569.

30. Fuxi Gan and Fengying Lin, *Physica Sinica* **8** (1979) 12.

31. Fuxi Gan and Fengying Lin, *Acta Optica Sinica* **1** (1981) 75.

32. He Deng and Fuxi Gan, Report of NBS **638** (1981) 568.

33. R. Fischer, *Sov. Quant. Electron.* **QE-7** (1977) 135.

8. Defects in Laser Materials

It is well known that the perfection of laser materials is one of the most important performances. The defects influence the optical homogeneity and losses, laser output efficiency and beam quality, as well as laser induced damage. Defects in crystals include constitutional supercooling, facets, scattering particles and dislocations. The inclusion particles and striae are the most harmful defects in glasses.

8.1 Characterization Techniques

8.1.1 *Optical Methods*

A. The conventional optical microscopy includes transmitted and reflected light (bright field) and differential interference contrast (both in transmitted and reflected light) for observing the inclusions, etch pit patterns of dislocations and other defects in crystals and glasses. The resolution is less than 1 μm.

B. Birefringence topography. Optical birefringence image (photoelastic pattern) induced by strain field of dislocations in isotropic crystals can be seen under cross polarizing microscope.

C. Ultramicroscopy.

The light scattering effect caused by small particles whose dimensions are less than the wavelength order of magnitude (Tyndall effect) can be observed by this method.

D. Laser light scattering tomography (LLST).

It is composed of ultramicroscopy, scanning technique and He-Ne laser. The resolution is 10.0 nm if the laser beam is focused into a very fine one (<50 μm) using LLST. The smaller particles, decorated dislocations and undecorated edge dislocations can be observed. Figure 8.1 shows the LLST apparatus.

E. Dark field microscopy (DFM)

Using the reflected and diffracted light from the surface of the object by tilted illumination the defects of crystals can be observed. The resolution is 4.0 nm.

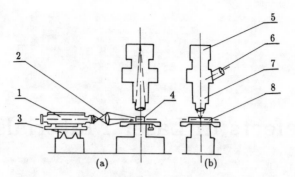

Fig. 8.1. Schematic diagram of the laser light scattering tomography apparatus. (a) side view; (b) . front view. 1, He-Ne laser; 2, focusing lens; 3, moving stage for beam scanning; 4, end of specimen; 5, reflex type camera; 6, eyepieces; 7, microscope; 8, side of specimen.

Table 8.1. X-ray absorption coefficient and sample thickness of four kinds of crystals.

	Molecular weight	Density (g/cm^2)	Mass absorption coefficient		Linear absorption coefficient	Sample thickness
			Target	μ_m	μ_l	(mm)
$Y_3Al_5O_{12}$	539.63	4.55	$CuK\alpha$	73.13	335.5	0.03
			$MoK\alpha$	49.61	225.7	0.04
			$AgK\alpha$	25.06	114.3	0.09
$BeAl_2O_4$	129.63	3.69	$CuK\alpha$	26.51	84.31	0.12
			$MoK\alpha$	3.03	11.8	0.89
MgF_2	62.30	3.18	$MoK\alpha$	2.88	9.18	1.10
Al_2O_3	101.94	3.98	$MoK\alpha$	3.50	13.93	0.70

8.1.2 *Chemical Etching Technique*

The etching method is always used for revealing the sites of emerging dislocations and defects in the crystals. Different kinds of etchants, such as boiling H_3PO_4. fused KOH, $KHSO_4$ etc., have been applied for different crystals and for different surfaces.

8.1.3 *X-ray transmission topography (XRMT)*

X-ray transmission topography (XRMT) is more convenient for mapping the distribution of defects in crystals. Table 8.1 shows that the X-ray absorption coefficient of Nd:YAG crystal is higher because it contains heavy element yttrium.

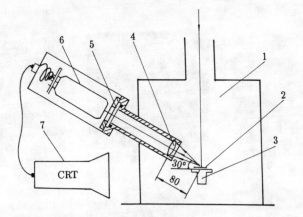

Fig. 8.2. Schematic of cathodluminescence analysis system. 1—SEM sample chamber; 2—sample; 3—sample stage; 4—focusing lens; 5—vacuum seal window plate; 6—photomultiplier (GDB-57); 7—kinescope.

8.1.4 *Scanning Electromicroscope (SEM), EDX and EPMA*

SEM, EDX and EPMA have been always used for observing and analysing the defects in crystals.

8.1.5 *Cathodoluminescent Technique (CL).*

Under high energy electron beam excitation, the light emission phenomenon produced by impurities and defects in materials is called cathodoluminescence (CL). It can be used for observing the luminescent image of defects in crystals.

Figure 8.2 is the schematic diagram of CL analysis system.

8.2 Defects in Nd:YAG Crystals

In literatures the macroscopic defects, such as facets, large size of scattering particles and constitutional supercooling in Nd:YAG crystals, have been reported numerously [1–5], but there are only a few works on microscopic defects such as dislocations. Some authors held that the dislocations do not form readily in Nd:YAG crystals due to their optical isotropy, unclearness of gliding direction and larger lattice parameter (1.2008 nm). However, the dislocations, which do occur in Nd:YAG crystals can be grouped into three categories: 1) Isolated straight lines of edge dislocations originated from seeds, inclusions and thermal stress induced by temperature fluctuation during growth. 2) Zigzag dislocation lines produced by jog motion from edge dislocations. 3) Dislocations helices and closed loops decorated by inclusions within

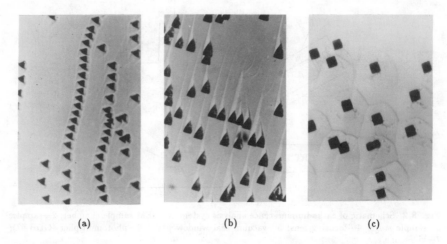

(a)   (b)   (c)

Fig. 8.3. (a) Etch pits formed on (111); (b) etch pits formed on (211); (c) etch pits formed on (100).

the melt during growth. The presence of dislocations in Nd:YAG crystals is still one of the principal sources leading to optical inhomogeneity of crystals. Also, the existence of inclusions in crystals effect strongly the quality of crystal, especially they can promote the climb motion of dislocations and result in the decoration of dislocations.

8.2.1 Dislocations in Nd:YAG Crystals

A. Observation of edge dislocations and mixed dislocations (zigzag dislocation) in Nd:YAG crystals.

a) Chemical etching technique. The crystal specimens were prepared using 85% H_3PO_4 for polishing and etching. The typical etch pit patterns of dislocations on (111), (211) and (100) faces in Nd:YAG crystals can be obtained (Fig. 8.3) using this technique.

b) Birefringence image of dislocation can be seen in Fig. 8.4. The strong birefringence image of dislocations in garnet type crystals can be produced because of their high hardness and larger Burgers vector.

c) Observation of undecorated edge dislocations by LLST. The theory of scattering factor of optical waves from an edge dislocation is similar to atomic scattering factor from an edge dislocation [6]. The scattering factor of optical waves is given by the same equation as that for X-ray scattering.

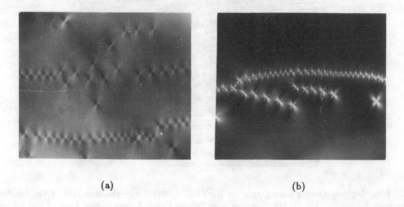

(a)   (b)

Fig. 8.4. (a) Rows of edge dislocation viewed end-on; (b) dislocation wall viewed end-on.

$$f = \int p(r)e^{i\frac{2\pi}{\lambda}gr}d\tau, \tag{8.1}$$

where $\vec{g}$—scattering vector, $\vec{g} = \vec{s} - \vec{s}_0$; $\vec{s}_0$—direction of incident light; $\vec{s}$—direction of scattering light; λ—the wavelength of laser beam.

The scattering power per unit length of the dislocation is

$$|F_2|^2 = |\frac{f_2}{2Z_0}|^2 = \frac{1}{2}\lambda^2 A^2 \cos^2 B = \frac{1}{8}\frac{b^2\lambda^2 k^2}{\pi^2}(\frac{1-2\mu}{1-\mu})\frac{g_y^2}{g_x^2 + g_y^2}, \tag{8.2}$$

where μ—Poisson ratio; g_x, g_y, g_z—the scattering vectors of edge dislocation in (x, y, z) coordinates; $2z_0$—the length of dislocation line resolved by the objective lens.

The scattering power of edge dislocations has a maximum at $g_x = 0$. The image caused by undecorated dislocations were clear lines and depended sharply on the direction of the incident beam. Figure 8.5 shows the light scattering image of edge dislocations parallel to $\langle 111 \rangle$ growth direction. The angle between the clearest and disappeared image of edge dislocations were 9° according to the calculation and experimental results.

Figure 8.6 shows that the scattering power of edge dislocations strongly depends on the direction of incident beam. The intensity of dislocations changed from strong to weak (B region) or in opposite (A region) when rotating the angle from zero to 9 degree [7].

B. Propagation of edge dislocations in Nd:YAG crystals [5]

Experiments show that the propagation of dislocations incorrelates with facets and it essentially depends on the shape of solid/liquid (S/L) interface. The propa-

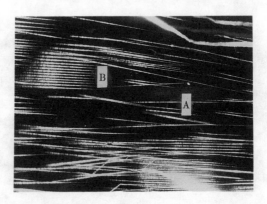

Fig. 8.5. Light-scattering tomography of Nd:YAG crystal. $\vec{s}_0$—the direction of the incident beam from bottom to top, ⟨111⟩ direction from left to right; A—individual lines of dislocation; B—equi-spaced multi-lines of dislocation.

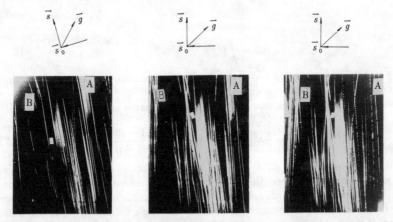

Fig. 8.6. Light-scattering intensities of dislocation lines changing with the direction of incident beam. $\vec{s}_0$—direction of incident beam; $\vec{s}$—direction of scattered light; $\vec{g}$—scattering vector.

gating direction of dislocations is always perpendicular to the S/L interface. Figure 8.7 shows the dislocation distribution patterns of crystal boule with three types of S/L interface. The dislocations tend to grow outward the crystal when S/L interface is convex toward the melt, and to focus toward the center of the crystal when S/L interface is concave. With flat S/L interface the propagating direction of dislocation lines is parallel to the growth axis and extend linearly almost to the end of the crystal. This rule can be explained with minimum energy principle, because the dislocations take the shortest way and locate in the lowest energy state only

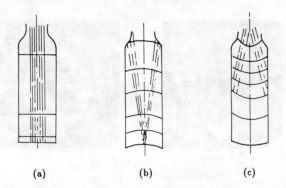

(a)   (b)   (c)

Fig. 8.7. Dislocation distributed patterns of crystal boules with three types of S/L interfaces. (a) Flat interface; (b) concave interface; (c) convex interface.

when they take such way after that, Schmidt [8] also found the similar rule in GGG crystal and calculated the propagating path of dislocations with Klapper theory [9]. He obtained good agreement with the experimental results and explained the causes of deviation from perpendicular propagating direction of a few edge dislocation lines.

Perfect Nd:YAG crystals can be obtained from CZ technique by reasonable technological process according to the above mentioned rule.

If the growth process is performed by growing the initial part of the crystal with high convex interface for eliminating dislocations remaining in the seeds and then change the S/L interface into flat for removing the facets, the dislocation free and unfacetted crystals can be got by the CZ technique.

C. Observation and identification of climb dislocations [10].

Generally, it is considered that when a crystal containing straight line dislocations whose density is of about 100 /cm^2 could be regarded as a perfect crystal or dislocation free crystal. But we found that the dislocation loops and helices existed in Nd:YAG crystals grown by different methods (flux. Czochralski pulling technique or temperature gradient technique) even in perfect crystals. Some are light and some are serious, they are formed by climb or jog-climb motion from edge dislocations.

We used the following methods to observe the different types of climb dislocations and their climb process.

a) Ultramicroscopy, light scattering tomography and dark field microscopy.

Because most of the dislocation loops and helices are decorated with impurities, they cause strong light scattering and can be seen clearly by the above mentioned techniques.

Figure 8.8 shows the different shapes of dislocation loops in Nd:YAG crystals grown by Czochralski method. They are all the climb dislocation loops. The climb

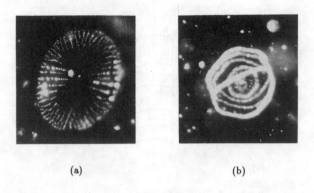

(a)   (b)

Fig. 8.8. (a) Photo of single loop (100×); (b) photo of multiple loop (100×).

Fig. 8.9. Ultramicroscopy photo of helix (100×).

plane of dislocation loops are (111), (110), (211) and (100) determined by X-ray Laue method.

Figure 8.9 shows the dislocation loops in Nd:YAG crystals grown by temperature gradient technique. Two types of climb dislocation helices have been found in Nd:YAG crystals. Type 1 helices are formed around inclusions and similar to those in GGG crystals, while type 2 helices are formed around the core composed of facets. The size of type 2 helical loops is 1.2~3 mm in diameter. This kind of giant helix has not previously been reported.

b) Birefringence topography. The climb dislocation loops and helices causes a strong strain field around them. The birefringence image of dislocation loops and helices can be seen in Fig. 8.10.

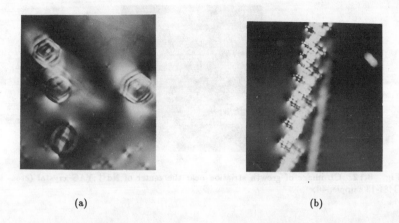

(a)   (b)

Fig. 8.10. (a) Birefringence image induced by dislocation loop (30×); (b) birefringence image induced by dislocation helix (30×).

Fig. 8.11. Etch pits formed where the turns of helix meet the surface of wafer (150×).

c) Chemical etching. Figure 8.11 demonstrates the etch pits formed when the turns of helices meet the surface of wafer.

d) Observation of constitutional supercooling, inclusions and growth striae and melt convection.

CL emission is a good technique for observing defects in crystals.

Under high energy electron beam excitation, strong CL image of Nd^{3+} ions in Nd:YAG crystals was induced, and it is similar to X-ray luminescent spectrum of Nd:YAG crystals. The luminescence emission in Nd:YAG crystals at 0.56∼0.58 μm is caused by transition between $^2F_{5/2}$–$^4G_{9/2}$, $^4G_{9/2}$. The variation of Nd^{3+}

Fig. 8.12. CL image of growth striation near the center of Nd^{3+}:YAG crystal (cross section) G481-13 sample, 40×.

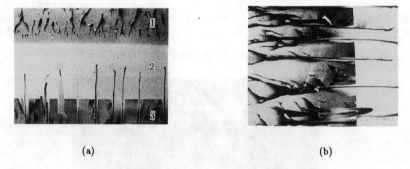

(a)   (b)

Fig. 8.13. CL image of the process of constitutional supercooling in Nd:YAG crystal. (a) 1—opaque layer; 2—transparent layer; 3—opaque layer; (b) a whole state of 3-areas.

concentration and decay of luminescence with defects in the crystals can be seen clearly [11].

Figure 8.12 shows the luminescent images of un-uniform growth striae and inclusions.

Figure 8.13 shows the luminescent images of formation process of constitutional supercooling in Nd:YAG crystals. The dark region shows the decay of CL emission caused by high Nd^{3+} concentration.

The inclusions in Nd:YAG crystals include solid inclusions (Ir, Mo and C particles), gas inclusions and liquid inclusions. There is a very important kind of inclusions found in Nd:YAG crystals. The sizes of this kind of inclusions are about 0.005~0.7 mm, and they are in definite shape and composed of very small particles. It can be seen using dark field microscopes and CL emission image.

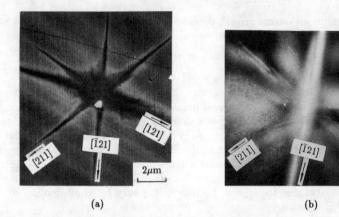

(a)   (b)

Fig. 8.14. CL and optical image of the triangle-like scattering micro-particle in Nd^{3+}:YAG crystal. (a) Optical image (dark field microscopy), YAG-2, 200×; (b) CL image YAG-2, 150×.

Figure 8.14(a) shows the nature of this type of inclusions. It can also be seen from the luminescent images as shown in Fig. 8.14(b).

8.2.2 *The Reason for the Formation of Dislocation Loops and Helices*

There are several reasons for the formation of dislocation loops and helices by climb motion in Nd:YAG crystals.

First, glide motion is usually easier to occur in crystals than climb motion.

The glide resistance of dislocations, F_s, is expressed by

$$F_s = \frac{2\nu b}{(1-\mu)} \cdot e^{\frac{-4\pi\rho}{b}}. \tag{8.3}$$

The climb resistance of dislocations, F_c, is expressed by

$$F_c = \frac{U_f}{b^2}, \tag{8.4}$$

where ρ—half width of dislocation; μ—Poisson ratio; ν—shearing modulus; U_f—formation energy of point defects.

Because Nd:YAG crystal is a high melting point (1970°C) crystal, it has a large Burgers vector and a small value of U_f at high temperatures. The glide resistance (F_s) is larger than the climb resistance (F_c), thus it is easier for climb motion to take place than glide motion in Nd:YAG crystals.

Second, the permeation force (f) of Nd:YAG crystals is large due to the formation of nonequilibrium and super saturation point defects under high melting and high doping conditions during growth process (interstitial atom excess and vacancy deficiency). The permeation force can be expressed by the following equation:

$$f = \frac{kT}{b^3}\ln\frac{c}{c_o} = \frac{kT}{b^3}\ln\alpha \simeq kT\frac{\sigma}{b^3},\tag{8.5}$$

where b—absolute value of Burger's vector; c_o— equilibrium concentration of point defects under given temperature; c—real point defect concentration at this temperature; σ—supersaturation degree of point defects; α— saturation ratio.

In this equation, the osmotic force, f, is proportional to σ, it means that the process of climb motion of dislocations enhances with the increase of the supersaturation degree of point defects. Because a large number of point defects are easy to occur in Nd:YAG crystals during growth at high temperature and high doping conditions, f is large due to large σ, therefore, it is easy for climb motion to take place in Nd:YAG crystals.

8.2.3 Interaction between Dislocations and Other Defects

We studied the interaction between dislocations and the other defects in Nd:YAG crystals by LLST [12]. The laser beam scans the whole crystal boule, light scattering patterns of the crystals can be taken layer by layer.

A poor quality Nd:YAG crystal boule containing different kinds of defects was chosen for investigation. The crystal was grown via Czochralski method parallel to (211) direction. Figure 8.15 demonstrates the formation of different defects during the crystal-growing process. At the initial stage of crystal growth, a large number of precipitated particles appeared in the as-grown crystal and many dislocation loops were formed due to precipitation of impurities. The stress field of the core attracted the dislocation loops arranged along two sides of core AA. In the region BB a strong stress field of growth striae is caused by temperature fluctuation. The edge dislocation lines CC are generated from region BB.

Figure 8.16 shows the interaction of dislocations with impurities, facets and growth striae. The strong stress field of growth striae promotes heavy precipitation of impurities and decoration, the shape of S/L interfaces can be clearly observed. The straight-line dislocation grows around the facet and forms a giant type 2 helix. The stress field of helix promotes the deposition of impurities on the dislocations and decorated dislocations can be observed. The large size deposits on the helix form a strong stress field and produce dislocation loop (H) again.

The properties and distribution of defects in Nd:YAG crystals were also successfully obtained by means of X-ray transmission topography with AgKα radiation [13]. Figures 8.17, 8.18 and 8.19 show the different kind of dislocations, such as edge, zigzag, jogged and helical dislocations, as well as mixed dislocations formed by dislocation motions in Nd:YAG crystals grown by CZ and TGT methods. Some experimental results are consistent with the results obtained by optical methods, however, X-ray method has its special advantage in discriminating the properties of dislocations.

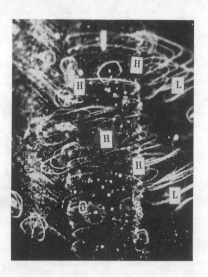

Fig. 8.15. Laser light scattering pattern of different defects during crystal growth process. AA—two rows of dislocation loops; BB—strong stress field of growth striation; CC—edge dislocation lines.

Fig. 8.16. Laser light scattering pattern of interaction of dislocations with impurities, facets and growth striae. L—larger turns of helix encircled facet; H—dislocation loop formed at large inclusions which deposited on the surface of facet.

From the above mentioned observations, it can be seen that besides the supersaturation degree of point defects, another important condition which induces the formation of dislocation loops, helices and decoration of defects is the precipitation of a large number of impurities from melt. Decorated dislocation loops and helices all induce the scattering loss and multiplication of dislocation in Nd:YAG crystals, in order to eliminate them the growth parameters, volatilization of crucible materials and segregation of dopant Nd_2O_3 should be controlled.

8.2.4 Effect of Dislocations on Laser Performances

It was well known that some macroscopic defects, such as facets, scattering particles and constitutional supercooling in Nd:YAG crystals have caused the optical inhomogeneity and deteriorated the laser output characteristics. It is interesting to study the influence of microscopic defects such as dislocations on laser properties. Badasarov [15] calculated the deviation of a straight line in the propagation of the light beam through a dislocation stress field in crystals. According to,

$$n(x) = n_0 + \delta n(x), \tag{8.6}$$

the light trajectory equation is expressed by

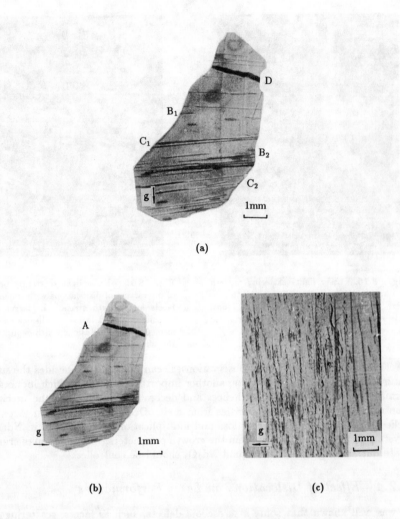

Fig. 8.17. X-ray topographs of different kinds of dislocations in Nd:YAG. (a) Diffraction vector $g=[\bar{4}4\bar{4}]$; B_1, B_2—edge dislocations; C_1, C_2—mixed dislocations; D—dislocation beam caused by constitutional supercooling. (b) Diffraction vector $g=[\bar{2}24]$; A—screw dislocations. (c) Zigzag-shaped dislocation lines.

$$\frac{d^2x}{dz^2} = \frac{1}{n_0} \cdot \frac{dn}{dx},\tag{8.7}$$

where n_0—refractive index of crystal; n_x—refractive index change due to the effect

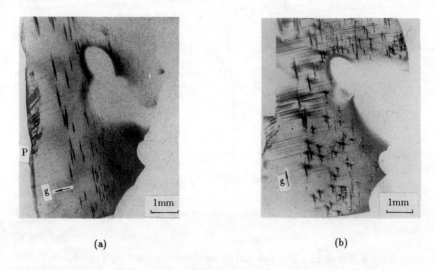

(a) (b)

Fig. 8.18. X-ray topographs of jogged dislocations in (111) slice. (a) Diffraction vector $g=[\bar{2}40]$; P, gliding bends. (b) Diffraction vector $g=[4\bar{4}\bar{4}]$.

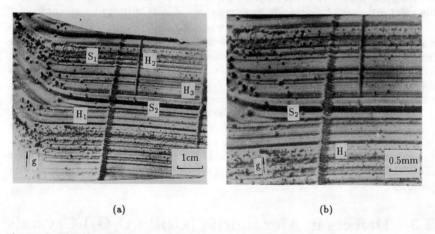

(a) (b)

Fig. 8.19. Distribution and stress pattern of defects in (211) slice obtained by X-ray topography and optical method. (a) Diffraction vector $g=[44\bar{4}]$; H_1, H_2, H_3: the helical dislocation; S_1, S_2: strong growth striation. (b) The magnification of helical dislocation H_1, H_2 and growth striation S_2.

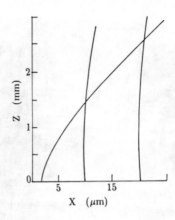

Fig. 8.20. Light trajectory calculated according to Eq. 8.7.

of dislocation stress field. The calculated results are shown in Fig. 8.20. It can be deduced that the presence of dislocations may have some effect on the laser performances of the crystal.

Table 8.2 shows the dislocation distribution and density, extinction ratio and laser output characteristics of six Nd:YAG rods. It is worth pointing out that there are great influence of dislocation properties on the laser output performances.

It demonstrates experimentally that when the dislocation density increases to a certain degree in the crystal, the stress field due to the dislocation will lead to intense birefringence, which gives rise to distorting the light wavefront, increase the interference fringes and decrease the extinction ratio. Hence, dislocation is one of the major sources of optical inhomogeneity of crystals. It can be seen from the near field pattern that at a given input energy, the areas with higher density of dislocations in the crystal will not oscillate due to the rise of the laser threshold, depolarization is produced due to the stress birefringent effect of dislocations; the frequency-doubling efficiency drops and the beam divergence increases. Furthermore, the laser output property suffers especially remarkable change when the decorated dislocations exist in the crystal. It is not only for breading of dislocation but also for many scattering particles.

8.3   Defects in Alexandrite (Cr:BeAl$_2$O$_4$) Crystals

Alexandrite crystal or chromium-doped chrysoberyl, is an important material for tunable solid state lasers. In order to improve the growth techniques for obtaining high quality crystals, the defects in crystals have been investigated by chemical

Table 8.2. Influence of defects on laser output performances of Nd:YAG rods.

No	Size of rods (mm)	Amplifier E_{out} (mj)	Gain G	Depolarization	Frequency doubling crystals			Defects in patterns	Near field	
					A	B	C		730 V Input	1000 V Input
1	$\phi9\times96$	573	7.5×	70/20	23	156	31.1%	Dislocation density 10^2 /cm²		
2	$\phi15\times95.5$	397	5.9×	74/16	54	92.5	23.2%	Local dislocation density 10^3 /cm²		
3	$\phi10\times81$	436	6.5×	60/20	85	85	19.4%	Local dislocation density 10^3–10^4 /cm²		
4	$\phi14\times86$	432	6.5×	60/26	54	92.5	21.0%	Local dislocation density 10^4 /cm²		
5	$\phi13\times81$	410	6.1×	68/20	56	67	16.0%	Decolated dislocation and scatter center		
6	$\phi9.8\times124$	567	8.5×	—	115	147	26.0%	Facets		

A—power density (MW/cm²); B—output energy (mj); C—efficiency.

etching, optical method, SEM, EPMA, EDX and X-ray transmission topography (XRMT) [16–18].

8.3.1 *Inclusions*

Under transmission and cross-polarized light, it is shown that the inclusions such as iridium particles (triangular and hexagonal in shape), bubbles, tunnels and secondary phase precipitated particles often appear in grown crystals.

The chemical composition of the inclusions caused by constitutional supercooling and some individual inclusion in crystals has been analysed by means of EPMA and EDX. The main impurities are Ca, Mg, Si, K and Na etc., which are introduced from raw materials and are shown in Tables 8.3 and 8.4 [19].

Inclusions and constitutional supercooling phenomena decrease obviously when the start materials are pre-purified.

8.3.2 *Spatial Distribution of Defects*

BeAl₂O₄:Cr crystal is composed of light elements Be, Al, O, so that Lang method is a good technique to show the spatial distribution of defects within the crystal.

Table 8.3. EPMA of various elements within inclusions of sample A.

Probe analysis area	Probe parameter ×10⁻⁹ nm, 20 kV	Counts of X-photon for various elements								
		Na	Mg	Al	Si	Cl	K	Ca	Cr	S
Tubular inclusion	5.02	206	116	38389	1078	113	178	227	13	75
Tubular inclusion	4.72	29	54	34774	147	25	—	46	—	21
Rod-like inclusion	4.78	58	2325	4381	12478	29	150	7681	—	124
Basis	4.90	—	—	36350	—	—	—	—	14	—

Table 8.4. EPMA of various elements within inclusions of sample B.

Probe analysis area	Probe parameter ×10⁻⁹ nm, 20 kV	Counts of X-photon for various elements								
		Na	Mg	Al	Si	S	Cl	K	Ca	Cr	Ba
Disk-like inclusion	4.02	—	25	260	106	16	76	26	235	—	—
Basis	4.02	—	—	20225	—	—	—	—	—	19	9
Rod-like inclusion	3.96	—	—	1641	15	(Fe) 24	13	25	—	13	58
Basis	3.96	—	—	19879	—	—	—	—	—	25	—

Figure 8.21(a) is an X-ray topograph of specimen 1, which is a (010) longitudinal slice. It contains two parts of crystal. The upper part is crystal seed and the lower part is grown crystal. It is shown that: (1) the quality of seed is very bad, it contains inclusions and dislocation lines, the dislocation lines propagate into grown crystal during growth from part A (right) and part B (left), (2) the inclusions in grown crystal can produce new dislocation lines. Figure 8.21(b) is an X-ray topograph taken with ⟨400⟩ of diffraction vector **g** of specimen 2, which is also a (010) slice. This photo shows that: (1) the bundle of dislocation lines at the two sides (A, B) of slice propagate from top to end, (2) dislocation densities in two sides A and B are very high, but in other part it is very low. Figure 8.21(c) shows the dislocations generated from the inclusions due to constitutional supercooling and they are also edge dislocation.

Therefore, the dislocations in CZ grown $BeAl_2O_4$:Cr crystal are generated from two sources. One generated from the seeds, which propagate into the grown crystal and considerably influence the further growth. The other is caused by inclusion. Most dislocations are edge ones [16].

8.3.3 *Dislocations*

Dislocation etch pit pattern can be formed after etching. The distribution of

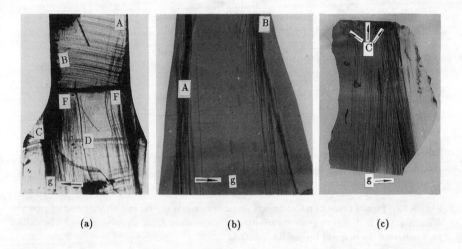

(a)   (b)   (c)

Fig. 8.21. (a) X-ray Lang topograph of 1# slice (400) diffraction. FF—two bundles of dislocation line come from defects A, B in seed; C—dislocation line generated at inclusions; D—inclusions. (b) X-ray Lang topograph of 2# slice (400) diffraction. A, B—two bundles dislocation come from seed at two sides of crystal. (c) X-ray Lang topograph of 3# slice (400) diffraction. C—dislocation lines generated from inclusions which induced by constitutional supercooling.

dislocation in alexandrite crystals is not uniform. The dislocation density is very high in some areas, up to 1×10^4–10^5 /cm^2. From dislocation etch pit pattern on (001) surface it indicates that the dislocation glide direction is $\langle 100 \rangle$ and the glide plane is (010).

The path of grown-in dislocation lines follows the minimum-energy theory (Klapper, Kuppers). The shape of the growing crystal, the traces of the growth interface at different stage of growth and the arrangement of grown-in dislocations are schematically shown in Fig. 8.22. They mainly originate from three sources: 1) dislocations continue from those already present in the seed, 2) dislocations nucleate at the seed-crystal interface, and 3) dislocations originate from the inclusion or gross imperfections. Most of the dislocations propagate along the paths normal or perpendicular to the local growth interface, which have the Burgers vectors parallel to the [100] direction. A few dislocations with different Burgers vectors form different angles with the local direction of the growth interface and proceed along different paths. They exhibit the propagation behaviour very similarly that observed in the crystal grown on planar faces. The dislocation density and arrangement closely relate to the seed quality.

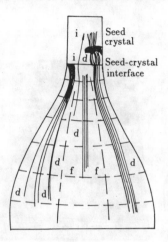

Fig. 8.22. Typical shape of a Czochralski alexandrite single crystal and arrangement of grown-in dislocations. i, Inclusion or gross imperfection; d, dislocation; f, facet growth region. The dashed lines indicated the shape of the growth interface.

8.4   Defects in Sapphire, Ruby and Ti:sapphire Crystals

Using X-ray Lang method we studied the defects in sapphire and ruby crystals made by different techniques, such as Verneuil, Czochralski, edge-defined film-fed growth and temperature gradient method [20]. There are many macroscopic defects such as mosaics, grain boundaries in ruby grown by Verneuil technique and edge-defined technique. The quality of ruby and sapphire by CZ are improved recently. There are no mosaic, glide and other macroscopic defects in sapphire grown by temperature gradient method.

It is found that high quality sapphire or ruby can be got by seed induced temperature gradient technique.

The geometry and distribution of dislocations, and the characteristic triangular cross-grid dislocation structure in sapphire and titanium doped sapphire single crystals grown by Temperature Gradient Technique (TGT) have been revealed by means of Lang X-ray transmission diffraction topography [21–23].

Figure 8.23 illustrates typical X-ray topographs which displayed the characteristic triangular cross-grid dislocation structure. Five pictures (Fig. 8.23a–e) were taken from the sample 1* by successive rotations of 30° in the plane of the wafer and oriented with the [1$\bar{1}$00] vertical, the corresponding diffraction vectors, g, were parallel to the (0001) basal plane. Figure 8.23f was also taken from sample 1* with the (0$\bar{1}$12) reflection, the corresponding diffraction vector, g, was at an about 30°

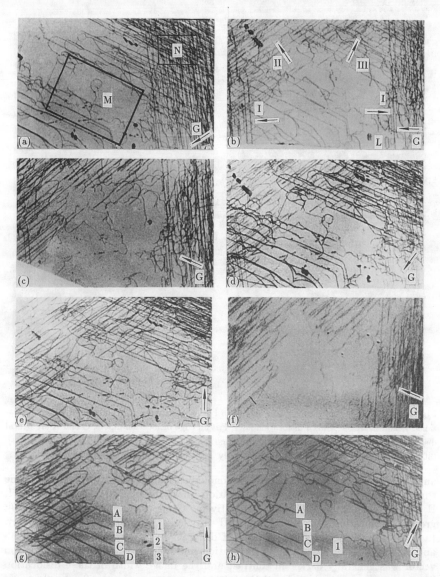

Fig. 8.23. Topographs of sample 1 showing the characteristic triangular cross-grid dislocation structure. (a), (b), (c), (d), and (e) are in the $(03\bar{3}0)$, $(11\bar{2}0)$, $(30\bar{3}0)$, $(\bar{1}2\bar{1}0)$, and $(3\bar{3}00)$ reflections, respectively; $t = 315$ μm. (f) is in the $(10\bar{1}4)$ reflection, and (h) and (g) are in the $(\bar{1}2\bar{1}0)$ and $(3\bar{3}00)$ reflections; $t = 203$ μm.

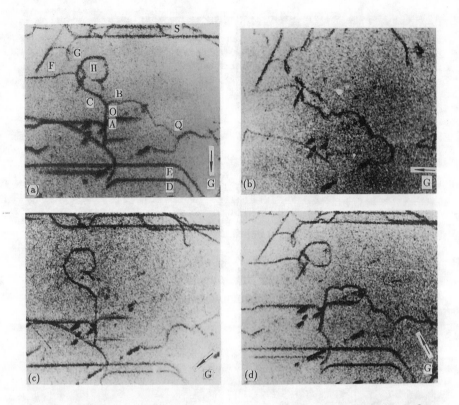

Fig. 8.24. Enlarged topographs of the area M in Fig. 8.23 showing different dislocation configurations. D and E are 60° kinked dislocation lines, F is an edge dislocation dipole, G and H are helical dislocations, and O and Q are dislocation reaction nodes. (a), (b), (c), and (d) are in the ($\bar{1}2\bar{1}0$), ($30\bar{3}0$), ($03\bar{3}0$), and ($3\bar{3}00$) reflections, respectively.

angle to the (0001) basal plane. The pronounced feature in the topographs is the three parallel groups of straight dislocation lines, three such systems marked I, II, and III are 60° or 120° apart, respectively. Group I system runs along [1$\bar{1}$00]; group II dislocation system and group III run along [10$\bar{1}$0] and [01$\bar{1}$0] respectively. In the 11$\bar{2}$0 reflections, three groups of dislocation lines appeared (Figs. 8.23b and 8.23d) in which the group of dislocation lines perpendicular to the diffraction vector was generally darker in contrast and the image width of these dislocation lines was approximately twice that of other two groups. In the 3$\bar{3}$00 reflections, only two groups of dislocation lines which display the same contrast and the same image width of dislocation appeared (Figs. 8.23a, 8.23c and 8.23e), the group of dislocation lines parallel to the diffraction vector vanished totally. The dependence of the contrast and image width of dislocation line on the diffraction vector implied that the three parallel groups of dislocation lines lay on the (0001) basal planes.

Further experimental evidence for dislocation distribution could be obtained by thinning the wafer again in molten borax for about 20 min. The prolonged polishing has led to a slightly beveled edge of the wafer. Figures 8.23g and 8.23h show the topographs. In comparison with Figs. 8.23d and 8.23e many dislocation lines vanished with the removal of the surfaces, and the segments of dislocation lines near the margin, labelled A, B and C disappeared due to the formation of the beveled edge. In addition, after the chemical etching relatively few dislocation etch pits visible in optical microscopy of the wafer surfaces, which correspond to dislocation outcrops, also confirmed that almost all dislocations lay on the (0001) basal planes. Also noticeable are the Pendellosung fringes, in the $(3\bar{3}00)$ reflection (Fig. 8.23g) it appeared as three fringes marked 1, 2 and 3 and in the $(1\bar{2}10)$ reflection (Fig. 8.23h) it appeared as only one fringe marked 1. This difference resulted from the different distinction distance, ξ_g, for different reflection planes. On the other hand the appearance of the Pendellosung fringes indicate the high degree of the perfection of the sapphire single crystal investigated.

Figure 8.24 shows the enlarged pictures of the area marked M in Fig. 8.23a. Different dislocation configurations, such as 60° kinked dislocation lines E, D, edge type of dislocation dipole F, spiral dislocations, G, H, and dislocation reaction nodes O, Q were clearly displayed. Some of these dislocations, i.e. the kinked dislocation lines D, P, have undergone glide along its Burgers vector direction due to the chemical stress during chemical polishing, and other dislocations, i.e. the spiral dislocation H, has climbed along the direction normal to (0001) basal plane.

The analyses of Burgers vector and the imaging width of dislocations had confirmed that three parallel groups of straight dislocation lines are all pure edge type of slip dislocations having $\langle 11\bar{2}0 \rangle$ type Burgers vector, and a few dislocation reactions observed are the type of $[11\bar{2}0]+[1\bar{2}10]+[\bar{2}110]=0$. Such a reaction represents the self-pinning of dislocations and influences the mobility of dislocations.

With the variation of the growth orientation the dislocation structure changes greatly, it appears as the tangle dislocation structure consisting of dense clusters of highly curved dislocations. The characteristic dislocation structure is closely related to the growth orientation of the crystal rather than the dopant or impurity.

8.5 Defects in Cr:forsterite Crystals

Chromium-doped forsterite ($Cr:Mg_2SiO_4$) is a newly developed tunable laser crystal. Three typical defects were found: dislocations, sub-structures and inclusions [24].

The most harmful defects in these crystals are inclusions. Figures. 8.25a and 8.25b show the inclusions exhibited on crystal cross-sections perpendicular to the growth directions along the a and b axes, respectively. From these figures, it can be seen that the inclusions are mainly located around the crystal core. The core on the (010) plane is round and that on the (100) plane exhibits four-fold symmetry,

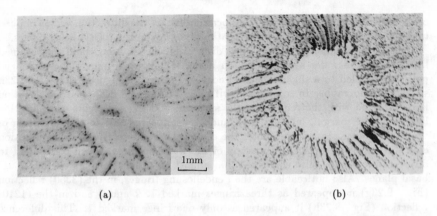

(a)   (b)

Fig. 8.25. (a) Inclusions and core on (100) surface; (b) inclusions and core on (010) surface.

Table 8.5. Analytical data of inclusions in forsterite measured by electron probe microanalysis.

Probe site	Probe parameters at 10^{-9} nm, 15 kV	Concentration of element (%)									
		SiO_2	Al_2O_3	MgO	Cr_2O_3	CuO	FeO	CaO	PtO_2	ZrO_2	Au_2O_3
Matrix	1.0	42.4618	0.0000	57.0053	0.0074	0.0000	0.0000	0.0000	0.0000	0.0000	0.1453
Matrix	1.0	42.1981	0.0305	57.3453	0.0419	0.0000	0.0000	0.0000	0.0738	0.0000	0.0442
Inclusion	3.0	57.0227	2.0654	38.6890	0.0100	0.0044	0.1082	1.7993	0.0000	0.0000	0.0000
Inclusion	3.0	58.0510	1.8326	38.7776	0.0687	0.0000	0.4402	0.0304	0.0000	0.0000	0.0000
Inclusion	3.0	54.3210	3.1239	36.7979	0.0000	0.0229	0.4390	0.9595	0.0000	0.0000	0.0000
Inclusion	3.0	84.0234	0.8063	8.1761	0.0000	0.0000	0.2981	0.5534	0.0000	0.0000	0.0094
Inclusion	3.0	83.5588	1.9943	10.6622	0.0147	0.0609	0.0622	0.0000	0.0000	0.0000	0.0000
Inclusion	3.0	81.1253	2.3265	10.5480	0.0123	0.0154	0.1820	0.0000	0.0210	0.0000	0.0132

implying that the core in forsterite is caused by facets on the solid-liquid interface. The results of the electron probe analysis on the crystal matrix and inclusions are listed in Table 8.5. From this table, it can be concluded that apart from some other impurities, almost every inclusion is rich in SiO_2. Based on this result, systematic experiments with 0.03, 0.06, 0.10 and 0.15 wt% excess of MgO over stoichiometry in the melt were tested. When the MgO excess reached 0.10 wt%, most of the inclusions disappeared, and the dimensions of those still present were greatly diminished. Therefore, this composition of the charge was used in subsequent crystal growth.

After chemical etching etch pits of dislocations appeared, which are shown in Fig. 8.26. The density of the dislocations at central area on the cross section perpendicular to the growth direction along an axis is about 1.0×10^3 cm^{-2}, which is at the edge, the density is about 2.7×10^4 cm^{-2}. The dislocation in the crystals grown along the b axis are of the same order of magnitude in density, and have almost the same distribution.

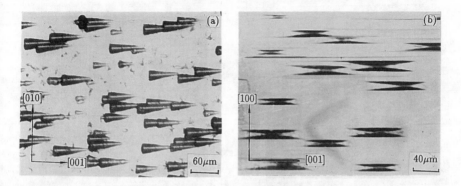

Fig. 8.26. (a) Etch pits on (100) surface and their orientation; (b) etch pits on (010) surface and their orientation.

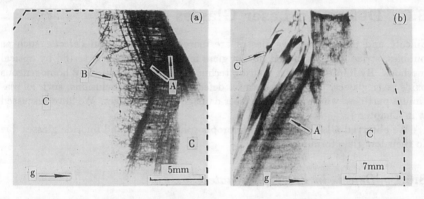

Fig. 8.27. X-ray topographs of (010) slices with thickness of 0.2 mm; Ag Kα radiation, g=002: (A) grown-in dislocations; (B) post-growth dislocations; (C) subgrain structure; dashed line shows missing parts of the samples.

Figures 8.27a and 8.27b are X-ray topographs. It can be seen that the misorientation among the subgrains permits only part of the sample to contribute to the Bragg scattering, while other subgrains show no image because they did not satisfy the reflection condition. The misorientation among the subgrains can be as large as 1°. Except for the subgrain structures, X-ray topographs also show that there are two different types of dislocations. One, labelled A, is extended along the crystal growth axis. The other, labelled B, is network-like. The former, we suggest, corresponds to the etch pits and is a grown-in dislocation; the latter may be caused in postgrowth cool-down. Moreover, the topographs verify the distribution of dislocations observed by etching.

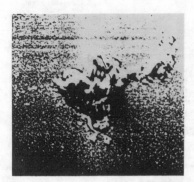

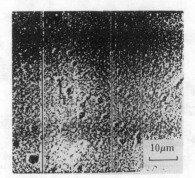

Fig. 8.28. Inclusion in fluorophosphate glass.

Fig. 8.29. Compositional variation in fluorophosphate glass.

8.6   Defects in Laser Glasses

Silicate and phosphate laser glasses are more stable, the main defects, such as bubbles, striae were happened during glass melting, they can be tested by optical methods. By improving glass melting technology we can obtain very homogeneous oxide laser glasses. The most harmful defects are metallic inclusions, such as platinium particles, which induce the laser damage in laser glasses. We have discussed it in Chapter 7.

We observed a lot of defects in fluorophosphate glasses and fluoride glasses due to their low glass forming ability.

8.6.1   Defects in Fluorophosphate Glasses

Many fluorophosphate glass samples were examined by optical and electron microscopy [25]. The bubbles in diameter ranging from 2 to 70 μm could be observed in many samples. Also, crystallites of various shapes and varied composition appeared in glasses. An example of which is shown in Fig. 8.28. The crystalline inclusion has more Na and less P than those in the host glass. A large compositional variation on a 1 μm scale could be also observed in fluorophosphate glasses. The sample shown in Fig. 8.29 has areas with high Fe content. The defects in fluorophosphate glasses are the origins of laser induced damage.

8.6.2   Defects in Fluoride Glasses

The microcrystallites are the main sources of optical losses in fluoride glasses. We have intensively studied the defects in ZrF_4-, AlF_3-, and InF_3-based fluoride glasses [26–28].

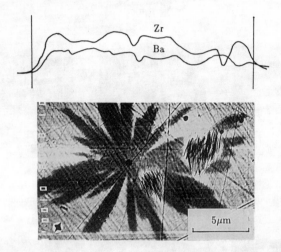

Fig. 8.30. Back-scattered electron image (bottom) and linear distribution (top) of the microcrystal.

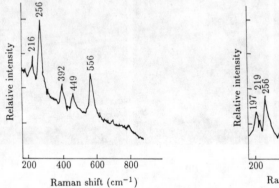

Fig. 8.31. Raman spectrum of the microcrystal.

Fig. 8.32. Raman spectrum of $2ZrF_4 \cdot BaF_2 \cdot LiF$.

Taking one type of micro-crystal in ZrF_4-based (ZBLALiY) glass as an example, Fig. 8.30 shows the back-scattered electron image and linear distribution of Zr and Ba of the micro-crystal, which composed of ZrF_4-BaF_4 confirmed by F line scanning, but it is difficult to identify the crystalline phase of this particle by EDS and WDS. Figure 8.31 shows the Raman spectrum of the micro-crystal. The typical Raman peaks of the micro-crystal are at 556, 449, 392, 256, 216 cm^{-1}. After heat treatment,

Fig. 8.33. Three kinds of crystalline particles commonly found in the fluoroaluminate glass.

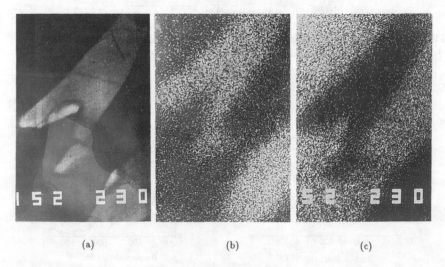

(a)   (b)   (c)

Fig. 8.34. (a) Back-scattered electron image of the crystallite in Fig. 8.33a. (b), (c) Two dimensional distribution of element Y and Al at the polished surface corresponding to this crystallite.

the crystallized product of ZBLALiY glass was $2ZrF_4 \cdot BaF_2 \cdot LiF$. Figure 8.32 demonstrates the Raman spectrum of $2ZrF_4 \cdot BaF_2 \cdot LiF$. The Raman peaks of micro-crystal are fitted with that of crystallized product. Therefore, it can be confirmed that the micro-crystal in the glass is $LiBaZr_2F_{11}$.

We paid more attention to study the original crystals in AlF_3-based and InF_3-based fluoride glasses. The phase analyses were based on a combination of optical

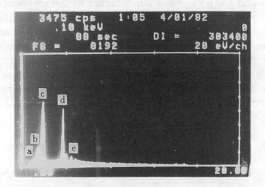

Fig. 8.35. EDS spectrum of the crystallite in Fig. 8.33 standing for Mg(K), Al(K), Sr,Y (L), Ca(K) and Ba(L) respectively.

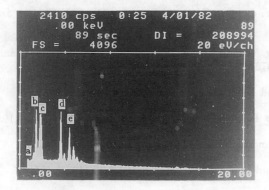

Fig. 8.36. EDS spectrum of the glass matrix.

microscope, SEM, EDS and WDS [29, 30].

Fluoroaluminate glasses are characterized by more ionic bonding than conventional glasses. As a result, such glasses are relatively unstable and tend to devitrify. Figure 8.33 exhibits the transmitted light micrographs of three kinds of crystalline particles which are commonly found in this glass under optical microscope.

The crystallite in Fig. 8.33a (crystal A) at the expolished surface was imaged by SEM (shown in Fig. 8.34a). Figures 8.35 and 8.36 where the EDS spectra corresponding to this crystallite and the glass matrix shown reveal striking difference of chemical composition between the crystallite and the matrix. It is clearly seen that the peaks belong to Mg(K), Al(K) and Ba(L) have a marked drop while the peaks of Sr, Y(L) and Ca(K) become much higher in the crystallite as compared to that of the glassy surrounding. This is consistent with the results of plane scan of every element contained in the glass. For example, Figs. 8.34b and 8.34c is the distribution of Y and Al. WDS analyses make clearly that the ascendent of Sr, Y(L) peak is due to the enrichment of element Y and the anion in this crystallite is F^-. Based on the above experimental results, crystal A may be preliminarily determined to be $CaF_2 \cdot 4YF_3$. The same techniques are used to examine the crystallite in Fig. 8.33b, where the crystallite phase may be $CaMg_2Al_2F_{12}$.

References

1. W. Bardsley and B. Cockayne, *Proceedings of International Conference on Crystal Growth*, (1966) 109.

2. B. Cockayne, *J. Crystal Growth* **3, 4** (1968) 60.

3. W. R. Wilcox, *J. Appl. Phys.* **36** (1965) 36.

4. Peizhen Deng, Jingwen Qiao and Zhenying Qian, *J. Chinese Silicate Soc.* **61** (1982) 8.

5. Peizhen Deng, Shoudu Zhang, Jingwen Qiao and Zhenying Qian, *Acta Physica Sinica* **25** (1976) 284.

6. K. Moriya and T. Ogawa, *J. Crystal Growth* **44** (1978) 53.

7. Jingwen Qiao, Peizhen Deng, *et al.*, *J. Synthetic Crystals* **16** (1987) 69.

8. W. Schmdit and R. Weiss, *J. Crystal Growth* **43** (1978) 515.

9. H. Klapper and H. Kuppers, *Acta Crystal* **A29** (1973) 495.

10. Peizhen Deng, Shoudu Zhang, Haobing Wang and Zhenying Qian, *J. Chinese Silicate Soc.* **73** (1978) 183.

11. Dequn Huang, Haobing Wang and Peizhen Deng, *Acta Optica Sinica* **7** (1987) 939.

12. Peizhen Deng and Jingwen Qiao, *J. Crystal Growth* **82** (1987) 579.

13. Peizhen Deng and Bing Hu, *Acta Optica Sinica* **8** (1988) 625.

14. Peizhen Deng, Jingwen Qiao and Zhenying Qian, *Acta Optica Sinica* **2** (1982) 259.

15. X. C. Badasarov, *Crystallography* (Russian) **2** (1970) 334.

16. Peizhen Deng, Zhenying Qian and Jingwen Qiao, *Acta Optica Sinica* **4** (1984) 922.

17. Qiang Zhang, Peizhen Deng and Fuxi Gan, *Materials Lett.* **9** (1989) 48.

18. Qiang Zhang, Peizhen Deng and Fuxi Gan, *Cryst. Res. Technol.* **25** (1990) 385.

19. Peizhen Deng, Dequn Huang and Haobing Wang, *J. Chinese Silicate Soc.* **13** (1985) 128.

20. Zhenying Qian, Yinchun Hou and Peizhen Deng, *J. Chinese Cer. Soc.* **8** (1980) 143.

21. Qiang Zhang and Peizhen Deng, *Materials Lett.* **8** (1989) 105.

22. Qiang Zhang, Peizhen Deng and Fuxi Gan, *J. Appl. Phys.* **67** (1990) 6159.

23. Qiang Zhang, Peizhen Deng and Fuxi Gan, *J. Cryst. Growth* **108** (1991) 377.

24. Bing Hu, Hongbi Zhu and Peizhen Deng, *J. Cryst. Growth* **128** (1993) 991.

25. S. Stokowski, M. Weber *et al.*, *Reports of LLNL*, Laser Program UCRL-50021-78, **2** (1978) 7–42.

26. Fuxi Gan, Quanqing Chen and Ruihua Li, *Mater. Sci. Forum* **32–33** (1988) 237.

27. Peizhen Deng, Ruihua Li, Po Zhang and Fuxi Gan, *J. Non-Cryst. Solids*, **140** (1992) 307.

28. Peizhen Deng, Ruihua Li and Fuxi Gan, *Proc. of XVI Intern. Congress on Glass*, Madrid, **2** (1992) 36.

29. Ruihua Li, Peizhen Deng and Haobing Wang, *Proc. SPIE* **1230** (1990) 565.

30. Peizhen Deng and Ruihua Li, *Proc. 8th Intern. Symp. on Halide Glasses*, Brittany, France, 1992, p. 74.

9. Crystal Growth and Structural Chemistry of Laser Crystals

Most laser crystals are inorganic dielectric monocrystals doped with active ions (RE and TM ions). High absorption at pumping wavelength, low loss at laser wavelength, high energy transfer efficiency, high homogeneous distribution of active ions and high perfection of the crystal hosts are the main requirements for laser crystals. The laser crystal quality is mostly ensured by crystal growth technique and suitable substitution of active ion in host crystal lattice. In this chapter the crystal growth and structural chemistry are briefly reviewed.

9.1 Crystal Growth Techniques of Laser Crystals

The laser crystals belong to high temperature oxide and fluoride single crystals, including simple and complex compounds, are most commonly grown from the melt. The main advantage of this method is high growth rate than that from solution or from vapor, but the material should not be decomposed at the melting point.

Depending on the melting temperature and growth atmosphere the principal container materials used in growing single crystals are platinum, rhodium, iridium and their alloys for oxidation atmosphere and graphite, molybdenum, tungsten and their alloys for reduction atmosphere. The basic requirement is mutual insolubility and the absence of chemical interaction between the melt and the container material. Sometimes, the container walls are coated with various materials for preventing reaction with the melt.

The principal methods of growing single laser crystals from melts are as follows. As for the in-depth discussion on crystal growth mechanisms and growth techniques, please refer to the relevant monographs [1, 2].

9.1.1 *Czochralski Method*

The Czochralski method is the main means for growing laser crystals. In Czochralski method the melt temperature is kept constant and the crystal is slowly pull out of the melt as it grows. This ensures a constant crystallization rate. The pulling rate depends on the physicochemical characteristics of the crystallizing material and

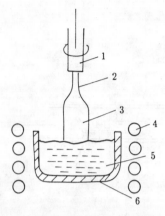

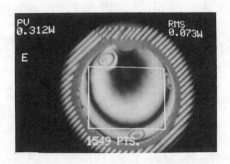

Fig. 9.1. Schematic diagram of Czochralski method. 1, Pulling bar; 2, seed; 3, growing crystal; 4, induction heating; 5, melt; 6, crucible.

Fig. 9.2. Interferometric pattern of a Nd:YAG crystal boule.

the size of the crystal. The schematic diagram of Czochralski method is shown in Fig. 9.1. The substance can be melted by high-frequency induction or resistance heating. It is worth pointing out the advantages of Czochralski method: there is no direct contact between the crucible walls and the crystal when helps in producing unstressed single crystals; the geometric shape of the crystal can be changed by varying the growth rate and melt temperature; high quality single crystal with few or no dislocation can be obtained by changing the solid-liquid interface during the growth. Due to the above advantages the Czochralski method has been widely used in growing single laser crystals, such as ruby, garnets, other oxides and fluorides.

Since 1965 we started to develop the Czochralski method for growing YAG crystals with a resistance heater or a radio-frequency generator. The equipment for crystal growth have been improved in the past 25 years, such as automatic temperature control and boule diameter control. Some experiments have been carried out to measure the crystal growth parameters, such as temperature distribution in the furnace, pulling rate and rotation rate, as well as melting flow in the crucible and the ambient gas situation for optimum growth condition. Combined crystal growth with defects study in crystals [3], by using suitable technology, we can get YAG crystals with not only decreased dislocations, but also with facets eliminated. Sometimes, large Nd:YAG crystals free of constitutional supercooling, with few scattering centers, and even free of dislocation can be obtained. Figure 9.2 shows the interferometric pattern of a Nd:YAG crystal boule.

Besides ruby and YAG, some other host crystals (e.g. $BeAl_2O_4$, Mg_2SiO_4, YAP) were grown by Czochralski method. Growth and perfection of the above mentioned laser crystals have also been studied for improving the crystal quality [4–6].

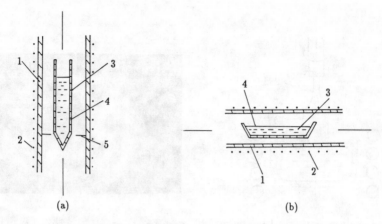

(a)   (b)

Fig. 9.3. Schematic diagram of Stockbarger-Bridgman method. 1, Furnace wall; 2, heater; 3, melt; 4, crucible; 5, thermal screen. (a) Vertical movement; (b) horizontal movement.

9.1.2   Stockbarger-Bridgman Method

Differing from Czochralski method, Stockbarger-Bridgman method is a unidirectional crystallization method. There are two modes of crystallization of melted substance in a container. In one mode the container is moved through the melting zone, and in the other, the temperature is reduced gradually at a constant temperature gradient. The more common of these is crystallization in which the container with substance is displaced through the melting zone. This method is technically simple, but here the container effect is not limited to possible contamination of the melt and the elastic interaction of the container wall with the crystal during cooling may cause stresses in the crystal boule. Figure 9.3 shows the schematic diagram of the Stockbarger-Bridgman method.

We have grown several laser crystals, such as Cr^{3+}:LiCaAlF$_6$, Sm^{3+}, Dy^{3+}, U^{3+}:CaF$_2$, by Vertical Bridgman method [7, 8]. Our crystal growth apparatus is simple and easily operated. The growth furnace is SiC resistance heating element which surrounds the SiO$_2$ glass growth chamber closely, and cylindrical high purity graphite crucibles of 20 mm diameter, 75 mm height and 25 mm wall thickness were employed, which make the boules easily detach from the crucible wall after growth. The temperature at the heating element is measured by a Pt–Rh thermal couple and controlled by a 702 temperature controller. The heating element is held by a platform raised or lowered by means of a precision mechanical device. During the growth, we keep the crucible in stationary state and only raise the heating element to move the thermal field up. Water cooler is also employed beneath the bottom of crucible to achieve a proper temperature field. In addition, for the purpose of protecting the melt and the crucible from contamination and oxidation, the flowing

Table 9.1. The crystal growth condition of laser crystals by flux growth methods.

Crystal	Solvent	Crystal growth condition	Growth method
$Cr:Al_2O_3$	PbF_2	T_c 1400, T_s 1000, D 1.5	ZA, ZC
$Cr:BaTiO_3$	$BaO-B_2O_3$	T_c 1150, T_s 1050, D 0.2, 90.1	ZA, SE
$Nd:CaWO_4$	$Na_2W_2O_7$	T_c (1100–1250), T_s 700, D 2.5	ZA
$Nd:GdAlO_3$	$PbO-PbF_2-B_2O_3$	T_c 1300, T_s 950, D 0.3–0.6	ZB
$Nd:YAG$	$PbO-PbF_2-B_2O_3$	T_c 1300, T_s 950, D 0.3–3	ZA
$Nd:YVO_4$	V_2O_5	T_c 1050, T_s 900, ΔT 4°/cm, P 0.005	SE

D—cooling rate °C/h, P—seed pulling rate mm/h, ΔT—temperature gradient °C/cm, T_c— maintain temperature °C, T_s—growth stopped temperature °C, ZA—spontaneous cooling, ZB—rotating crucible, ZC—evaporation, SE—pulling seed.

atmosphere of Ar gas with purity of 99.995% is used during the melting stage before crystal growth.

For some volatile and hydrolytic materials, such as fluorides, We also adopted closed crucibles and liquid sealing techniques to avoid volatilization of melt. In detail, after loading the starting material into the crucible, put a graphite plug above them, and fill the rest space with the same material, then screw up the crucible cover. At the growth temperature, the melts below and above the graphite plug are simultaneously melted, so the melt above the plug isolates the melt below from the outside, which can protect it from contamination of O_2 and H_2O in the atmosphere.

9.1.3 *Flux Growth Method*

Flux growth method has been widely used for growing single crystals of complex multicomponent materials. This method takes advantage of high solubility of high-melting compounds in liquid inorganic salts and oxides. The successful use of flux growth method depends on the choice of solvent. In Table 9.1 summarized the crystal growth condition of laser crystals by flux growth method.

The crystal is grown from flux by the supersaturation, which can be achieved either by reducing the temperature (ZA, ZB) or by evaporating the solvent, the former is more popular. The crystallization can take place spontaneously or on a seed (SE): Large and perfect single laser crystals can be obtained by crystallization on a seed, which is placed at the bottom of the crucible or pulled from the solid-liquid interface above. Figure 9.4 shows the schematic diagram of flux growth method by seeded nucleation.

Since 1964, Nd:YAG crystals were grown by flux method in SIOFM. We have obtained large, transparent crystals with good shape. The crystal grown by seeded nucleation were larger than 4 cm in length. The weight of crystal was up to 500 g [8, 9]. A series of experiments to determine the optimum condition for growth of large $NdLiP_4O_{12}$ crystals by flux method. The crystals grown by spontaneous nucleation have dimensions of $9\times5\times2$ mm^3 and the largest crystals by seeded growth were up

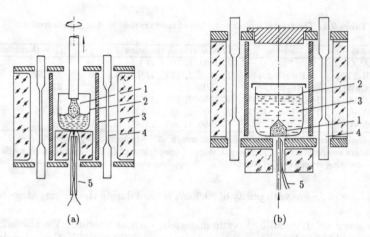

(a)   (b)

Fig. 9.4. Schematic diagram of flux growth by seeded nucleation. 1, Growing crystal; 2, crucible; 3, solution; 4, heater; 5, thermocouple. (a) By pulling; (b) at bottom.

to 20 mm in length [10].

In recent years two new laser crystal growth methods were developed in SIOFM, named the Temperature Gradient Technique (TGT) and the Induction Field Upshift Method (IFSM).

9.1.4   Temperature Gradient Technique (TGT) [11–14]

The heat exchanger method (HEM), vertical solidification of the melt method (VSOM) and vertical gradient freeze method (VGF) of crystal growth techniques, which were developed by Khattak, Powell and Struss respectively, have been successful in producing large diameter single crystals. However, the apparatus of these growth techniques are complicated in structure, which make it difficult to establish the suitable temperature gradient field in growth system. We modified the above mentioned methods to grow single laser crystals using the temperature gradient technique [11].

The TGT growth apparatus is shown in Fig. 9.5. The furnace is resistance heated by means of a cylindrical plumbago element. The crucible is insulated by multilayer molybdenum heat shields which consist of upper, lower and round shields. There are molybdenum support. ZrO_2 ceramic, and water cooling stainless steel shaft at the bottom of molybdenum support. We increase the resistance of cylindrical plumbago element from the bottom to the top by means of making appropriate holes on the plumbago designed by ourselves. A vertical temperature gradient is imposed on the crucible by cylindrical plumbago element and cooling the crucible from the bottom, which contact with the water cooling stainless steel shaft. We

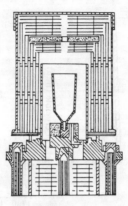

Fig. 9.5. Growth apparatus of TGT method.

make a favorable temperature field which is in the opposite direction to the gravity field as shown in Fig. 9.6. Large and perfect single crystals of sapphire (Al_2O_3), titanium doped sapphire (Ti:Al_2O_3) and neodynium doped garnet (Nd:YAG) have been successfully grown by TGT. The interference patterns of a sapphire single crystal sample with 50 mm in diameter and 30 mm in thickness are shown in Fig. 9.7.

The growth technology and perfection of laser crystals grown by TGT have been studied in detail [13, 14].

9.1.5 *Induction Field Up-shift Method (IFSM) [15, 16]*

We have grown laser crystals Ti:Al_2O_3 first by IFSM.

In ordinary Stockbarger growth technique, some factors limit this technique in production of large diameter and laser quality single crystals with higher melting point, such as Al_2O_3 and Al_2O_3:Ti^{3+}. First the furnace temperature is lower. This problem can be solved by using induction heating method and molybdenum crucible, which allows growing crystals that have both higher melting point and high specific gravity. Next the propagation of the growth interface during crystal growth is achieved by moving the crucible downwards, it can be easily result in vibration and thermal fluctuation, and produce thermal stress in the cooling crystal, such stress not only decreases the optical quality of crystal but generally also results in microcracks, particular for Al_2O_3:Ti^{3+} single crystal growth the larger vibration and thermal fluctuation can cause severe constitutional supercooling and other growth defects due to smaller distribution coefficient of Ti^{3+} ion in grown crystals, which are detrimental to the production of laser quality single crystals. So the crystals

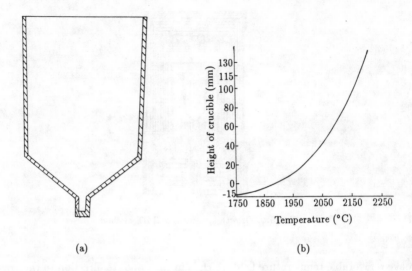

(a)   (b)

Fig. 9.6. The crucible shape (a) and the distribution curve of the temperature gradient in crucible (b).

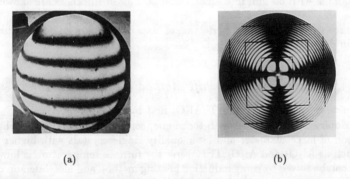

(a)   (b)

Fig. 9.7. Interference pattern (a) and interference figure (under conoscopy) (b) of a sapphire single crystal.

should be grown in a vibration-free environment with a planar or slightly convex growth interface. This problem can be solved by keeping the crucible in stationary state and only raising the heating coil to realize the moving of thermal field up, although the problem discussed above cannot be eliminated entirely, this effect can be significantly minimized by use of the IFSM technique.

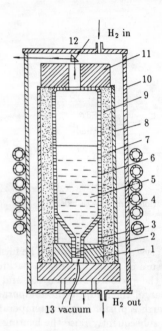

Fig. 9.9. Interference pattern of 55×25 mm Al_2O_3:Ti^{3+} by Mark III interferometer (double beam path). The optical homogeneity of crystal is high except for a region A.

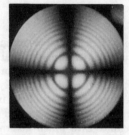

Fig. 9.8. Schematic diagram of apparatus for induction field up-shift method. 1, ZrO_2 pad; 2, seed crystal; 3, lower after heater (Mo); 4, coil; 5, melt; 6, molybdenum crucible; 7, ZrO_2 powder; 8, aluminue tube; 9, upper after heater (Mo); 10, quartz tube; 11, ZrO_2 cover; 12, prism; 13, thermocouple.

Fig. 9.10. Interference figure of 55×25 mm Al_2O_3:Ti^{3+} crystal (under conoscopy), very uniform concentric circles shows the crystal with good quality and less stress.

A schematic diagram of IFSM crystal growth apparatus is shown in Fig. 9.8. The starting material consisting of highly purified γ-Al_2O_3 and TiO_2 powders are loaded into a cylindrical container with the diameter of the required crystal size. An oriented sapphire seed crystal is placed at the center of the bottom of the crucible. In order to facilitate removal of the as-grown crystal from the crucible after growth run, the crucible is tapered at an angle of about 2°. In this technique, the crucible is put on a ZrO_2 pad in a quartz tube which serves as the growth chamber, and surrounded by ZrO_2 powder insulating material to protect the quartz tube from the high crucible temperature. A radio frequency (R. F.) induction furnace at 2.5 MHz is used to heat the crucible, the heating coil surrounds the quartz tube while the coil is raised or lowered verticallly by means of a precise mechanical device. The temperature is measured by a W/W–Re thermocouple located at the seed which monitors the growth process.

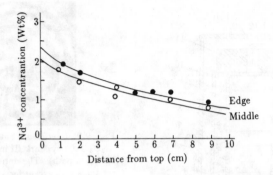

Fig. 9.11. Nd^{3+} concentration distribution along growth axis.

The system operations for growing Ti:Al$_2$O$_3$ crystals are as follows: The furnace is evacuated with a mechanical pump to a pressure of about 0.1333 Pa, backfilled the flowing growth ambient of purified H$_2$ at the pressure between 0.1 and 0.4 kg/cm^2, and heated sufficiently to melt the charge, then adjust the temperature by observing the image of the flowing melt through a prism, keep the temperature constant for several hours, and stop flowing H$_2$, subsequently raise the heating coil at a definite rate which determines the driving force for crystal growth and the shape of growth interface. After the melt is condensed totally stop moving the coil and then gradually cool the grown crystal to room temperature. The initial cooling rate is small, after the top of the crystal reaches a temperature that is well below the melting point of crystal, the rate is increased greatly.

With the gradual modification in technological parameters of crystal growth, the quality and perfection of crystals have been improved greatly, in which the main macrodefect and sub-microdefect, such as mosaic structure and grain boundary etc., have been eliminated in most of as-grown crystals. Figure 9.9 is the interference pattern of a crystal with Zygo Mark III interferometer, which shows that except for the small region marked with A in this figure, the crystal has high optical homogenity, the refractive index difference was calculated to be $\Delta n = 1.5 \times 10^{-5}$. Figure 9.10 is the interference figure of Al$_2$O$_3$:Ti^{3+} crystal with the incident light along the c (optical) axis, the appearance of uniform concentrical circles shows the crystal tested with high optical quality and less interior stress. A number of high quality laser rods have been cut and fabricated from these crystal boules.

One disadvantage of both IFSM and TGT is the large concentration gradient of doped ions along the growth axis.

The axial concentration of Nd^{3+} in the Nd:YAG boule were measured using spectrophotometry. Figure 9.11 shows the axial distribution of Nd^{3+} concentration in the boule. Curve I of Fig. 9.11 shows the axial concentration along the radial

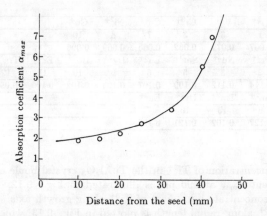

Fig. 9.12. Peak Ti^{3+} absorption coefficient α_{max} vs distance from the seed.

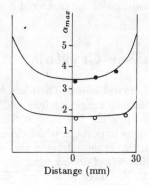

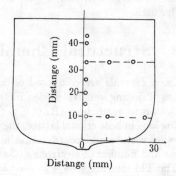

Fig. 9.13. Peak Ti^{3+} absorption coefficient α_{max} vs distance from the axis center.

Fig. 9.14. Positions of the measured points in Fig. 9.13 in crystal. The sample is $(1\bar{1}00)$ slice with the thickness of 1.2 cm.

centre of the boule and curve II shows the axial concentration along the edge of the boule. The Nd^{3+} concentration varied from 2.3 wt% at the edge of the top to 0.7 wt% at the centre of the boule bottom. Fluorescence lifetimes of the top and bottom parts of the boule were measured and found to be 160 and 235 μs respectively. In the crystal the Nd^{3+} doping concentration was up to 2.3 wt% and the crystalline perfection and optical homogeneity were still excellent as determined by transmitted microscope light of a Leitz wide field microscope.

Table 9.2. Ionic radius (r) and number of electrons (N) at $3d$ and $4f$ orbit.

Ion	Ti^{3+}	V^{2+}	Cr^{3+}	Co^{2+}	Ni^{2+}	Cu^+		
N_{3d}	1	3	3	7	8	10		
r (nm)	0.067	0.079	0.062	0.065	0.069	0.096		
Ion	Pr^{3+}	Nd^{3+}	Sm^{3+}	Eu^{3+}	Dy^{3+}	Ho^{3+}	Er^{3+}	Tm^{3+}	Yb^{3+}
N_{4f}	2	3	5	6	9	10	11	12	13
r (nm)	0.114	0.112	0.109	0.107	0.103	0.102	0.100	0.099	0.098
Ion	Sm^{2+}	Tm^{2+}	Dy^{2+}	Eu^{2+}				
N_{4f}	6	13	10	7				
r (nm)	0.127	0.102	0.121	0.125				

The axial concentration of Ti^{3+} in the $Ti:Al_2O_3$ crystal boule expressed as absorption coefficient α_{max} at 490 nm is illustrated in Fig. 9.12. There exists a relatively large concentration gradient of Ti^{3+} along growth axis. The absorption coefficient of Ti^{3+} along radial length is plotted in Fig. 9.13, which displayed the crystal growth with planar or slightly convex growth interface. It can be seen that the radial concentration distribution is rather flat at central part of the boule. Therefore it is benificial to cut the laser crystal rod or slab along the plane perpendicular to the growth axis. Figure 9.14 shows the positions of the measured points in Fig. 9.13.

9.2   Structural Chemistry of Laser Crystals

Laser crystals are composed of doping ions and crystal hosts. There are four groups of doping ions: transition metal ions (TM), trivalent rare earth ions (RE^{3+}), bivalent rare earth ions (RE^{2+}), actinouranium group ions. For substitution of doping ions in host crystal lattice, the ionic radius is more important, which depends on valence state and coordination number (C. N.). In Table 9.2 listed the ionic radius and number of electrons at $3d$ or $4f$ electronic orbit of doping ions.

The TM ions are always located at tetrahedon (C. N. $\sim$4) or octahedron (C. N. $\sim$6) and the RE-ions possess higher coordination number (C. N. $\sim$8).

For laser crystal hosts there are five groups: simple oxides, oxide compounds, oxygen acid salts, simple fluorides and mixed fluorides. The crystal symmetry of the hosts and the site symmetry of the doping ions are the most important structural parameters for laser crystals. Table 9.3 lists these parameters for the most important laser crystals.

The crystal structure type, substitutional site, optical class and common growth method of several laser crystals are shown in Table 9.4.

Table 9.3. Structural parameters of laser crystals.

Crystal	Host	Cation in host crystal			Active ion (I. R.)	
	Space group	Cation	Site symmetry	Ionic radius (C. N.)	3d	4f
I. Simple oxide						
Al_2O_3	D_{3d}^6–$R\bar{3}C$	Al^{3+}	C_3	0.53 (6)	Cr^{3+} (0.62), Ti^{3+} (0.67)	Eu^{3+} (1.05)
$BeAl_2O_4$	D_{2h}^{16}–P_{nma}	Be^{2+}, Al^{3+}	C_5, C_i	0.27 (4), 0.53 (6)	Cr^{3+}, Ti^{3+}	
$YAlO_3$	D_{2h}^{16}–P_{bnm}	Y^{3+}, Al^{3+}	C_s, C_i	1.12 (12), 0.53 (6)	Ti^{3+}, Cr^{3+}	Nd^{3+} (1.12), Er^{3+}, Tm^{3+}
$Y_3Al_5O_{12}$	O_h^{10}–I_{a3d}	Y^{3+}, Al^{3+}	D_2, C_{3i}	1.02 (8), 0.53 (6)	Cr^{3+} (0.53), Cr^{4+} (0.44)	Nd^{3+}, Gd^{3+}, Er^{3+}, Ho^{3+}, Tm^{3+}, Yb^{3+}
$Gd_3Ga_3O_{12}$	O_h^{10}–I_{a3d}	Gd^{3+}, Ga^{3+}	D_2, S_4	1.06 (8), 0.62 (6)	Cr^{3+}	Nd^{3+}
$Gd_3Sc_2Al_3O_{12}$	O_h^{10}–I_{a3d}	Gd^{3+}, Sc^{3+}, Al^{3+}	C_{3i}, D_2, C_{3i}, S_4	1.06 (8), 0.75 (6), 0.39 (4)	Cr^{3+}, Cr^{4+}	Nd^{3+}
Mg_2SiO_4	D_{2h}^{16}–P_{nma}	Mg^{2+}, Si^{4+}	C_s, C_i	0.72 (6), 0.26 (4)	Cr^{4+}	
YVO_4	D_{4h}^{19}–$I4/amd$	V^{5+}, Y^{3+}	D_{2d}	0.36 (4), 1.02 (8)		Nd^{3+}, Eu^{3+}, Tm^{3+}
$CaWO_4$	C_{4h}^6–$I4_1/a$	Ca^{2+}, W^{6+}	S_4	1.00 (6), 0.42 (4)		Pr^{3+}, Nd^{3+}, Ho^{3+}, Er^{3+}, Tm^{3+}
$KY(WO_4)_2$	C_{2h}^6–$C2/c$	W^{6+}, Y^{3+}		0.42 (4), 1.02 (8)		Nd^{3+}
$Ca_5(PO_4)_3F$	C_{6h}^2–$P6_3/m$	Ca^{2+}, P^{5+}	C_s, C_3	1.00 (6), 0.17 (4)		Nd^{3+}, Ho^{3+}
$LiNbO_3$	C_{3v}^6–R_{3c}	Li^+, Nb^{5+}	C_3	0.59 (4), 0.64 (6)		Nd^{3+}, Ho^{3+}, Tm^{3+}
NdP_5O_{14}	C_{2h}^5–$P2_1/C$	Nd^{3+}		1.12 (8)		Nd^{3+}

Table 9.3 *(to be continued)*

Table 9.3 (*Continued*)

Host		Cation in host crystal			Active ion (I. R.)	
Crystal	Space group	Cation	Site symmetry	Ionic radius (C. N.)	3d	4f
II. Mixed oxide						
CaY₄(SiO₄)₃O	C²₆ₕ–P6₃/m	Y³⁺		1.02 (8)		Nd³⁺, Ho³⁺
ZrO₂–Er₂O₃	O⁵ₕ–Fm3m	Ca²⁺		1.00 (6)		Nd³⁺, Ho³⁺, Er³⁺, Tm³⁺
		Er³⁺		1.07 (8)		
		Zr⁴⁺		0.72 (6)		
III. Simple fluoride						
MgF₂	D¹⁴₄ₕ–P4/mnm	Mg²⁺	D₂ₕ	0.86 (6)	Co²⁺ (0.89), V²⁺ (0.93) Ni²⁺ (0.83)	
LiYF₄	C⁶₄ₕ–I4₁/a	Li		0.73 (4)		Tm³⁺, Gd³⁺
		Y³⁺	S₄	1.16 (8)		
LiCaAlF₆	D²₃d–P3̄1C	Ca²⁺	C₃	1.14 (6)	Cr³⁺ (0.76)	Nd³⁺ (1.26), Ce³⁺ (1.28), Er³⁺ (1.00)
		Al³⁺		0.67 (6)		
LiSrAlF₆	D²₃d–P3̄1C	Sr²⁺	C₃	1.27 (6)	Cr³⁺	
		Al³⁺		0.67 (6)		
KMgF₃	O¹ₕ–Pm3m	K⁺		1.65 (8)	Co²⁺, V²⁺	
		Mg²⁺		0.86 (6)		
CaF₂	O⁵ₕ–Fm3m	Ca²⁺	C₄ᵥ, C₃ᵥ Oₕ	1.14 (6)		Nd³⁺, Ho³⁺, Er³⁺, U³⁺ Sm²⁺, Dy²⁺, Tm²⁺
LaF₃	D⁴₃d–P3̄C1	La³⁺	C₂	1.32 (8)		Nd³⁺, Er³⁺
IV. Mixed fluoride						
CaF₂–YF₃	O⁵ₕ–Fm3m	Ca²⁺	C₃	1.14 (6)		Nd³⁺, Ho³⁺, Er³⁺
		Y³⁺	S₄	1.16 (8)		
CaF₂–ErF₃	O⁵ₕ–Fm3m	Ca²⁺	C₃	1.14 (6)		Er³⁺, Ho³⁺, Tm³⁺
		Er³⁺	S₄	1.14 (8)		

Table 9.4. Some characteristics of several laser crystals.

Host	Growth method	Optical class	Crystal structure type	Substitutional site. coord. No.
LiYF$_4$ (YLF)	CZ	Uniaxial	Scheelite	Y, 8
LaF$_3$	ZM	Uniaxial	Tysonite	La, 9
SrF$_2$	BM	Isotropoc	Fluorite	Sr, 8
BaF$_2$	BM	Isotropic	Fluorite	Ba, 8
KCaF$_3$	BM	Biaxial	Perovskite (D)	Ca, 6
KY$_3$F$_{10}$	ZM	Isotropic	—	Y, 8
Rb$_2$NaYF$_6$	ZM	Isotropic	Elpasolite	Y, 6
BaY$_2$F$_8$	ZM	Biaxial	—	Y, 8
Y$_2$SiO$_5$(YOS)	CZ	Biaxial	—	Y, 6
Y$_3$Al$_5$O$_{12}$ (YAG)	CZ	Isotropic	Garnet	Y, 8
YAlO$_3$ (YALO)	CZ	Biaxial	Perovskite (D)	Y, 12
Ca$_5$(PO$_4$)$_3$F (FAP)	CZ	Uniaxial	Apatite	Ca$_{II}$, 7
LuPO$_4$	Flux	Uniaxial	Zircon	Lu, 8
LiYO$_2$	CZ	Biaxial	Rock salt (D)	Y, 6
ScBO$_3$	CZ	Uniaxial	Calcite	Sc, 6

ZM=zone-melting; CZ=Czochralski; BM=Bridgman; D=distorted version of structure type.

References

1. A. A. Chernov, *Modern Crystallography III, Crystal Growth* (Springer Series in Solid-State Sciences, Springer-Verlag, Berlin, 1984).

2. *Crystal Growth*, eds. Kezong Zhang and Laowei Zhang (Science Press, Beijing, 1981) (in Chinese).

3. Peizhen Deng, Jingwen Qiao and Zhengying Qian, *Acta Optica Sinica* **2** (1982) 251.

4. Laser Crystal Research Group, SIOFM, *J. Lasers* **6** (1979) 48.

5. Qiang Zhang, Peizhen Deng and Fuxi Gan, *Cryst. Res. Technol.* **25** (1990) 385.

6. Bing Hu, Hongbi Zhu and Peizhen Deng, *J. Cryst. Growth* **128** (1993) 991.

7. Xiaodong Liu, Peizhen Deng, *et al.*, *Proc. SPIE* **1863** (1993) 513.

8. Shanshan Zhang, Peizhen Deng *et al.*, *Laser Materials and Components*, in *Research Reports of SIOFM*, Vol. 8, 1980, p. 12.

9. Flux Growth Research Group, SIOFM, *J. Chinese Cer. Soc.* **8** (1980) 137.

10. Baosheng Lu, Huanchu Chen and Fusheng Chen, *Acta Phys. Sinica* **27** (1978) 609.

11. Yongzong Zhou, Peizhen Deng and Jingwen Qiao, *J. Chinese Cer. Soc.* **11** (1983) 357.

12. Yongzong Zhou, *J. Cryst. Growth* **78** (1986) 31.

13. Peizhen Deng, Jingwen Qiao, Bing Hu and Yongzong Zhou, *J. Cryst. Growth* **92** (1988) 276.

14. Yongzong Zhou, Peizhen Deng, *et al.*, *Proc. SPIE* **1627** (1992) 230.

15. Yao Chai, Shenghui Yan, Peicong Pan and Peizhen Deng, *J. Chinese Cer. Soc.* **19** (1991) 338.

16. Peizhen Deng, Yao Chai, *et al.*, *Proc. SPIE* **1338** (1990) 207.

10. Rare Earth Ions Doped Laser Crystals

Rare earth (RE) ions are generally introduced into the host crystals as a substitutional impurities in concentration in the order of 1%. Laser oscillations have also been obtained with rare earth ions as the stoichiometric components of the host crystals. Although as many as two hundreds of RE ions doped laser crystals have been emerged, only a few of them have practical applications, which we will discuss in this chapter.

10.1 Laser Emissions of Rare Earth Ions Doped Crystals

There are hundreds of laser emission lines of RE ions doped in different oxide and fluoride crystals. In Fig. 10.1 the shaded areas show the band of optical transparency of fluoride and oxide crystals and the vertical lines indicate the wavelengths of laser emission of RE^{3+} ions in these crystals. The $4f$–$4f$ intermanifold transitions and $4f$–$5d$ transitions of RE^{3+} ions cover wide spectral regions from UV ($\sim$0.18 μm) to IR ($\sim$5 μm), as shown in Fig. 10.2. The shortest wavelength (0.172 μm) of laser emission was achieved in $4f$–$5d$ transition of Nd^{3+} ions by laser pumping at 300 K [1], and the longest wavelength (5.15 μm) of laser emission was also performed by Nd^{3+} ions using cascade transition by lamp pumping at 300 K [2]. The most popular ions are still Nd^{3+} ion, followed by Ho^{3+}, Tm^{3+} and Er^{3+} ions in term of their applicability. The richest by the number of laser emission wavelengths is Ho^{3+}, which has 12 laser channels, Er^{3+} has 11 and Pr^{3+}—8.

10.2 Nd^{3+} Ions Doped Laser Crystals

As many as a hundred Nd^{3+}-doped laser crystals have been emerged, but from application point of view only a few of them are useful.

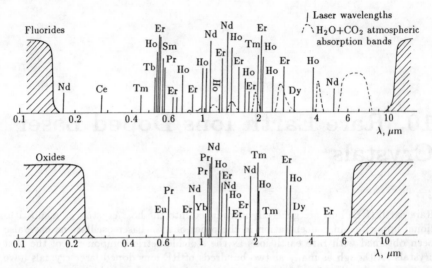

Fig. 10.1. Laser emission bands and optical transparency of the laser crystals doped with RE^{3+} ions.

10.2.1 *Nd:YAG and Nd:YAP Laser Crystals*

Although Nd:YAG and Nd:YAP laser crystals have been already commercialized, researches for improving the crystal quality and laser performances are still going on.

The laser performances of Nd:YAG crystal rods are dependent on crystal quality greatly. The optical inhomogeneity, scattering loss and dislocations in crystal all influence the laser output performances. Table 10.1 shows this relationship, the Nd:YAG crystals were grown by Czochralski method in the Shanghai Institute of Optics and Fine Mechanics (SIOFM) with decreasing light loss coefficient, the slope efficiency and laser output energy increase obviously in the case of CW operation [3].

The main laser crystals for high power laser output (CW or high repetition rate) are Nd:YAG and Nd:YAP crystals. 300 W laser output power at 1.06 μm can be obtained in Nd:YAG and Nd:YAP single crystal rod of 8 mm in diameter and 120–140 mm in length (Table 10.2). We put more emphasis on the growth method (temperature gradient technique, TGT). Figure 9.2 shows the interferometric pattern of a Nd:YAG disk, indicating that there is only a little facets at the edge of crystal and a large high quality region within the crystal [4]. Using a slab Nd:YAG crystal (6×20×94 mm³ size), 230 W output power can be got at 50 Hz repetition rate. The quality of output laser is greatly improved, and the divergence of the laser beam is about 3/8 mrad [5]. 300 W average laser power in multimode were generated

Fig. 10.2. Wavelengths at which various transitions of RE ions are observed.

Table 10.1. Quality of Nd:YAG rods.

Growth technique	Diameter (mm)	Length (mm)	ϵ (dB/5cm)	α (%cm^{-1})	β (%cm^{-1})	Pulsed opertion			CW operation		
						E_{th} (j)	$\tan\theta$ (%)	E_{out} (j)	W_{th} (W)	$\tan\theta$ (%)	W_{out} (W)
Czochralski	ϕ5.4	79	25–29	1.0	0.25	1.0	0.9	0.7	990	2.05	35.2
induction	ϕ5.5	75	27	1.4	0.21	1.8	1.56	0.98	966	1.40	29.7
heating											
Czochralski	ϕ5.5	71	26–30	1.8	0.19	2.5	1.54	1.0	1070	1.9	24.9
resistance	ϕ5.6	71	31	1.1	0.3	1.3	1.52	0.96	1080	1.72	28
heating											

ϵ, Extinction ratio; α, light loss coefficient; β, large-angle scattering coefficient; E_{th}, threshold energy; $\tan\theta$, slope efficiency; E_{out}, output energy; W_{th}, threshold power; W_{out}, output power.

Table 10.2. Laser performance of Nd:YAG and Nd:YAP rods.

Rod size (mm)	High repetition rate						CW operation
	Repetition rate (pps)	E_{max} (j)	P_{max} (W)	θ (mrad)	Power stability	E-O efficient	P_{out} (W)
1	20–50	23	500				400
2	0–40	34	400	10	2–5%	5.5%	240
3	0–20	40	700	20	2–5%	4.5%	300

1, ϕ7×100×2; 2, ϕ8×110; 3, ϕ8×140 YAP.

in free running repetitively pulsed (30 pps) operation and the slope efficiency was 3.3%. The size of the slab-crystal Nd:YAG was 6×15×100 mm^3 [6].

For material processing industrial Nd:YAG lasers over 1 kW power in CW and quasi-CW mode have been developed. Figure 10.3 shows typical output characteristics of CW output laser versus input power of the lamp. At 40 kW input power the output of the laser can be more than 1.4 kW. The overall efficiency of the system (four laser heads are connected) was 2.8% [7].

Using KTP crystal for frequency doubling, 6.5 W CW and 33 W quasi-CW (an acousto-optical Q-switch operating at 25 kHz) green lasers at 532 nm have been also achieved [8].

Nd:YAG crystal can also be operated at 1.3 μm by transition $^4F_{3/2}\rightarrow^4I_{132}$. Due to low gain at 1.3 μm in comparison with that at 1.06 μm, the depression of gain at 1.06 μm is necessary. Figure 10.4 shows the laser output energy vs flash lamp input energy, the slope efficiency is 0.7%. The Nd:YAG crystal rod was 5 mm in diameter and 75 mm in length [9].

The effective peak cross-section at 1.06 μm and 1.33 μm for several laser crystals with Nd^{3+}ions at 300 K was measured as shown in Table 10.3.

The gain coefficient at 1.3 μm of Nd:YAP (YAlO$_3$) is larger than that of Nd:YAG,

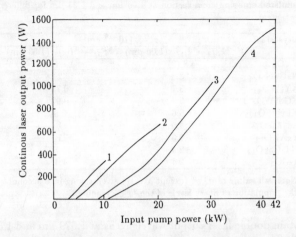

Fig. 10.3. Laser output characteristic of Nd:YAG laser for up to 4 modules within the resonator. 1, One module; 2, two modules; 3, three modules; 4, four modules.

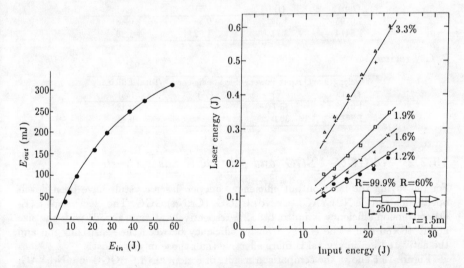

Fig. 10.4. Laser output character of Nd:YAG crystal at 1.3 μm wavelength.

Fig. 10.5. Free running laser efficiency for various Nd,Cr:GGG laser rods of different Cr-doping as well as Nd:YAG and Nd:YAP for comparison. Rod size φ5 mm×60 mm.

Table 10.3. Stimulated emission cross-section at 1.06 μm and 1.33 μm for different Nd^{3+}-doped crystals.

Crystal	$\sigma_e^{\text{eff,p}}$ $(10^{-19}$ $cm^2)$	
	$^4F_{3/2}\rightarrow^4I_{11/2}$ (1.06 μm)	$^4F_{3/2}\rightarrow^4I_{13/2}$ (1.33 μm)
YVO_4	8 (?)	1.8 (?)
$NaGdGeO_4$	3.8	2
$KY(WO_4)_2$	3.7	0.73
$KGd(WO_4)_2$	3.7	0.92
$KLu(WO_4)_2$	3.6	0.9
$Lu_3Al_5O_{12}$	3.5	1
$Y_3Al_5O_{12}$	3.3	1
$Gd_3Ga_5O_{12}$	2.9	1.5
$YAlO_3$	2.4	1.1

Note: all values of $\sigma_e^{\text{eff,p}}$ (except YVO_4) given taking into account the population of Stark levels of the $^4F_{3/2}$ manifold.

it is easy to obtain double laser output wavelengths at 1.079 μm and 1.341 μm from Nd^{3+}:YAP [10]. The laser output performances are shown as follows:

1. Pulsed regime

Wavelength (μm)	Output energy (j)	Slope efficiency (%)	Total efficiency (%)
1.0795	1.39	0.59	0.48
1.3414	3.71	1.62	1.29

2. CW regime

Wavelength (μm)	Output Power (W)	Slope efficiency (%)	Total Efficiency (%)
1.0795	33.7	—	0.48
1.3414	30.0	—	0.43

10.2.2 *Nd:GGG, YSGG, and Nd:GSGG Laser Crystals*

For improving the laser output efficiency codoped laser crystals have been developed, such as (Ce,Nd):YAG, (Cr,Nd):GSGG, (Cr,Nd):GGG. The electro-optic (E-O) conversion efficiency is up to 4, 6, 2% respectively at crystal rods with the size about ϕ6×80 mm. The energy transfer efficiency between the sensitizing ion and the activated ion in crystal is more effective than those in glass host.

Figure 10.5 shows the comparison results of codoped Nd,Cr:GGG and Nd:YAG, Nd:YAP. With increasing codoping concentration of Cr in Nd,Cr:GGG crystals, the output efficiency is raised obviously. The optimizing free running efficiency for a Nd,Cr:GGG and Nd,Cr:YSGG laser rod can be up to 4.2%, as shown in Fig. 10.6 [11].

Due to high energy transfer efficiency from Cr^{3+} to Nd^{3+} in GSGG crystal the

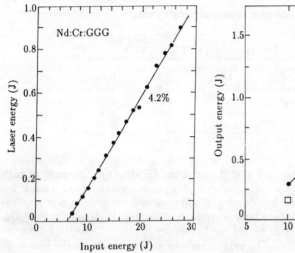

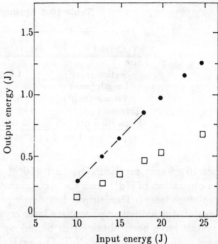

Fig. 10.6. Optimized free running laser efficiency for a Nd,Cr:GGG laser rod (2.4×10^{20} Nd-at/cm³ and 3.7×10^{13} Cr-at/cm³). Rod size $\phi 5$ mm×75 mm.

Fig. 10.7. Laser output energy vs input energy (capacitor bank) for Cr,Nd:GSGG and Nd:YAG in the same free-running laser resonator. Output-coupler reflectivity was 49%. •, Cr,Nd:GSGG; □, Nd:YAG.

intrinsic lasing efficiency of Cr,Nd:GSGG laser rods exceeded that of the best Nd:YAG rods by more than a factor of 2, although the Cr,Nd:GSGG crystals had significantly higher internal losses than commercial YAG laser rods, which is understandable in view of the fact that GSGG crystal-growth technology has had relatively little time to develop. Figure 10.7 shows the comparison results of Cr,Nd:GSGG with Nd:YAG, the slope efficiency of laser output vs input energy is about 7% for Cr,Nd:GSGG crystals [12].

An actively and passively mode locked Nd,Cr:GSGG laser has been constructed, which delivers 25 ps single pulses with energy up to 30 mj at 1.06 μm [13].

10.2.3 *Nd³⁺ Ions Doped Phosphate, Borate, Aluminate and Vanadate Crystals*

The high Nd³⁺ concentration is present in stoichiometric phosphate crystals, such as NdP_5O_{14}, $LiNdP_4O_{12}$, $NaNdP_4O_{12}$, $KNdP_4O_{12}$ etc. The concentration quenching behaviour observed for concentrated Nd phosphate crytals in which Nd³⁺ ions are separated by phosphate cross-chains is not as serious as those in Nd:YAG, Nd:YAP crystals. For example, in $La_{1-x}Nd_xP_5O_{14}$ the fluorescence lifetime was only reduced from 310 μs to 120 μs as X increased to unity. Reported cross sections range from

Table 10.4. Quality of NdP_5O_{11} rods.

Crystal	From abroad	Shandong Univ.	
		Rod No. 1	Rod No. 2
Size			
Section (mm^2)	1.2	9.6	5.75
Length (mm)	7	5.2	10.7
Volume (mm^3)	8.4	49.9	61.5
Threshold (j)	0.16–0.18	0.7	0.36
Slope efficiency (%)	0.52–0.78 *	0.4	0.95

Appl. Phys. Lett. **34** (1979) 387.

1.8×10^{-19} cm^2 for NdP_5O_{14} to 9×10^{-19} cm^2 from $LiNdP_4O_{12}$. Because of high concentration of Nd^{3+} ions the stoichiometric phosphate crystals are of interest for miniature lasers. Danielmyer has developed and reviewed the physics and applications of stoichiometric phosphate laser crystals in 1970s [14]. High quality NdP_5O_{11} crystals have been obtained in Shandong University, China. Table 10.4 lists the laser performances of the NdP_5O_{11} crystal rods in comparison with the literature data.

Due to the narrow fluorescence linewidth of Nd^{3+}-doped calcium fluorophosphate [FAP, $Ca_5F(PO_4)_3$] the stimulated emission cross section σ_p is high, thus low laser threshold can be expected. Flash lamp pumped Nd:FAP lasers operating at 1.0623 μm have been reported in early time [15, 16]. The experimental results of an efficient CW Nd:FAP laser pumped by a dye (LD-700) laser, operating at the strong line 1.0629 μm [transition from $R_1(^4F_{3/2})$ to $Y_1(^4I_{11/2})$] and the weak line 1.1259 μm (transition from $R_1 \rightarrow Y_5$) are shown in Fig. 10.8. In comparison with Nd:YAG, it can be found that the laser threshold is much lower in Nd:FAP [17].

Low threshold, high efficiency laser at 1.059 μm and 1.328 μm was reported for strontium fluorapatite, $Sr_3(PO_4)_3F$ or SFAP, doped with Nd^{3+}. The threshold energy and power were as low as 5 μj and 4 mW respectively, and the slope efficiency of output was around 40–60% for pulsed and CW operation [42].

Neodynium aluminum borate [$NdAl_3(BO_3)_4$] (NAB) is a typical stoichiometric borate with high Nd-doping concentration (5×10^{21} cm^{-3}) and stimulated cross section (8×10^{-19} cm^2), low laser output threshold and high gain. High quality and large size (up to $\phi 3.2 \times 23.7$ mm^3) NAB crystals have been obtained by flux growth in China [18]. The maximum laser output of a NAB rod with dimensions of $\phi 3.2$ mm $\times 23.7$ mm was 422 mj a pulse when it was pumped with a single flashlamp. The laser threshold of a NAB rod with dimensions of $\phi 1.8$ mm $\times 2.4$ mm was determined to be 67 mj. By using BDN dye for switching, the laser pulse width of a NAB rod with dimensions of $\phi 1.8$ mm $\times 8.4$ mm was 8 ns. The output laser beam was linearly polarized with beam divergence angle of about 2 mrad and wavelength of $\lambda = 1.063$ μm [19]. Figure 10.9 shows the single pass gain G_0 and gain coefficient g_0 of a NAB crystal rod with 1.9 mm in diameter and 8.4 mm in length with flash lamp pumping.

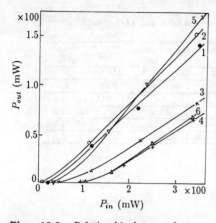

Fig. 10.8. Relationship between laser output power and pumping power of Nd:FAP and Nd:YAG. Nd:FAP: 1, 1.0629 μm ($\pi-$); 2, 1.0629 μm ($\sigma-$); 3, 1.1259 μm ($\sigma-$); 4, 1.1259 μm ($\pi-$). Nd:YAG: 5, 1.0641 μm; 6, 1.1225 μm.

Fig. 10.9. Relation between single pass gain G_0, gain coefficient g_0 and pumping energy E_{in} of a NAB crystal rod with dimensions of $\phi 1.9 \times 8.4$ mm. 1, g_0; 2, G_0.

Table 10.5. Luminescence and laser characteristics of Nd^{3+}-doped YAG, YSGG, LSB.

Crystal	N_{Nd} ($\times 10^{20}$ cm^{-3})	N_{Cr} ($\times 10^{20}$ cm^{-3})	τ_L (μs)	σ_p ($\times 10^{-19}$ cm^2)	Loss (%)	Laser slope eff. (%)
YAG	1.4	—	210	3.69	3.34	0.76
YSGG	3	2	240	1.38	5.68	2.6
LSG	10.25	0.6	75	2.2	0.87	5.4

La$_{1-x}$Nd$_x$(Sc$_{1-y}$Cr$_y$)$_3$(BO$_3$)$_4$ (LSB) is another new stoichiometric borate laser crystal. The Nd^{3+} doping concentration can reach up to 5×10^{21} cm^{-3} and the fluorescence lifetime can keep to 70 μs. Cr^{3+} ions substitute Sc sites in LSB crystal, the quantum efficiency of energy transfer Cr$^{3+} \rightarrow$Nd^{3+} is about 0.8. Table 10.5 lists the luminescence and laser characteristics of LSB crystal in comparison with those of Y$_{3-x}$Nd$_x$Al$_5$O$_{12}$ (YAG) and (Y$_{1-x}$Nd$_x$)(Sc$_{1-y-z}$Ga$_y$)$_z$Ga$_3$O$_{12}$ (YSGG) [20].

LSG has a broad FWHM absorption peak (3 nm) that is four times that of a Nd^{3+}:YAG crystal (0.8 nm). When pumped by 1.2 W from a 808 nm diode laser, Nd:LSG produced 0.74 W output at 1.063 μm, the slope efficiency was near 70%.

The crystal—La$_{1-x}$MgNd$_x$Al$_{11}$O$_{19}$ (LNA) has been prepared first by Kahn in 1981 [21], and 5 W CW laser output with Kr lamp pumping was obtained in 1988. LNA single crystals were grown by Czochralski method in SIOFM, large crystals of LNA (up to 120 mm long, 20 mm in diameter) have been obtained, the optical

Table 10.6. Spectral properties of Nd^{3+} in LNA and YAG crystal.

Crystal	$^4F_{3/2}$ $\rightarrow$	Peak (μm)	$\Delta\lambda_{flu}$ (nm)	τ_{flu} (μs)	τ_0 (μs)	η	N_{Nd} at%
Nd^{3+}:LaMgAlO	$^4F_{11/2}$	1.054	1.2	315	338	0.93~1	5
Nd^{3+}:YAG	$^4I_{11/2}$	1.064	0.65	240	420	0.56	1

Crystals	$\sigma =$ $\times10^{19}$ cm^2	β_{ij} 9/2	11/2	13/2	15/2	$\Omega_{(2,4,6)}$ Ω_2	Ω_4	Ω_6
Nd^{3+}:LaMgAlO	3.2	0.35	0.54	0.15	0.009	0.21	0.48	1.5
Nd^{3+}:YAG	4.5	0.32	0.58	0.11	0.003	0.2	2.7	5.0

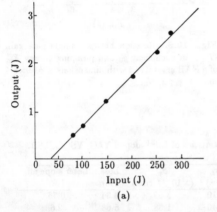
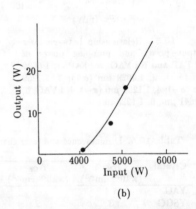

(a)   (b)

Fig. 10.10. Laser output characteristics of LNA laser crystal rod with dimensions of $\phi6.2$ mm$\times110$ mm. (a) Pulsed operation, (b) CW operation.

quality of obtained LNA crystal is similar to that of Nd:YAG [22].

As shown in Table 10.6, some spectral parameters of Nd^{3+} in LNA crystal are better than those in YAG crystal, the fluorescence lifetime is longer and the quantum efficiency is higher. The concentration quenching effect is weaker in LNA crystal.

The laser output characteristic curves are shown in Fig. 10.10. The lasing experiments were carried out using a $\phi6\times110$ mm laser rod oriented along the "a"-axis. The pump source was a pulse xenon flashlamp. The laser rod was put in a single elliptical cavity. The double lamp resonator consists of two plane mirrors with transmission $T_1 =0$ and $T_2 =40\%$ respectively at the lasing wavelength. When the repetition frequency is 5 Hz, the lasing threshold is $P_{th} =30$ J. The output energy $E =2.65$ J/pulse at 1.054 μm, slope efficiency $\eta =1\%$.

CW lasing experiments were carried out using $\phi6\times110$ mm rod with "a"-axis parallel to the rod length, the pump source was two krypton lamps, the resonator

Table 10.7. Nd density, absorption coefficient (α) at the pumping wavelength (809 nm) and absorption depths (l_a) of the laser materials.

Laser material	Nd density (at%)	α (cm^{-1})	l_a (mm)
α-Cut Nd:YVO$_4$			
(π polarization)	2.02	72.4	0.14
	1.1	28.8	0.35
Nd:YAG	1.1	7.1	1.41

consists of two plane mirrors with the transmission of $T_1 = 0$ and $T_2 = 5\%$ or 10% respectively at 1.054 μm. When the input power is 5 kW and $T_2 = 10\%$, the output power is 16 W. The output power is limited by the thermal focusing effect because the temperature coefficient of refractive index of LNA crystal is larger than that of Nd:YAG ($dn/dT = 18 \times 10^{-6}$ K^{-1} in LNA, 9.86×10^{-6} K^{-1} in YAG).

Nd^{3+} ions possess rather high absorption coefficient and broad band at the absorption wavelength (809 nm) in YVO$_4$ crystal host. Table 10.7 lists Nd density, absorption coefficient and absorption length of Nd:YAG and Nd:YVO$_4$ crystals, it can be obviously found, the absorption of Nd:YVO$_4$ is 4 times that of Nd:YAG. Therefore, Nd^{3+}:YVO$_4$ crystal has been popularly used as a diode pumped miniature laser crystal.

For example, an end-pumped, simple standing-wave-type laser resonator was constructed by Sasaki *et al.* [23]. A 1-W single- and broad-stripe laser diode (Sony SLD 304V or Mitsubishi M4049E) was used as the pumping source. Three *a*-axis-cut Nd:YVO$_4$ crystal plates of different Nd concentrations, which had parallel flat surfaces with a thickness of 1 mm, were used as gain medium and one of the resonator mirrors. The Nd concentration and the absorption coefficient of Nd:YVO$_4$ used are listed in Table 10.7. A Nd:YAG crystal was also given for comparison. The focusing spot size of the pumping diode laser was approximately 230 μm $\times$ 100 μm at 760 mW output. Figure 10.11 shows the output power of Nd:YVO$_4$ and Nd:YAG crystals at 1.064 μm as a function of absorbed pumping power. The maximum output power of 325 mW was achieved for 1.1 at% Nd:YVO$_4$ at a pumping power of 760 mW. The slope efficiency was 54.6%. They also found that the smaller the waist size, the larger the absorption coefficient, which, therefore, gives a high output power. An intracavity SHG experiment was carried out using a type II KTP of 7-mm thickness with the 2.02% Nd:YVO$_4$. A green light output power of as much as 16 mW was obtained in a single longitudinal mode at a pumping power of 620 mW. The conversion efficiency was 2.6% [23].

T. Taira reported a highly efficient and compact Nd:YVO$_4$ laser. In CW operation, a single-mode output of 95 mW and a multimode output of 435 mW were obtained with a 1 W laser-diode pump. Using a KTP crystal in the laser cavity, 105 mW of green light was generated. A Q-switched pulse with a peak power of 230 mW and pulse width of 8 ns was obtained with the intracavity KTP crystal used as both the Q-switch and the frequency doubler [40].

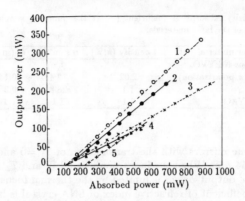

Fig. 10.11. Output power of Nd:YVO$_4$ and Nd:YAG lasers at 1.064 μm as a function of absorbed pumping power. 1, YVO$_4$, 2.02 at%, multimode; 2, YVO$_4$, 2.02 at%, single longitudinal mode; 3, YVO$_4$, 1.1 at%, multimode; 4, YVO$_4$, 1.1 at%, single longitudinal mode; 5, YAG, 1.1 at%, multimode.

10.3   Er^{3+}, Tm^{3+}, Ho^{3+} Ions Doped Laser Crystals

It can be seen from the energy diagram (Fig. 10.12), there are many radiative transition channels of Er^{3+}, Tm^{3+} and Ho^{3+} ions in dielectric crystals. People more interested in near infrared laser emissions of Er^{3+}, Ho^{3+} and Tm^{3+} ions by cascade and cross-cascade energy transfer from upper to lower levels as follows:

Ho^{3+}:

1) $^5S_2 \rightarrow {}^5I_5 \rightsquigarrow {}^5I_6 \rightarrow {}^5I_8$;

2) $^5S_2 \rightarrow {}^5I_5 \rightarrow {}^5I_7$;

3) $^5S_2 \rightarrow {}^5I_5 \rightarrow {}^5I_6$;

4) $^5I_6 \rightarrow {}^5I_7 \rightarrow {}^5I_8$;

Er^{3+}:

5) $^4S_{3/2} \rightarrow {}^4I_{11/2} \rightarrow {}^4I_{13/2}$;

6) $^4S_{3/2} \rightarrow {}^4I_{13/3} \rightarrow {}^4I_{15/2}$;

7) $^4S_{3/2} \rightarrow {}^4I_{9/2} \rightarrow {}^4I_{11/2}$;

8) $^4S_{3/2} \rightarrow {}^4I_{9/2} \rightsquigarrow {}^4I_{11/2} \rightarrow {}^4I_{13/2}$;

9) $^4I_{11/2} \rightarrow {}^4I_{13/2} \rightarrow {}^4I_{15/2}$;

Tm^{3+}:

10) $^3F_4 \rightarrow {}^3H_5 \rightsquigarrow {}^3H_4 \rightarrow {}^3H_6$;

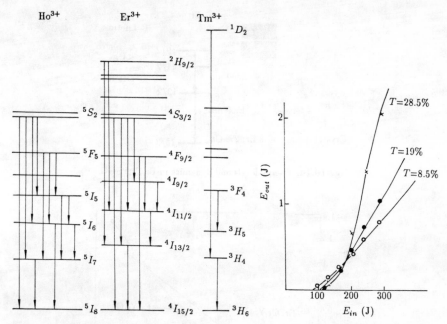

Fig. 10.12. Energy level diagram and laser channels of Ho^{3+}, Er^{3+} and Tm^{3+} in inorganic crystals.

Fig. 10.13. Laser output character curve of Er^{3+}:YAG crystal ($\phi 6 \times 96$ mm, 300 K).

Table 10.8. Low-temperature laser channels of high-concentration erbium crystals.

Crystal (Er^{3+} ion concentration)	Laser channel; wavelength μm; (threshold, J)			
	$^4S_{3/2} \rightarrow ^4I_{13/2}$	$^4S_{3/2} \rightarrow ^4I_{11/2}$	$^4S_{3/2} \rightarrow ^4I_{9/2}$	$^4F_{9/2} \rightarrow ^4I_{11/2}$
$LiErF_4$ ($1.4 \cdot 10^{22}$ cm^{-3})	0.8540 (7.5)	1.2288 (13)	1.7042 (15)	2.0005 (70)
$BaEr_2F_8$ ($1.3 \cdot 10^{22}$ cm^{-3})	0.8425 (8)	1.2320 (10)	1.6455 (7)	1.9975 (25)
	0.8543 (5)		1.7355 (7.5)	
$Er_3Al_5O_{12}$ ($1.4 \cdot 10^{22}$ cm^{-3})	—	—	1.7762 (70)	—
$Y_{1.5}Er_{1.5}Al_5O_{12}$ ($0.7 \cdot 10^{22}$ cm^{-3})	0.8627 (57)	—	1.7767 (40)	—
$KEr(WO_4)_2$ ($0.65 \cdot 10^{22}$ cm^{-3})	0.8621 (5.1)	1.2460 (10)	1.7370 (1.9)	—
$ErAlO_3$ ($2 \cdot 10^{22}$ cm^{-3})	—	—	1.6632 (22)	—

10.3.1 Er^{3+} Ions Doped Laser Crystals

Kaminskii summarized the low-temperature laser channels of high-concentration erbium crystals, as listed in Table 10.8 in the early years [24]. More useful laser wavelengths for medical applications (specially for eye-safe) of erbium crystals are at 1.54 μm and 2.94 μm, which originated from transitions $^4I_{13/2}$–$^4I_{15/2}$ and $^4I_{11/2}$ –$^4I_{13/2}$ respectively. High efficiency Er^{3+}:YAG lasers have been emerged. Figure

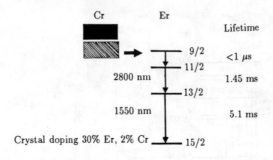

Fig. 10.14. Energy levels and dynamics in Er,Cr:YSGG.

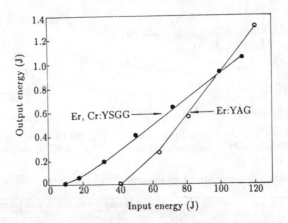

Fig. 10.15. Er,Cr:YSGG vs Er:YAG laser performance.

10.13 shows the laser output character at 2.94 μm of Er^{3+}:YAG crystal developed in SIOFM [25]. In the fluoride crystal (Er:YLF) the laser wavelength is 2.81 μm.

Recently, the E-O conversion efficiency was up to 1–2%. It has been done by Cr^{3+} sensitization, the energy levels and dynamics in Er,Cr:YSGG are shown in Fig. 10.14. The laser output energy versus input energy curve was shown in Fig. 10.15, in comparison with that of Er:YAG, the threshold energy of Er,Cr:YSGG was much lower and the E-O efficiency was higher than those of Er:YAG.

10.3.2 *Tm^{3+} Ions Doped Laser Crystals*

Laser emission has been obtained from two levels of Tm^{3+}—$^3H_4$ and $^3F_4$. Both pulsed and CW laser action were observed from $^3H_4 \rightarrow {}^3H_6$ transition at wavelength

Table 10.9. Laser output characteristics of Tm^{3+}-doped crystals.

Host crystal	Sensitizer ion	Wavelength (μm)	Slope eff. (%)	Pumping source	Crystal size (mm)	Operation temp. (K)	Ref.
YAG	—	1.833	—	F·L	$l = 25$	77	26
YAG	—	1.87–2.16	30	Ti:Al$_2$O$_3$	$l = 3.2$	300	27
YVO$_4$	—	1.94	25	Ti:Al$_2$O$_3$	1.5×1.5×2.5	300	28
YAG	Cr	2.014	4.5	F·L	$\phi 5 \times 76.3$	300	29

F·L: Flash lamp.

of $\sim$1.9 to 2.1 μm [26]. The broad absorption band at 780 nm is useful for Ti:Al$_2$O$_3$ and diode laser pumping. There are 9 Stark sublevels of $^3F_4$ and 13 Stark sublevels of $^3H_6$, it results in that the laser emission wavelength around 2 μm can be tuned continuously from 1.85 to 2.14 μm. Tm^{3+} has only a few absorption band in the visible range and the energy cascade is inefficient due to the large energy gaps between J states. Er^{3+} is an effective sensitizer for $^3H_4$ laser action, and Cr^{3+} has been used as a sensitizer for $^3F_4$–$^3H_5$ laser action for flash lamp pumping. Table 10.9 summarized the laser output characteristics of thulium laser crystals with different pumping.

10.3.3 Ho^{3+} Ions Doped Laser Crystals

In 1965 Johnson first reported the laser emission of Ho^{3+} ions in YAG at 2.0975 μm [26]. Due to the high threshold energy and the operation at liquid nitrogen temperature, the practical application has been delayed. Great improvement has been achieved for holmium crystal laser operated at room temperature by sensitization. The main energy donor ions are Tm^{3+} ions, the nonradiative energy transfer takes place between $^3H_4$ (T_m^{3+}) and $^5I_7$ (Ho^{3+}). The energy can be also transferred by cascade process, such as $Er^{3+}\leadsto Tm^{3+}\leadsto Ho^{3+}$, $Cr^{3+}\leadsto Tm^{3+}\leadsto Ho^{3+}$. Figure 10.16 shows the energy transfer in Ho, Cr, Tm system. The comparison of laser output performances of Ho, Cr, Tm:YSGG and Ho, Cr, Tm:YAG crystals is shown in Fig. 10.17. Table 10.10 summarizes the recent research results of codoped holmium laser crystals.

By transition $^5I_6$–$^5I_7$ laser emission of Ho^{3+} can be achieved at 2.9054 μm in BaYb$_2$F$_8$ fluoride crystal [32].

10.4 Pr^{3+} and Yb^{3+} Ions Doped Laser Crystals

The short wavelength (0.5 μm) Pr^{3+} doped laser demonstrated by German et al. [33] is due to the $^3P_o \rightarrow ^3H_4$ transition, hence the threshold was high and the laser operation was at low temperature, there was no further development. More attention has been paid on laser emission channels from $^3P_0$ to ^3F sublevels. Kaminski

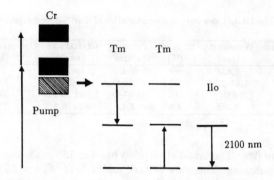

Fig. 10.16. Energy transfer in Ho, Cr, Tm systems.

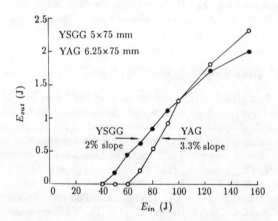

Fig. 10.17. Comparison of Ho,Cr,Tm:YSGG and Ho,Cr,Tm:YAG.

Table 10.10. Laser output characteristics of codoped holmium laser crystals (at $\sim 2\ \mu$m).

Sensitizer ions	Host crystal	Slope eff. (%)	Crystal size (mm)	Temp. (K)	Pumping source	Ref.
Er^{3+}, Tm^{3+}	YLF	1.3	$\phi 4 \times 25$	300	F·L	30
Tm^{3+}	YAG	30	3.2	300	$Ti:Al_2O_3$	27
	YSGG					
Cr^{3+}, Tm^{3+}	YAG	5.1	$\phi 5 \times 7$	300	F·L	31
	YSGG					

F·L: Flash lamp.

Table 10.11. Laser performances of Pr^{3+}-doped laser crystals.

Crystal host	Laser channel	Laser wave-length (μm)	E_{th} (j)	Operation temp. (K)	Pumping source	Term. level energy (cm^{-1})
YAlO$_3$	$^3P_0-^3F_3$	0.7195	25	110–250	F·L	6460
LaF$_3$	$^3P_0-^3F_4$	0.7198	20	300	F·L	7030
LiYF$_4$	$^3P_0-^3F_2$	0.6395	10	300	F·L	5220

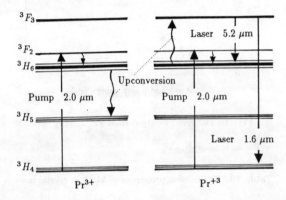

Fig. 10.18. Energy level diagram of Pr^{3+}:LaCl$_3$ showing the observed laser and pump transitions as well as one possible upconversion process.

reported the laser performances of Pr^{3+}:LaF$_3$ and Pr^{3+}:YAlO$_3$ in 1980s [34–36], as shown in Table 10.11.

New stimulated emissions of Pr^{3+}-doped LaCl$_3$ crystal have been observed recently, 1.6 μm laser action was caused by transition $^3F_3\rightarrow^3H_4$ and $^3F_3\rightarrow^3H_6$ transition resulted in 5.242 μm laser emission at temperature of 130 K. The pumping source was a pulsed Tm:YAG laser, which gave the laser output at 2.02 μm. The possible process of energy transfer is shown in Fig. 10.18. The pump energy at 2.02 μm directly excites the $^3F_2$ level, which rapidly thermalizes with the $^3H_6$ level. Excitation of the higher lying $^3F_3$ level then occurs rapidly as a result of a quadratic upconversion process:

$$^3H_6 +^3 H_6\rightarrow^3F_3 +^3 H_5. \tag{10.1}$$

Figure 10.19 shows the energy output from Pr^{3+}:LaCl$_3$ laser crystal at the laser wavelength of 5.24 μm [41].

There is only one excited level $^2F_{5/2}$, the transition $^2F_{5/2}\rightarrow^2F_{7/2}$ is laser action of Yb^{3+} ions. The laser threshold for oscillation is high for flashlamp pumping since the laser emission terminates on the Stark level of the ground state manifold, it needs low temperature for laser action. The diode laser pumping source is

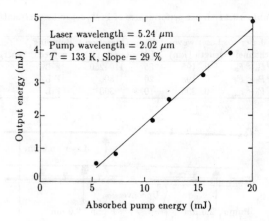

Fig. 10.19. Energy output from Pr^{3+}:$LaCl_3$ laser.

Table 10.12. Emission properties of Yb^{3+} in crystal hosts.

Host	τ_{em} (ms) Measured	τ_{rad} (ms) Calculated	τ_s (ms) Selected	λ_{ext} (nm)	ΔE (cm^{-1})	β_{min}	σ_{ext} (10^{-20} cm^2)	Concentration (10^{20} Yb^{3+}/cm^3)	Density (g/cm^3)
$LiYF_4$	2.16	2.27	2.21	1020	480	0.098	0.81	1.44	3.77
LaF_3	2.22	2.10	2.16	1009	353	0.113	0.36	0.22	5.94
SrF_2	9.72	(8.6)	9.2	1025	593	0.054	0.16	0.10	4.24
BaF_2	8.20	(7.8)	8.0	1024	578	0.058	0.14	0.16	4.83
$KCaF_3$	2.7	(4.0)	2.7	1031	(593)	0.070	0.22	1.40	2.67
KY_3F_{10}	2.08	1.66	1.87	1011	369	0.118	0.44	1.45	4.31
Rb_2NaYF_6	10.84	—	10.84	1012	(372)	0.140	0.10	1.03	~3.4
BaY_2F_8	2.04	—	2.04	1018	(458)	0.097	0.67	1.40	4.97
Y_2SiO_5	1.04	—	1.04	1042	(617)	0.047	0.33	1.83	4.44
$Y_3Al_5O_{12}$	1.08	0.93	1.01	1031	628	0.055	2.03	1.36	4.55
$YAlO_3$	0.72	0.42 [a]	0.42 [a]	1014	353	0.141	1.31	6.82	5.35
$Ca_5(PO_4)_3F$	1.08	1.30	1.2	1043	603	0.046	5.90	0.36	3.19
$LuPO_4$	0.83	—	0.83	1011	349	0.160	0.53	0.80	5.53
$LiYO_2$	1.13	—	1.13	1020	(474)	0.090	0.56	2.27	4.12
$ScBO_3$	4.80	—	4.80	1022	(472)	0.091	0.19	1.32	3.45

[a] Only one of the three polarizations was accounted for.
(τ_{em}, Measured emission lifetime; τ_{rad}, lifetime calculated by reciprocity; τ_s, average or selected lifetime; $\Delta E = E_{ZL} - h\nu$; β_{min}, minimum fractional population inversion required to have transparency at the extraction wavelength; σ_{ext}, emission cross section at extraction wavelength judged to be optimal; parentheses suggest greater uncertainty in the magnitude of the quantity.)

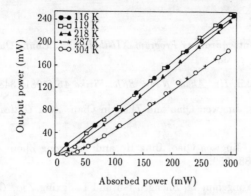

Fig. 10.20. Output power of Ti:sapphire laser-pumped Yb:YAG laser at 1030 nm versus the absorbed pump power at 969 nm at various temperatures.

suitable for Yb^{3+}-doped laser crystals, which are rather useful for high power lasers. Because of the long radiative lifetime (1∼2 ms) of the $^2F_{5/2}$ state, Yb^{3+} provides good energy storage.

Deloach *et al.* measured the spectral properties of Yb^{3+} ions doped in different fluoride and oxide hosts, the experimental results are listed in Table 10.12 [37].

In the framework of requirements for an effective diode-pumped Yb^{3+} laser system, Yb^{3+}:YAG and Yb^{3+}:Ca$_5$(PO$_4$)$_3$F exhibit the most useful properties because of high stimulated emission cross section σ_{ext}.

The first efficient room temperature diode pumped Yb:YAG crystal laser was demonstrated by Lacovara *et al.* [38]. Quite recently Giesen *et al.* determined the laser threshold and efficiency of Yb:YAG for crystal temperatures between 100 K and 340 K. A longitudinal pumping scheme with an Ar$^+$-pumped Ti:sapphire laser was chosen because of its excellent beam quality, narrow tunable linewidth and good overlap between the pump beam and the TEM$_{000}$ laser mode, which gives the best performance possible. The 0.7 mm thick crystal with a doping level of about 19 at% was AR/HR-coated on the front side and AR/AR-coated on the backside for the pump beam (940 nm) and the laser beam (1030 nm), respectively. A typical plot of the laser output power at 1030 nm versus absorbed pump power for various temperature is shown in Fig. 10.20. The slope efficiencies are 85% for 969 nm pumping and 83% for 940 nm between 100 K and 210 K. The extrapolated threshold is 40 mW, the real 25 mW one was at room temperature [39].

The main problem of Yb:YAG crystal laser is the thermal population of the lower laser level which leads to the high pumping power density in the range of 10 kW/cm^2, which results in the thermal lensing and efficiency reduction by the subsequent increase of the lower level population.

References

1. R. W. Waynant, *Advances Program XII-th Intern. Conf. Quantum Electron.*, Munich, 1982, p. 33.

2. A. A. Kaminskii, *Izv. Akad. Nauk. SSR, Fizika* **45** (1981) 348.

3. Peizhen Deng, Jingwen Qiao and Zhenyin Qian, *Acta Optica Sinica* **2** (1982) 259.

4. Peizhen Deng, Jingwen Qiao, Bing Hu and Yongzong Zhou, *Chinese Physics: Lasers* **15** (1988) 659.

5. Huijuan He, Yongchun Li, Shengru Gu and Longxing Zao, *Chinese J. Lasers* **17** (1990) 513.

6. H. P. von Arb, Ch. Lüchinger and F. Studer, *Proc. SPIE* **1021** (1988) 24.

7. H. Takenaka, K. Okino, M. Tani and J. Buchholz, *Proc. SPIE* **1021** (1988) 153.

8. Chaoen Huang, Pengdi Han, *et al.*, *Chinese J. Lasers* **15** (1988) 586.

9. Seisong Zhou and Kie Huan, *Chinese J. Lasers* **14** (1987) 576.

10. Hongyuan Shen, R. R. Zeng, Yuping Zhou *et al.*, *J. Quantum Electron.* **27** (1991) 2315.

11. V. V. Raptev *et al.*, *Sov. J. Quant. Electron.* **18** (1991) 579.

12. J. A. Caird, W. F. Krupke, M. D. Shinn, L. K. Smith, and R. E. Wilder, Digest of Technical Papers, 1985 Conf. on Lasers and Optoelectronics, Washington D.C., 1985, p. 232.

13. H. J. Eichler and B. Liu, *Opt. Mater.* **1** (1992) 21.

14. H. G. Danielmeyer, *Stoichiometric Laser Materials*, in *Festkörperprobleme, Vol. XV*, ed. H. J. Queisser (Pergamon/Vieweg, Braunschweig, 1975) pp. 253–277.

15. O. Deutschbein *et al.*, *Appl. Opt.* **17** (1978) 2228.

16. W. E. Martin, *IEEE J. Quant. Electron.* **QE-18** (1982) 1155.

17. Zhongxing Shao and Yi Chen, *Chinese J. Lasers* **17** (1990) 46.

18. Baolin Wang *et al.*, *J. Chinese Cer. Soc.* **12** (1984) 259.

19. Yichuan Huang, Minwang Qiu *et al.*, *Chinese J. Lasers* **14** (1987) 524.

20. S. A. Kutovoe *et al.*, *Sov. J. Quant. Electron.* **18** (1991) 149.

21. A. Kahn *et al.*, *J. Appl. Phys.* **52** (1981) 6864.

22. Xiurong Zhang, Ximin Zhang, Jun Xu *et al.*, *Chinese J. Lasers* **B2** (1993) 221.

23. T. Sasaki, T. Kojima *et al.*, *Optics Letters* **16** (1991) 1665.

24. A. A. Kaminski, *Izv. Akad. Nauk SSR, Neorganicheski Material* **20** (1984) 901.

25. Xiurong Zhang, Guangzhao Wu *et al.*, *Chinese J. Lasers* **18** (1991) 630.

26. L. F. Johnson *et al.*, *Appl. Phys. Lett.* **7** (1965) 127; **8** (1966) 200.

27. R. C. Stoneman, *Proc. SPIE* **1223** (1990) 231; *Opt. Lett.* **15** (1990) 486.

28. H. Sasito, *et al.*, *Optics Lett.* **17** (1992) 189.

29. G. J. Quarlis *et al.*, *Optics Lett.* **15** (1990) 42.

30. E. P. Chichlis, *et al.*, *Appl. Phys. Lett.* **19** (1971) 119.

31. A. A. Allatev, *Sov. J. Quant. Electron.* **18** (1988) 167.

32. A. A. Kaminski, *et al.*, *Izv. Akad. Nauk SSR, Neorganicheski Mater.* **18** (1982) 482.

33. K. R. German, A. Kiel and H. Guggenheim, *Appl. Phys. Lett.* **22** (1973) 87.

34. A. A. Kaminski, A. G. Retroyan, K. L. Ovanesyan and M. I. Chertanov, *Phys. Stat. Sol.* **77** (1983) KI73.

35. A. A. Kaminski, *Izv. Akad. Nauk. SSR., Neorgan. Mater.* **17** (1981) 185.

36. A. A. Kaminski, *Dok. Akad. Nauk. SSR* **271** (1983) 1357.

37. L. D. Deloach, S. A. Payne, *et al.*, *IEEE J. Quant. Electron.* **29** (1993) 1179.

38. P. Laovara, C. A. Wang, *et al.*, *Opt. Lett.* **16** (1991) 1089.

39. A. Giesen, H. Hügel, A. Voss and K. Wittig, U. Brauch and H. Opower, *Symp. Advanced Solid State Lasers*, Salt Lake City, Utah, 1994, PD4-1 .

40 T. Taira and T. Kobayashi, *Proc. Symp. Compact Blue-green Lasers*, Salt Lake City, Utah, 1994, PD8-1.

41 S. R. Bowman, J. Ganem, B. J. Feldman, *Proc. Symp. Advanced Solid State Lasers*, Salt Lake City, Utah, 1994, PD1-1.

42 X. X. Zhang, P. Hong, *et al.*, *Proc. Symp. Advanced Solid State Lasers*, Salt Lake City, Utah, 1994, PD7-1.

11. Transition Metal Ions Doped Laser Crystals

As we have discussed in Chapter 2, due to stronger interaction between transition metal (TM) ions and crystal host, differing from rare earth (RE) ions, the spectral behaviours of TM ions are characterized by strong absorption in visible and near infrared regions, which is a great advantage for selecting light pumping sources, and the broad phonon side band is suitable for laser wavelength tuning. The ruby ($Cr^{3+}:Al_2O_3$) laser was the first solid state laser emerged in 1960. A great number of TM ions doped laser crystals are still playing important role in laser science and technology. In this chapter we will introduce the TM ions doped laser crystals in general, and discuss the tunable laser crystals in detail in the next chapter.

11.1 Laser Emissions of Transition Metal Ions Doped Crystals

The laser actions were performed mainly from TM ions of $3d$ electron configuration, such as Ti^{3+}, Cr^{4+}, Cr^{3+}, V^{2+}, Ni^{2+} and Co^{2+}. According to the ionic radius, the coordination number of $3d$ TM ions is from 4 to 6, therefore the TM ions substitute the sites of Al^{3+}, Ga^{3+}, Mg^{2+}, Ca^{2+}, Zn^{2+} in the crystal lattice. Table 11.1 lists the ionic radius, number of $3d$ electrons, valence state and coordination number of laser active TM ions.

Table 11.1. Structural parameter of TM active ions.

Ion	Ti^{3+}	Cr^{4+}	Cr^{3+}	V^{2+}	Ni^{2+}	Co^{2+}
Ionic radius (nm)	0.067	0.044	0.062	0.079	0.069	0.067 (0.057)
Number of $3d$ electrons	1	2	3	3	5	7
Coordination state	6	4	6	6	6	6 (4)

230

Table 11.2. Laser performances of TM ions doped crystals.

Ion	Crystal host	Laser wavelength (nm)	Temp. (K)	Laser operation	Pumping source	Transition	Emission lifetime (μs)
d^3 Ti^{3+}	Al$_2$O$_3$	680~1178	300	P., CW	F. L, L	^2E_g-^2T	3.2
	BeAl$_2$O$_4$	780-820	300	P.	L.		4.9
	YAlO$_3$	610-630	300	P.			14
d^2 Cr^{4+}	Mg$_2$SiO$_4$	1130-1367	300	P., CW	Nd:YAG L	$^3T_{2g}$-$^3A_{2g}$	25
	YAG	1196-1303	300	P.	F. L		
	YAG	1420-1500	300	P., CW	Nd:YAG L		
d^3 Cr^{3+}	Al$_2$O$_3$	694.3	300	P., CW	F. L	$^2E \rightarrow {}^4A_2$	3500
	BeAl$_2$O$_4$	700-830	300	P. CW	F. L		260
	Be$_3$Al$_2$Si$_6$O$_{12}$	751-759	80	P.	F. L		65
	KZnF$_3$	758-875	300				180
	LiCaAlF$_6$	720-840	300	P. CW	L.		175
	ZnWO$_4$	980-1090	77	P.	Kr F.		8.6
	Y$_3$Ga$_5$O$_{12}$	740	300	CW	L.	$^4T_2$-$^4A_2$	241
	Gd$_3$Ga$_5$O$_{12}$	760	300	CW	L.		159
	Gd$_3$Sc$_2$Al$_3$O$_{12}$	765-801	300	CW	L.		–
	Y$_3$Sc$_2$Ga$_2$O$_{12}$	730	300	CW	L.		139
	Gd$_3$Sc$_2$Ga$_3$O$_{12}$	745-820	300	P.	L.		115
	La$_3$Lu$_2$Ga$_3$O$_{12}$	820	300	P., CW	L., F. L		–
	SrAlF$_5$	852-1005	300	CW	Kr, L		80
	ScBO$_3$	780-890	300	P.	L.		–
d^3 V^{2+}	CsCaF$_3$	1240-1330	80	CW	Kr, L	$^4T_2$-$^4A_2$	2500
	MgF$_2$	1050-1300	200	CW	Kr, L		2300
d^8 Ni^{2+}	MgO	1310-1410	77	P., CW	Nd:YAG L		3600
	CaY$_2$Mg$_2$Ge$_3$O$_{12}$	1460	80	P.	L.		–
	KMgF$_3$	1591	77-300	P.	L.	$^3T_2 \rightarrow {}^3A_2$	11400
	MgF$_2$	1610-1740	80-200	CW	Nd:YAG L		12800
	MnF$_2$	1920-1940	77-85	P., CW	L.		–
d^7 Co^{2+}	MgF$_2$	1630-2450	80-225	CW, P.	Nd:YAG L	$^4T_2 \rightarrow {}^4T_1$	1200
	KMgF$_3$	1620-1900	80	P.	L.		3100
	KZnF$_3$	1650-2070	80-200	CW	L.		–
	ZnF$_2$	2165	77	CW	Ar L		400

P., Pulsed; CW, continuous wave; L., laser; F. L, flash lamp.

Table 11.3. Groups of Cr^{3+}-doped laser crystals.

Crystal field	D_q/B	Δ_g	Transition and emission line shape	Crystal host	Emission wavelength (nm)
High	72.3	>0	$^2E \rightarrow {}^4A_2$ Sharp line	Al_2O_3	694
Middle	~2.3	~0	Mixed emission $^2E \rightarrow {}^4A$ $^4T_2 \rightarrow {}^4A_2$	MgO, $BeAl_2O_4$ $Y_3Ga_5O_{12}$, $Be_3Al_2(SiO_3)_6$	720, 750 750, 730
Weak	<2.3	<0	$^4T_2 \rightarrow {}^4A_2$ Broad band	K_2LiScF_6, $Al(PO_3)_3$ K_2NaAlF_6 Cs_2NaYF_6	750, 785 765 1050

Owing to broader fluorescence linewidth the stimulated emissions are not so easy to obtain from the TM ions doped inorganic crystals in comparison with that from RE ions doped crystals. In Table 11.2 we summarized the laser performances of TM ions doped laser crystals.

11.2 Cr^{3+} and V^{2+} Ions Doped Laser Crystals

From Cr^{3+} energy level diagram (Fig. 1.3) it can be seen that the energy gap between $^4T_2$ and ^2E depends on the crystal field strength D_q. At high crystal field $\Delta_g = E(^2E) - E(^4T_2) > 0$ the transition $^2E \rightarrow {}^4A_2$ takes place, because this is the spin-forbidden transition, the narrow linewidth emission (R-line) can be obtained. At low crystal field ($\Delta_g < 0$), a broad band emission from $^4T_2 \rightarrow {}^4A_2$ transition can be observed [1]. Therefore, the Cr^{3+}-doped laser crystals are classified as three groups shown in Table 11.3.

11.2.1   Ruby (Cr^{3+}:Al_2O_3) Laser Crystals

It has already been stated that the first laser built used a ruby operating at 694.3 nm. The concentration of Cr_2O_3 is around 0.05 wt%, corresponding 1.2 cm^{-1} of the absorption coefficient at 400 nm. The crystal has almost cubic symmetry, with a distortion along one of the body diagonals, therefore, the true symmetry of ruby is rhombohedral. The essential features of the energy diagram are the two wide (absorption) bands $^4F_1$ and $^4F_2$ and the split ^2E level ($\delta = 29$ cm^{-1}) as shown in Fig. 11.1. Fluorescence of ruby consists of R_1 and R_2 lines, but ordinarily laser action will take place only at R_1 line (694.3 nm) with light propagating along c-axis, the peak absorption coefficient for R_1 line at room temperature is $R = 0.4$ cm^{-1}, the lifetime of R_1 transition is 3 ms at room temperature.

Ruby crystal is first grown by Verneuil method, which is also called flame fusion

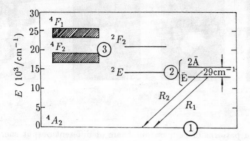

Fig. 11.1. Energy level diagram of ruby crystal.

Fig. 11.2. "Star" image of ruby crystal rod (a) at focusing point; (b) in front of focusing point; (c) behind focusing point.

method. Powdered alumina with Cr_2O_3 raw material falls from a bin, passes through a gas burner, and reaches the upper fused end face of the single crystal ruby seed. Dropping through a hydrogen-oxygen flame, the charge particles partially fuse and fall into a thin film of the melt. Since the seed is lowered slowly, the melt film crystallizes at the seed surface. The main advantages of Verneuil method are rather simple equipment, no container, crystallization conducted at 2000 °C in air, but the quality of ruby crystals obtained by this method is poor, such as high internal stresses, low optical homogeneity and impurities came from the working gas and the surrounding. We have prepared ruby laser crystal rods with large size (30 mm in diameter and 1 m in length) in 1960s, the crystal boules were grown by Verneuil method. For improving the ruby crystal quality, the Czochralski method was introduced for crystal growth in 1970s.

Although the threshold energy of ruby laser was high due to the laser action by three levels mechanism, but the ruby laser was the only visible laser among the crystal lasers, hence it has been used for a long time. A lot of research work on ruby crystals and lasers have been done since 1960 [2, 3]. High energy experiments of ruby laser have been carried out in the Shanghai Institute of Optics and Fine

Fig. 11.3. Near field patterns of laser output beam at different output energy (a) and interferometric pattern of ruby rod (b).

Table 11.4. Change of laser output energy before and after repairing optical path length of ruby rod.

Size of ruby rod (mm)	Before repairing		After repairing	
	Laser output (j)	Displacement of focusing plane (mm)	Laser output (j)	Displacement of focusing plane (mm)
$\phi 10 \times 259$	16.0	+18.2	61.3	+0.1
$\phi 12 \times 284$	22.2	+13.2	47.9	−0.5
$\phi 11.5 \times 300$	11.8	+15.0	22.2	−0.2

Mechanics (SIOFM) [4]. Due to optical inhomogeniety of ruby crystal the light passing through ruby rod was diverged. As shown in Fig. 11.2, the "Star" image after the focusing point was enlarged greatly. Therefore, the optical path length of ruby rods should be repaired. In Table 11.4 the change results of laser output energy for repairing optical path length have been listed. After repairing the optical path length of ruby rods the laser output energy were increased by 4~5 times. The near field pattern of laser output beam corresponded to the interferometric pattern of the ruby rod, as shown in Fig. 11.3.

Using two ruby rods with the dimensions of $\phi 18$ mm×540 mm and $\phi 18$ mm×575 mm, as the oscillator and amplifier, and pumped by two Xe flash lamps with the dimensions of $\phi 21$ mm×1200 mm, the laser output energy was 1500 j, and the E-O efficiency was 0.4% in a F-P cavity with the length of 1.7 m. 3000 j laser energy has been reached using two ruby rods with the dimensions of $\phi 28$ mm×86 mm and $\phi 28$ mm×930 mm. The experiments have been done in 1965–1966.

11.2.2 Cr^{3+} Ions Doped Oxide Laser Crystals

As mentioned above R-line lasing has been demonstrated in Cr:YAG [5], alexandrite [6, 7] and emerald [8]. In GSGG and GSAG the R-lines do not appear at room temperature, because the inversion-site symmetry does not allow for a strong electric dipole transition. Except for ruby and Cr:YAG all Cr^{3+}-doped oxide crystal produce vibronic lasing from transition $^4T_2 \rightarrow {}^4A_2$.

Table 11.5. R-line spectral characteristics of Cr^{3+}-doped crystals.

	Al_2O_3	$MgAl_2O_4$	$Y_3Al_5O_{12}$	$BeAl_2O_4$
Peak position of				
absorption band (cm^{-1})	18400	18520	16700	16393
R_1 line (cm^{-1})	14400	14660	14530	14706
Lifetime of level	3.5	5.0, 8.0	1.5	0.3
^2E (τ_{300}) (ms)				
τ_{77}/τ_{300} (K)	1.14	2.6, 4.5	6.0	9.3
σ $(\times 10^{-19}$ cm$^2)$	1.2	0.23	0.35	$k = 12$ cm^{-1}
σ_{R_1}, $(\times 10^{-20}$ cm$^2)$	2.5	—	0.26	$k = 7.5$ cm^{-1}

Table 11.6. Laser performances of Cr^{3+}-doped laser crystals.

Parameter	Host material			
	$BeAl_2O_4$	GSGG	$Na_3Ga_2Li_3F_{12}$	$SrAlF_5$
Free-lasing wavelength,				
λ_{free} (nm)	753	785	791	932
Tuning range (nm)	704–801	750–842	741–841	825–1011
σ_e at λ_{free} $(\times 10^{-20}$ cm$^2)$	0.55	0.92	0.61	2.1
Maximum slope efficiency (%)	71 (752 nm)	28 (785 nm)	20 (783 nm)	15.5 [a] (930 nm)
Output coupling, C_{out} (%)	5.1	1.6	4.3	2.32
Round-trip loss, L (%)	1.1	3.1	13.5	8.1
Crystal thickness, t (cm)	0.36	0.136	0.45	0.196
Loss coefficient,				
$L/2t$ (%cm^{-1})	1.5	11.2	15.0	20.6
Measured G_p (λ) (%W^{-1})	—	5.65 (780 nm)	3.42 (780 nm)	2.60 (910 nm)
Predicted $G_p(\lambda)$ (%W^{-1})	—	5.78	11.0	9.44

[a] Tuner-plate output coupling.

Table 11.5 lists the spectral characteristics of R-line emission of several Cr^{3+}-doped crystals.

Table 11.6 lists the vibronic laser output performances of Cr^{3+}:$BeAl_2O_4$ and Cr^{3+}:GSGG crystals, where $G_p(\lambda)$ is the gain coefficient at the peak wavelength [9]. As shown in Fig. 11.4, the laser-pumped slope efficiency of Sr^{3+}:$BeAl_2O_4$ is much higher than that of Cr^{3+}:$ScBO_3$. The pumping source was Kr ion laser. For most of Cr^{3+}-doped crystals the slope efficiency for conversion of pump power to output power was unable to achieve values higher than 30%. In alexandrite (Cr^{3+}:$BeAl_2O_4$) a slope efficiency of 71% has been achieved.

11.2.3 Cr^{3+} Ions Doped Fluoride Laser Crystals

All Cr^{3+}-doped fluoride crystals possess vibronic lasing. The first tunable fluoride laser to operate at room temperature was Cr^{3+}:$KZnF_3$ [10]. The slope efficiency was 14% by Kr ion laser pumping. In recent years Cr^{3+}-doped new fluoride crystals,

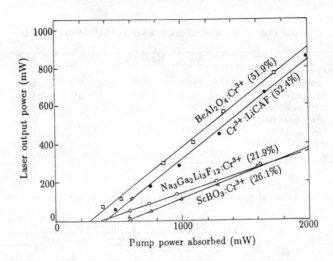

Fig. 11.4. Laser output power versus pump power absorbed by Cr^{3+}-doped crystals pumped by Kr ion laser.

such as $LiCaAlF_6$, $LiSrAlF_6$, $LiBaAlF_6$, exhibit excellent laser performances [11], which are demonstrated in Table 11.6 and Fig. 11.4. Due to the low nonlinear refractive index it is beneficial in reducing the self-focusing and the laser-induced damage.

In Table 11.7 the measured efficiencies of Cr^{3+} vibronic lasers utilizing the laser pumped laser configuration have been summarized [11]. The intrinsic quantum efficiencies of Cr^{3+}-doped crystals are shown in Table 11.8. It can be seen that the $LiCaAlF_6$ and emerald are the most pronounced ones.

11.2.4 V^{2+} Ions Doped Laser Crystals

V^{2+} ion has similar electronic configuration ($3d^3$) to Cr^{3+} ion, but due to the low valence state it has a weaker crystal field interaction in comparison with Cr^{3+}, therefore V^{2+}-doped crystals can operate at longer wavelength. Figure 11.5 shows the relative gain cross-section and CW power output versus wavelength for V^{2+}:MgF_2 laser crystal at 77 K. The main tuning range is from 1.05 μm to 1.25 μm. Table 11.9 summarizes the laser performances of V^{2+}-doped laser crystals.

As mentioned in Table 11.2, the long fluorescence lifetime, wide absorption band and high laser-induced damage threshold of V^{2+}-doped fluoride crystals attracted great attention for long-pulsed flash lamp operated lasers and amplifiers in the past ten years. However, it was found that the excited state absorption of V^{2+} ions was so strong, which decreased the laser efficiency greatly and prevented the laser operation at low temperature.

Table 11.7. Measured efficiencies of Cr^{3+} vibronic lasers utilizing the laser-pumped laser configuration (as of October, 1988).

Host material	Name	Peak lasing wavelength (nm)	Slope efficiency
Be$_3$Al$_2$(SiO$_3$)$_6$	Emerald	768	64
LiCaAlF$_6$	LiCAF	780	54
BeAl$_2$O$_4$	Alexandrite	752	51
LiSrAlF$_6$	LiSAF	825	36
ScBeAlO$_4$	Scalexandrite	792	30
ScBO$_3$	Borate	843	29
Gd$_3$Sc$_2$Ga$_3$O$_{12}$	GSGG	785	28
Na$_3$Ga$_2$Li$_3$F$_{12}$	GFG	791	23
Y$_3$Sc$_2$Al$_3$O$_{12}$	YSAG	767	22
Gd$_3$Sc$_2$Al$_3$O$_{12}$	GSAG	784	19
SrAlF$_5$	Pentafluoride	932	15
KZnF$_3$	Perovskite	820	14
ZnWO$_4$	Tungstate	1035	13
La$_3$Ga$_5$SiO$_{14}$	LGS	968	10
Gd$_3$Ga$_5$O$_{12}$	GGG	769	10
Ca$_3$Ga$_{5.5}$Nb$_{0.5}$O$_{14}$	Niobate	1040	5
Y$_3$Ga$_5$O$_{12}$	YGG	(740)	5
Y$_3$Sc$_2$Ga$_3$O$_{12}$	YSGG	(750)	5
La$_3$Lu$_2$Ga$_3$O$_{12}$	LLGG	830	3
Mg$_2$SiO$_4$	Forsterite	1235	1
Al$_2$(WO$_4$)$_3$	Tungstate	810	?
BeAl$_6$O$_{10}$	Aluminate	834	?

Table 11.8. Intrinsic quantum efficiencies of Cr-doped laser materials.

Host	Name	λ (nm)	n_q
LiCaAlF$_6$	LiCAF	720–840	0.86
Be$_3$Al$_2$(SiO$_3$)	Emerald	720–842	0.86
BeAl$_2$O$_4$	Alexandrite	701–850	0.76
LiSrAlF$_6$	LiSAF	760–920	0.67
BeScAlO$_4$	Scalexandrite	792	0.40
Na$_3$Ga$_2$Li$_3$F$_{15}$	GFG	786	0.34
ScBO$_3$	—	787–892	0.34
Y$_3$Sc$_2$Al$_3$O$_{12}$	YSAG	760	0.29

11.3 Ti^{3+} Ions Doped Laser Crystals

Ti^{3+} ions have lased in sapphire (Al$_2$O$_3$), yttrium orthoaluminate (YAlO$_3$) and BeAl$_2$O$_4$, where Ti^{3+} substitutes for Al^{3+} in the crystal lattice. The broad band absorption at 450–550 nm is beneficial for Ar ion laser and frequency doubled Nd:YAG (DNYL) laser pumping. In Table 11.10 the laser performances of Ti^{3+}-doped crystals were summarized.

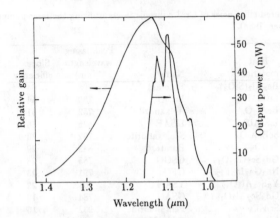

Fig. 11.5. Relative gain cross-section predicted from fluorescence spectrum and CW power output versus wavelength for a V:MgF$_2$ laser at a crystal temperature of 80 K.

The first Ti^{3+}-doped sapphire laser was demonstrated by D. F. Moulton in 1982 [14], since then it has been developed rapidly, we will introduce it in detail in the next chapter. The Ti^{3+}:YAlO$_3$ exhibits longer fluorescence lifetime than that of Ti^{3+}:Al$_2$O$_3$ and is possible to lase at shorter wavelength (600 nm), but this material is difficult to grow.

The Ti:BeAl$_2$O$_4$ was demonstrated to lase by flash lamp pumping. The slope efficiency is rather low (0.013%). The tuning range was found to cover 753 nm–949 nm, which was narrower than that of Ti:sapphire [24].

11.4 Co^{2+} and Ni^{2+} Ions Doped Laser Crystals

The Ni^{2+} ions were among the first explored for tunable solid state lasers. Ni^{2+} doped MgF$_2$ crystal was cooled by cryogenic liquids and pumped by flash lamp or tungsten-filament lamp [15]. The laser performance was substantially improved later on by Nd:YAG laser laser pumping at 1.32 μm [16]. As shown in Fig. 11.6, nearly 2 W of CW output power and continuous tuning from 1.6 μm to 1.75 μm were obtained with a slope efficiency of 28% at 1.67 μm for a Ni:MgF$_2$ crystal at 77 K.

From fluoride to oxide hosts the absorption and fluorescence bands of Ni^{2+} ions are moved to shorter wavelengths. The laser output peak wavelength of Ni^{2+}:MgO is at 1.318 μm. The laser action can be performed by Nd:YAG pumping at 1.06 μm, which is more efficient than Nd:YAG at 1.3 μm. Figure 11.7 shows the laser output behaviours of Ni:MgO laser crystal. High slope efficiency has been achieved at 80 K.

The behaviour of laser output of Ni^{2+}:GGG is characterized between Ni:MgF$_2$

Table 11.9. Laser performances of V^{2+}-doped laser crystals.

Host crystal	Activator concentration (%)	Laser transition	Wavelength or tuning range (μm)	Upper level lifetime (μs)	Temperature (K)	Crystal size (mm)	Mode of operation	Optical pump	Threshold	Efficiency (%)	Ref.
$CsCaF_3$	2	$^4T_2 \rightarrow {}^4A_2$	1.24–1.33	2500	80	2.6	CW	KrL	110 mW	0.06	12
MgF_2^*	0.5	$^4T_2 \rightarrow {}^4A_2$	1.07–1.16	2400	80	20	CW	ArL	1 W	16	13

Table 11.10. Laser performances of Ti^{3+}-doped laser crystals.

Host crystal	Activator concentration (%)	Laser transition	Wavelength or tuning range (μm)	Upper level lifetime (μs)	Temperature (K)	Crystal size (mm)	Mode of operation	Optical pump	Threshold	Efficiency (%)
Al_2O_3	0.02–0.15	$^2E \rightarrow {}^2T_2$	0.66–1.06	3.15	300	2–21	CW	ArL	0.1–2.2 W	3–19
				3.87	80	21	CW	ArL	4.8 W	8
				3.15	300	2–50	P	dye	0.5–600 mJ 75–150 kW/cm²	13–52
					300	6–40	P	DNYL	0.5–30 mJ	14–60
				0.2	510	6	P	DNYL	—	14
				3.15	300	7×49—6.3×120	P	Xe	12–150 J	0.02–0.5
					300	27	Q	DNYL	9.6 mJ	12
					300	17	AML 200 ps	ArL	4.4 W	14
					300	—	SML 10 ps	DNYP	4 mJ	10
$BeAl_2O_4$	0.05–0.1	$^2E \rightarrow {}^2T_2$	0.73–0.95	4.9	300	10.5	CW	DNYL	0.3 W	15
					300	10–27	P	DNYL	0.3–15 mJ	2.4
$YAlO_3$	0.08	$^2E \rightarrow {}^2T_2$	0.6116	14	300	3.5	P	DNYP	45 mJ	0.004

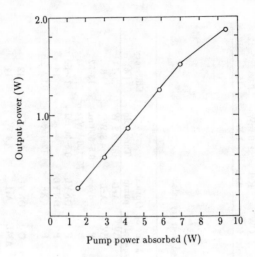

Fig. 11.6. CW output at 1.67 μm versus absorbed 1.32 μm pump power of Ni^{2+}:MgF$_2$ laser crystal.

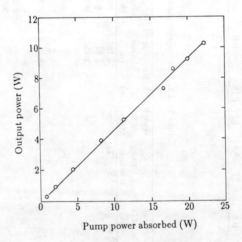

Fig. 11.7. CW output at 1.318 μm versus absorbed 1.065 μm pump power of Ni^{2+}:MgO laser crystal.

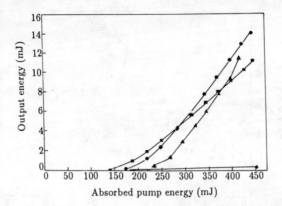

Fig. 11.8. Input-output curves of Ni:GGG laser crystal for different transmissions of the output coupler. —■—, 2%, η =3.8%; —●—, 5%, η =5.5%; —▲—, 10%, η =5.7% at λ_{pump} =1.32 μm and λ_{laser} =1.47 μm.

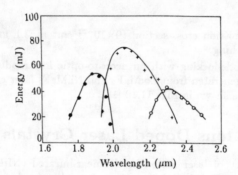

Fig. 11.9. Tuning curves (using three mirror sets with nominal 2% output coupling) at 299 K for normal-mode operation of the Co:MgF₂ laser at an absorbed pump-pulse energy of 211 mJ.

and Ni:MgO. The input-output curves are shown in Fig. 11.8. The laser ouput wavelength is at 1.47 μm by Nd:YAG laser pumping at 1.32 μm. The laser output efficiency is around 5% [19].

During the past few years there has been remarkable progress in room temperature operation of Co²⁺-doped laser crystals. A normal-mode, pulsed Co:MgF₂ laser was operated at room temperature for the first time by Welford and Moulton in 1988 [17]. Continuous tuning from 1.75 μm to 2.5 μm with pulse energy up to 70 mj and 46% slope efficiency was obtained with Nd:YAG pumping at 1.34 μm. Figure 11.9 shows the tuning curves of Co:MgF₂ at 299 K with different tuning components. The main problems of Co:MgF₂ laser are low gain ($1-2\times10^{-21}$ cm⁻¹)

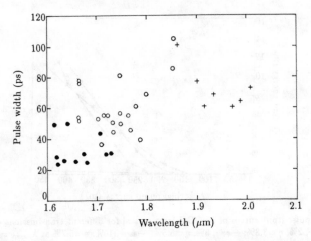

Fig. 11.10. Mode-locked pulse width versus tuning wavelength of Co:MgF$_2$ (o and +) and Ni:MgF$_2$ (•).

due to the low emission cross-section (9×10^{-22} cm^2 at 2.1 μm) and the strong temperature quenching.

By actively mode-locking with an acousto-optic loss modulator shorter laser pulses have been generated from Co:MgF$_2$ and Ni:MgF$_2$ laser crystals, the experimental results were shown in Fig. 11.10 [18].

11.5   Cr^{4+} Ions Doped Laser Crystals

Since the discovery of laser action in the near-infrared (NIR) 1.2 μm band in Cr:Mg$_2$SiO$_4$ (forsterite) [19], a lot of research work were dedicated to determine the valence state and site structure of chromium ions in dielectric crystals [20, 21]. The valence state of Cr^{4+} is unstable in comparison with Cr^{3+}. Cr^{3+} ions are always located at octahedral site, whereas Cr^{4+} ions at tetrahedral site.

Figure 11.11 shows the absorption spectra of Cr-doped forsterite grown from different atmospheres. The main absorption peaks around 500 nm and 700 nm belong to Cr^{3+} ions, and the absorption peak at 1100 nm is due to Cr^{4+} ions formed in pure O$_2$ atmosphere [22]. Cr^{4+} ions are in distorted tetrahedral site with local symmetry of D_{2d}.

Laser action of Cr^{4+} ions has been observed in Mg$_2$SiO$_4$ [19], YAG [20] and Y$_2$SiO$_5$ [23]. The useful pumping source is Nd:YAG laser at 1.06 μm or 0.532 μm (frequency doubling). Table 11.11 shows the first laser experimental result of Cr:Mg$_2$SiO$_4$ pumped by Nd:YAG at 1.06 μm and 0.532 μm.

Laser performances of Cr:Mg$_2$SiO$_4$ and Cr:YAG crystals have been much

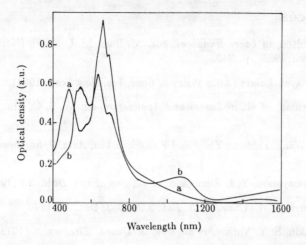

Fig. 11.11. Absorption spectrum of Cr-doped forsterite grown in pure N_2 atmosphere (a) and in pure O_2 atmosphere (b).

Table 11.11. Properties of laser emission of Cr:forsterite for two excitation wavelengths.

Properties	Value at the excitation wavelength	
	1064 nm	532 nm
Lasing threshold (absorbed energy)	1.25 mJ	1.37 mJ
Slope efficiency	1.8%	1.4%
Spectral bandwidth (FWHM)	25 nm	22 nm
Center wavelength	1235 nm	1235 nm

Table 11.12. Laser performance characteristics of the free-running Cr^{4+}:Y_2SiO_5 laser at 77 K.

Pump wavelength (nm)	1064	532	840
Threshold energy (mJ)	0.32	1.5	0.93
Maximum slope efficiency (%)	6.3	12.4	8.2
Peak laser wavelength (nm)	1270	1225	1225
Spectral bandwidth (nm)	10	23	30

improved recently. The maximum slope efficiency of pulsed operation at room temperature was over 20% pumped by Nd:YAG laser at 1.06 μm. We will discuss it in detail in the next chapter. Although the chromium ions in Y_2SiO_5 are all in tetrahedral site and at tetravalent state, but the laser output performances of Cr^{4+}:Y_2SiO_5 crystals are still poor (see Table 11.12), Cr^{4+}:Y_2SiO_5 crystal can only work at liquid nitrogen temperature.

References

1. P. F. Moulton, in *Laser Handbook*, eds. M. Bass, M. L. Stitch (North-Holland, Amsterdam, 1985), p. 243.

2. B. A. Lengyel, *Lasers* (John Wiley & Sons, Inc, New York, 1964).

3. W. Ruderman, *et al.*, in *Lasers and Application*, ed. W. S. C. Chang (1963) p. 20.

4. Zhijiang Wang, Fuzheng Zhou and Kangchun Lin, *Acta Optica Sinica* **3** (1983) 521.

5. B. K. Sevastyanov, Y. I. Remigailo, *et al.*, *Sov. Phys. Dokl.* **26** (1981) 62.

6. R. C. Morris, C. F. Cline, *U. S. Pat.* 3,997,853, Dec. 14, 1976.

7. G. V. Bukin, S. Y. Volkov, *et al.*, *Sov. J. Quant. Electron.* **8** (1978) 671.

8. J. Buchert, A. Katz, R. R. Alfano, in *Proc. Int. Conf. Lasers 82*, ed. R. Powell (STS Press, Mclean, Va, 1983) pp. 791–798.

9. S. A. Payne, L. L. Chase, H. W. Newkirk, *et al.*, J. Quant. Electron. **QE-24** (1988) 2243.

10. U. Brauch and U. Durr, *Opt. Lett.* **9** (1984) 441; *Opt. Commun.* **49** (1984) 61; **55** (1985) 35.

11. S. A. Panyne, L. L. Chase, *et al.*, *J. Appl. Phys.* **66** (1989) 1051.

12. P. Moulton, in *Proc. Int. Conf. Lasers 81*, ed. C. B. Collins (STS Press, Mclean, Va 1981).

13. U. Brauch and U. Durr, *Opt. Commun.* **49** (1984) 61.

14. P. F. Moulton, *Opt. News*, No. 8 (1982) 9.

15. L. F. Johnson, R. E. Dietz and H. J. Guggenheim, *Phys. Rev. Lett.* **11** (1963) 318; *Phys. Rev.* **149** (1966) 179.

16. P. F. Moulton and A. Mooradian, *Appl. Phys. Lett.* **35** (1979) 838.

17. D. Welford and P. F. Moulton, *Optics Lett.* **13** (1988) 975.

18. B. C. Johnson, P. F. Moulton and A. Mooradian, *Opt. Lett.* **9** (1984) 116.

19. V. Petricevic, S. K. Gayen, R. R. Alfano, *Appl. Opt.* **27** (1988) 4162.

20. G. M. Zverev, A. V. Shestakov, OSA Proc. Tunable Solid State Lasers, **5** (1989) 66.

21. T. H. Alik, B. H. T. Chai, L. D. Merkle, OSA Proc. Tunable Solid State Lasers, **10** (1991) 84.

22. Peicong Pan, Hongbin Zhu, *et al.*, *J. Crystal Growth* **121** (1992) 141.

23. C. Deka, B. H. T. Chai, *et al.*, *Appl. Phys. Lett.* **61** (1992) 2141.

24. A. Sugimoto, Y. Segawa, *et al.*, *Japan J. Appl. Phys.* **29** (1990) L1136.

12. Tunable Laser Crystals

The tunable lasers give widespread use in scientific field, such as spectroscopy and photochemistry, and important practical applications, such as remote sensing, pollution detecting and isotope separation. The recent advances in tunable laser crystals have allowed solid-state tunable lasers to play a more important role than liquid dye lasers. Most tunable laser crystals are doped with paramagnetic ions (the most of transition metal (TM) ions). Due to random phonon-electron interaction and high nonradiative transition probability the TM ion doped glasses are excluded from tunable laser materials. This chapter discusses the class of tunable laser crystals containing paramagnetic ions from the first or $3d$ transition-metal group in the Periodic Table and low valent rare-earth ions.

12.1   Tunable Laser Emissions

L. F. Johnson and co-workers successfully obtained flashlamp pumped laser operation from Ni^{2+}-doped MgF_2 in 1.63 μm wavelength range in 1963 [1]. Subsequently the laser operation was performed from Co^{2+} and V^{2+} ions doped crystals [2, 3]. The Sm^{2+}-doped CaF_2 crystal was another type of tunable laser crystal tuned at 0.7 μm [4]. The early tunable laser crystals did not have practical application due to the need for cryogenic cooling of the crystals.

The important development in tunable laser crystals was the demonstration of tunable laser operation from alexandrite crystal (Cr^{3+}:$BeAl_2O_4$), which could be operated at room temperature around 0.75 μm efficiently pumped by flashlamp [5]. More recently a series of Cr^{3+}-doped tunable laser crystals have been developed [6–9]. Further major advances in tunable laser crystals involved the emergence of Ti^{3+}-sapphire (Al_2O_3) [10] and Cr^{4+}-forsterite (Mg_2SiO_4) [11], which show wide tuning range in near infrared wavelength range (0.65–1.02 μm and 0.9–1.3 μm respectively) and become the most prospective tunable laser crystals. Attraction has been paid on Ce^{3+}-doped fluoride crystals with the potential for tunable near-UV output around 0.3 μm [12].

Figure 12.1 shows the wavelength tuning range of different TM ions doped laser crystals.

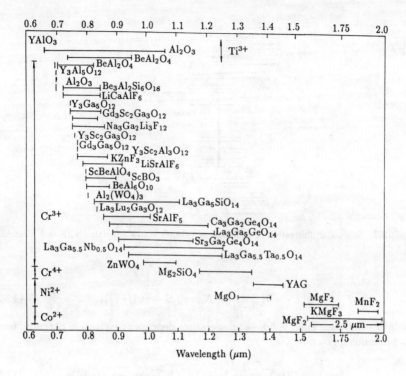

Fig. 12.1. Wavelength tuning range of TM ions doped crystals.

12.2 Vibronic Laser Gain

As we discussed in Chapter 4, the gain coefficient $g(E)$ (laser gain per unit length) in the case of two nondegenerated electronic levels can be expressed as

$$g(E) = \sigma(E)(N_j - N_i), \qquad (12.1)$$

where N_j and N_i—ion density in the upper energy level and lower energy level, $\sigma(E)$—emission cross section.

With the addition of phonon coupling to the electronic system, the vibronic levels associated with each electronic level represent the many phonon energy. The vibronic levels can be divided into independent manifolds, which are then involved into upper and lower electronic levels to form "j" and "i" manifolds, as shown in Fig. 12.2.

Considering the thermal distribution of the ground state level, Eq. 12.1 can be written as [13]

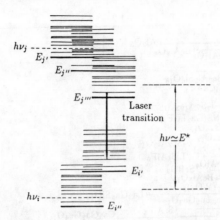

Fig. 12.2. Schematic illustration of manifolds of vibronic states associated with each electronic state.

$$g(E) = \sigma(E)\{N_j - N_i\exp[(E - h\upsilon)/kT]\}, \tag{12.2}$$

where E—radiation energy, $h\upsilon$—a temperature-dependent excitation potential from i to j manifolds given by

$$h\upsilon = -kT\{\ln\sum_j\exp(-h\upsilon_j/kT) - \ln\sum_i\exp(-h\upsilon_i/kT)\}. \tag{12.3}$$

To express the vibronic laser rate equation, the value g_j/g_i, normally representing the upper and lower electronic level degeneracy can be replaced by $\exp[(E-h\upsilon)/kT]$.

12.3   Ti:sapphire Crystals

Ti:sapphire (Ti^{3+}:Al_2O_3) single crystal is one of the most attractive broadly tunable solid state laser material [14]. Both CW and pulsed lasers have been demonstrated with very high efficiency over the tuning range in excess of 500 nm centered at 800 nm. In order to develop this material to meet the needs of current commercial laser systems a variety of growth techniques have been used to grow large diameter, high purity and laser quality Ti^{3+}:Al_2O_3 single crystals. However, laser performance has generally been limited by the infrared absorption band present in the laser output wavelength region which is primarily due to the presence of Ti^{3+}–Ti^{4+} ion pairs [15]. A lot of researches have been done for improving the crystal quality and the laser performances recently [16–18].

12.3.1   Crystal Growth Techniques and Crystal Quality

In Shanghai Institute of Optics and Fine Mechanics (SIOFM), Academia Sinica, two new techniques have been invented to grow high quality $Ti:Al_2O_3$ laser crystals, they are named by Induction Field Up-Shift method (IFUS) [18, 19] and Temperature Gradient Technique (TGT) [20], which differ from Czochralski method (CZ) and Heat Exchange method (HEM). The merits of $Ti:Al_2O_3$ crystals grown by IFUS and TGT are as follows: 1) High perfection and low defects in grown crystals [21, 22]. The dislocation defect density is lower than 10^3–10^4 /cm^3. 2) High optical homogeneity: $\Delta n \leq 1$ λ per inch of crystal boules and $\Delta n \leq 1/2$ λ per inch of laser rods measured by Zygo III interferometer. 3) High titanium doping level: 0.02–0.45 wt% Ti_2O_3. The absorption coefficient of crystals at 490 nm can be reached to 7.0 cm^{-1}, which is higher than that of CZ and HEM grown crystals. 4) High Figure of Merit value (FOM) of $Ti:Al_2O_3$ crystals before annealing can be obtained, because of reduction atmosphere applied during crystal growth processes. 5) Large size of crystal boule: 55 mm and 100 mm in diameter by IFUS and TGT techniques respectively [28]. The properties and specifications of Ti:sapphire laser crystal are summarized in Table 12.1.

The annealing is an important process for improving the quality of as-grown $Ti:Al_2O_3$ crystal. According to our research results the role of high temperature annealing can be considered as follows: first, to change the valence state of titanium (from Ti^{4+} to Ti^{3+}), thus increase the FOM value; second, to improve the perfection of the crystal. Figure 12.3 shows the high resolution image of Ti:sapphire crystal, it illustrates the high perfection of lattice image after annealing. Figure 12.4 shows the laser slope efficiency of Ti:sapphire crystal rod before and after annealing. The laser efficiency increased about 50% through the annealing process.

12.3.2   The Laser Characteristics of $Ti:Al_2O_3$ Grown by IFUS and TGT

The laser efficiency of $Ti:Al_2O_3$ rods are demonstrated by pulsed, Quasi-CW and CW laser operation with different pumping sources, such as Q-switched, frequency-doubled Nd:YAG laser, copper vapor laser, argon ion laser and excimer laser.

    I. Pulsed laser operation [23]

        1. Pumping source: Q-switched frequency-doubled Nd:YAG laser.

            Wavelength: 532 nm

            Pulse width: 10 ns

            Repetition rate: 5–10 Hz

            Mode: TEM$_{00}$

            Laser performances: $Ti:Al_2O_3$ rod, 8.7×8.3×18.5 mm^3 with α_{490} =1.9 cm^{-1}.

                Threshold energy: 0.5 mJ

                Slope efficiency: 55%

                Energy conversion efficiency: 40%

Table 12.1. IFUS and TGT Ti-sapphire properties and specifications.

Chemical formula	Ti:Al$_2$O$_3$
Crystal structure	Hexagonal
Refractive index	1.76
Hardness	Mohs 9
Dopant level	0.02–0.45 wt% Ti$_2$O$_3$
Absorption coefficient at 490 nm (α_{490})	0.5–7.0 cm^{-1}
Absorption band	400–600 nm
Tunable range	650–1100 nm
Figure of merit (FOM)	80–200

(low FOM at high Ti^{3+} concentration crystals)
End faces : flat/parallel 15 arcsec; Brewster's angle.
Crystal orientation : c axis normal to rod axis.

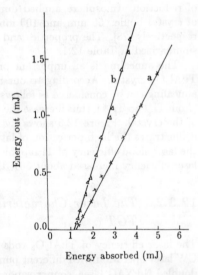

Fig. 12.3. High resolution lattice image of high perfection Ti:sapphire crystal B=[1100].

Fig. 12.4. Laser slope efficiency curve of IFUS Ti:sapphire rods pumped by 532 nm laser light. (a) Before annealing=36.1%; (b) after annealing=58.2%.

Maximum slope quantum efficiency: 82%
Tuning range: 668–938 nm (two set of mirrors)
2. Pumping source: excimer laser [24].
    Wavelength: 499 nm
    Pulse width: 40 ns
    Repetition rate: 2 Hz
    Mode: multimode

Laser performances: Ti:Al$_2$O$_3$ rod, 4×4×20 mm^3 with α_{490}=1.98 cm^{-1}.

> Threshold energy: 0.54 mJ
> Energy conversion efficiency: 42%
> Slope efficiency: 59%
> Maximum quantum efficiency: 87%
> Tuning range: 662–833 nm (a set of mirrors)

II. Quasi-CW laser operation

1. Pumping source: Internal cavity Q-switched frequency-doubled Nd:YAG laser [25].

> Wavelength: 532 nm
> Frequency: 1–10 kHz
> Pulse width: 200 ns
> Input power: 0–2 W tune

Laser performances: Ti:Al$_2$O$_3$ rod, 4×4×10 mm^3 with α_{490}=2.8 cm^{-1} (unannealed rod)

> Threshold energy: 0.9 W
> Slope efficiency: 35%
> Laser output power: 300 mW
> Tuning range: 685–809 nm (a set of mirrors)

2. Pumping source: copper vapor laser [26].

> Wavelength: 510.2 nm
> Pulse width: 80 ns
> Frequency: 5 KHz
> Input power: 1–8 W

Laser performances: Ti:Al$_2$O$_3$ rod, 5×5×10 mm^3 with α_{490}=1.5 cm^{-1} (unannealed rod)

> Threshold energy: 0.5–0.8 W
> Slope efficiency: 31–35%
> Laser output energy: 2.5 W
> Tuning range: 750–820 nm

III. CW laser operation

> Pumping source: argon ion laser [27].
> Wavelength: 488 nm
> Laser input power: 5–10 W

Laser performances: Ti:Al$_2$O$_3$ rod, 6×8×14 mm^3 with α_{490}=2.0 cm^{-1}.

> Threshold energy: less than 2.5 W
> Conversion efficiency: 14.5%
> Slope efficiency: 41%
> Laser output energy: 1030 mW
> Tuning range: 680–1010 nm

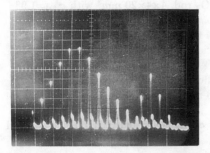 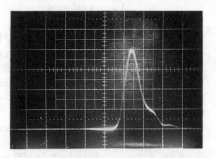

Fig. 12.5. A pulse train of synchronous mode-locking of the Ti:sapphire laser (right) and the pump light (left) (20 ns/div).

Fig. 12.6. Temporal profile of the shortest output pulse (2 ns/div).

12.3.3 Short Laser Pulse Generation of Ti:sapphire Crystal

The Ti:sapphire crystal has a wide absorption band (extends about 200 nm) centered at near 490 nm and broad gain profile (extends about 400 nm) at near infrared spectral region. Thus the Ti:sapphire laser has the potential to produce ultrashort pulses and is now competing with the dye lasers in tunability at the near infrared spectral range.

Pumped by frequency doubled Q-switched or mode-locked Nd:YAG laser, the experimental results showed that the temporal profiles of the output pulses were strongly correlative with both the density of the pump light power and the position of the Ti:sapphire crystal in the F-P cavity [29]. For mode-lock pumping, with the density of pump power increased, a giant pulse with modulation was produced. As soon as the density of pump power was higher than the threshold value, a pulse train of synchronous mode-locking was generated, as shown in Fig. 12.5. For Q-switched pumping with the density of pump power increased, the pulse width decreased. When the length of the Ti:sapphire laser cavity was 35 mm, the pump energy 14.5 mJ, and the output coupler reflectivity 50% at 780 nm, the minimum laser pulse of 1.2 ns was obtained. The temporal profile of this output pulse is shown in Fig. 12.6.

At the early stage passive mode locking with a saturable absorber was observed in a flashlamp-pumped Ti:sapphire crystal, the pulse durations of the mode-locked laser were 100 ps [30]. By CW Ar ion laser pumping through active mode-locking the pulse of 80 ps duration was obtained. Using nonlinear external cavity feedback the pulse has been compressed to less than 80 fs [31]. Later on, stable pulses of less than 150 fs were generated directly from a tunable CW passively mode-locked Ti:sapphire laser, through a balance of self-phase modulation in the Ti:sapphire rod and negative group-velocity dispersion produced by a prism pair. After external fiber compression, 50 fs pulses were obtained at approximately 750 nm [32]. The schematic diagram of passively mode-locked Ti:sapphire laser is shown in Fig. 12.7.

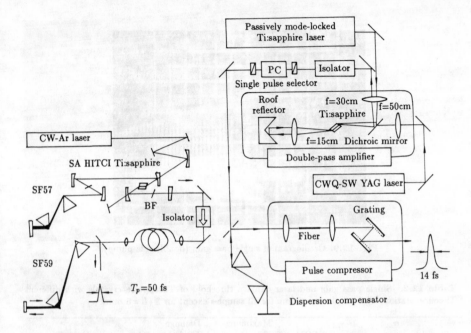

Fig. 12.7. Schematic of the femtosecond CW passively mode-locked Ti:sapphire laser system with a fiber compressor. SA, Saturable-absorbed dye jet; BF, birefringent filter.

Fig. 12.8. A Ti:sapphire laser system, consisting of a femtosecond Ti:sapphire oscillator, a double pass amplifier pumped by a CW Q-switched YAG laser and a pulse compressor.

Combining with this oscillator and a double pass Ti:sapphire amplifier pumped by a CW Q-switched YAG laser (3 kHz) enabled fiber-grating pulse compression down to 14 fs [33]. The experimental arrangement is shown in Fig. 12.8. More recent result of ultra-short pulse of 8.5 fs has been obtained from Ti:sapphire crystal [34].

12.3.4 *Amplification Gain*

Because of its large stimulated emission cross section, high saturation energy and wide gain bandwidth, Ti:sapphire crystal is an important amplification medium. For high power laser system, some amplification characteristics of a Ti:sapphire crystal with higher Ti-doped concentration grown by IFUM and TGT methods was investigated. The absorption coefficient at 532 nm of the crystal was 2.45 cm^{-1} and at 800 nm was 0.044 cm^{-1}. In the experiments, 50 times of single pass gain (double side pump) were observed, it was shown in Fig. 12.9.

Using these crystals with different Ti-concentration, the single pass gain, and the damage threshold have been measured under 532 nm pulse pumping (Table

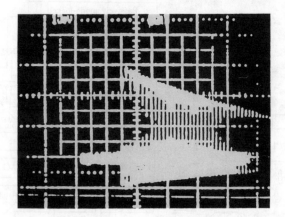

Fig. 12.9. Oscillogram of single pass gain (double side pump).

Table 12.2. Single pass gain and laser damage threshold of Ti:sapphire crystals with different Ti-concentration. The length was 20 mm for all samples except for 3 ($L =6$ mm).

Sample	α_{488} (cm^{-1})	α_{532} (cm^{-1})	FOM	Maximum single pass gain (dB)	Damage threshold (J/cm^2)	Comments
1	2.9	2.5	100	30.4	8.7	
2	4.2	3.3	150	32.0	13.9	
3	5.8	4.4	150	22.7	11.6	
4		0.5	200	16.1	7.9	Crystal from other supplier
5					9.6	Un-doped sapphire

12.2). The maximum single pass gain was 32 dB ($\times$1600) for single side pumping. It can be enhanced to 45.5 dB ($\times$35300) in double side pumping, and the gain duration was reduced to less than 100 ns. This single pass gain is the highest ever observed in Ti:sapphire [35]. The gain is limited by the absorption-independent surface damage, and the damage threshold was similar to that of undoped sapphire even in the highest concentration crystal (Table 12.2).

Furthermore, in constructing a confocal double pass amplifier [36], which is consisted of a 2 cm Ti:sapphire (α_{532}=3.3 cm^{-1}) located at the beam waist of a confocal lens pair, two dichroic mirrors to overlap the input beams with independently focused pumping beams, and a rotating prism for vertical displacement (Fig. 12.10), 1.5 ps and 770 nm selected pulses were amplified to 2 mJ with 62 dB gain. Up to this level, no beam-break-up effect was observed owing to high single pass gain even without the chirped pulse amplification scheme. The advantages compared with a

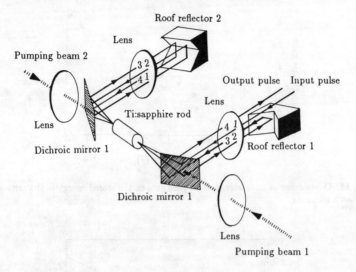

Fig. 12.10. Experimental layout for studying four pass gain.

regenerative amplification are the easy tuning and operation due to its construction from passive optical components, and elimination of timing requirements [37].

Chirped pulse amplification (CPA) is always used in Ti:sapphire high power laser system. Since the emission lifetime of Ti:sapphire is low, it makes flashlamp pumping difficult; most Ti:sapphire-based CPA laser systems use laser-pumped amplifiers. In the regenerative amplifiers the laser gain can reach high value, for example after approximately 50 round trips in the cavity, a 75 fs and 100 pJ laser pulse has depleted the gain, with an output pulse energy of 2.5 μJ with 10 ps duration [38]. However, flashlamp pumping of Ti-sapphire amplifiers has advantages over laser-pumping ones in terms of simplicity of configuration and scalability to high output energy. The chirped pulse amplification using a dye converter-free, flashlamp pumped Ti:sapphire crystal amplifier was studied by I. Shimada etc. [39]. The small signal single pass gain and saturation behaviour of flashlamp pumped Ti:sapphire amplifier are shown in Figs. 12.11 and 12.12 respectively. The crystal used in the amplifier is 1.35 cm in diameter and 21 cm in length with 0.15 wt% doping. An output energy of 500~600 mJ/pulse is achieved with an input energy of 170 mJ with 125 fs pulse duration.

The amplification in Ti:sapphire beyond 1 μm is interested for high power laser system, the final amplifiers are Nd^{3+}-, Yb^{3+}-doped laser media. Figure 12.13 shows the gain cross section of Ti:sapphire beyond 1 μm measured by single pass gain determination [40]. The value of 2.8×10^{-20} cm^2 at 1053 nm was obtained.

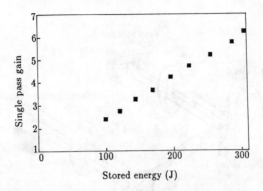

Fig. 12.11. Dependence of small signal single pass gain on the stored energy in the capacitors used for flashlamp pumping.

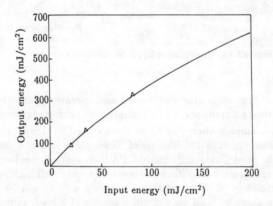

Fig. 12.12. Measured output energy versus input energy.

N. Blanchot *et al.* demonstrated a generation of 10 J, 250 fs pulses in a Ti:sapphire-mixed Nd glasses system [41]. After the output of a Ti:sapphire regenerative amplifier, the 1.058 μm, 0.5 mJ, chirped pulse is amplified in a series of alternated six phosphate and six silicate glass laser heads.

12.4 Cr:forsterite and Cr:YAG Crystals

Chromium doped forsterite ($Cr^{4+}:Mg_2SiO_4$) is a newly developed tunable laser crystal. The wavelength tuning range of Cr:forsterite extends to longer wavelength (0.9~1.5 μm) than that of Ti:sapphire in near IR region. The lasing centers in this

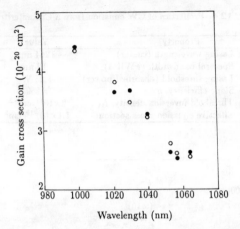

Fig. 12.13. Measured Ti:sapphire gain cross section by single pass gain determination at pump fluences of 1.8 J/cm^2 ($\bullet$) and 2.8 J/cm^2 ($\circ$).

Table 12.3. Characteristics of Cr:forsterite.

a) Large crystal boule 20 mm in diameter and 100 mm in length.
b) No visible inhomogeneities, such as gas bubble and crack.
c) Average dislocation density $10^3/cm^2$.
d) Cr_2O_3 doping content 0.01–0.06 wt%.
e) Figure of merit ($\alpha_{1064}/\alpha_{1250}$) 30–40.
f) Effective distribution coefficient K =0.15.
   There is inhomogeneous concentration distribution of Cr not only in axis but also in radial direction.

crystal are attributed to Cr^{4+} substituting for tetrahedrally coordinated Si^{4+}. This crystal was grown by Czochralski method with RF heating generator and Ir crucible of size ϕ50 mm$\times$40 mm in SIOFM. The crystal growth and perfection study have been reported before [42–44]. In order to obtain Cr^{4+} ions in crystal the strong oxidation atmosphere is necessary during crystal growth. Figure 11.11 shows the absorption spectra of Cr-doped forsterite in different atmosphere. The absorption peak around 1.1 μm is characterized for Cr^{4+} ions. The characteristics of as-grown crystal are listed in Table 12.3.

The quality improved Cr:forsterite crystal was tested in a flat-concave cavity pumped by a Nd:YAG laser (1.064 μm, 10 Hz and 80 ns) [45]. Tens of millijoules of laser output could be achieved at room temperature. The slope efficiency was about 20%, the overall E-O conversion efficiency was 13.2%. The central laser wavelength was at 1.23 μm with linewidth of 24 nm. The crystal size was 6$\times$7$\times$26 mm^3 with

Table 12.4. Properties of CW emission from a Cr:forsterite laser.

Property	Value
Lasing wavelength (center)	1244 nm
Spectral bandwidth (FWHM)	12 nm
Lasing threshold (absorbed power)	1.25 W
Slope efficiency, η	6.8%
Threshold inversion density, N_t	2×10^{17} cm^{-3}
Effective emission cross section	1.1×10^{-19} cm^2

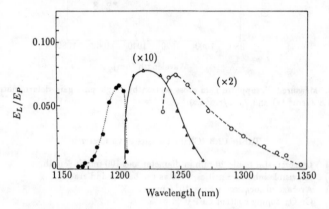

Fig. 12.14. Ratio of Cr:forsterite laser output (E_L) to the absorbed pump energy (E_p) as a function of wavelength.

the FOM value of 33.

Petricevic *et al.* firstly performed CW laser operation of Cr:forsterite [46]. Table 12.4 listed the properties of CW emission of Cr:forsterite laser. The pump source was CW Nd:YAG laser at 1.064 μm. Figure 12.14 shows the wavelength tuning curves of Cr:forsterite using three coupling mirrors. The CW laser operation of Cr:forsterite has been improved later. Carrig *et al.* obtained over 1.8 W of output power from a CW Cr:forsterite laser operated at 77 K when it was pumped by a 7.3 W Nd:YAG laser at 1.06 μm (the slope efficiency was around 30%) [47] with increasing temperature the output power decreases as shown in Fig. 12.15.

Short pulse generation of Cr:forsterite has been studied in recent years. The typical Q-switched pulse was 60 ns to 80 ns in width, varied with wavelength [48]. Produced pulse as short as 31 ps tunable between 1204 nm and 1277 nm was obtained by active mode locking [49]. With optimizing the dispersion of a self-mode-locked Cr:forsterite laser shortest pulses of 25 fs in duration were achieved [50].

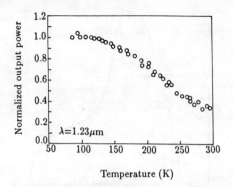

Fig. 12.15. Normalized output power versus temperature for the free-running Cr:forsterite laser.

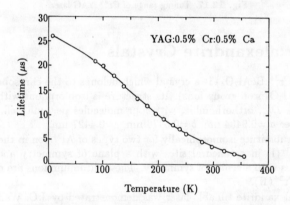

Fig. 12.16. Temperature dependence of the Cr^{4+} lifetime.

Cr^{4+} ions enter the YAG host with different Ca co-dopant concentration, the Cr^{4+} ions substitute the Al^{3+} ions in tetrahedral sites with Ca^{2+} charge compensation. Cr^{4+} ions are unable to dope into $YAlO_3$ host because in $YAlO_3$ no tetrahedral sites exist. At low Cr^{4+} concentration the emission lifetime of Cr^{4+}:YAG is 3.6 μs and drops with raising temperature, as shown in Fig. 12.16. The wavelength tuning range of Cr^{4+}:YAG further moves to longer wavelength in near IR in comparison with that of Cr^{4+}:forsterite (see Fig. 12.17).

The pulsed and CW laser operations of Cr^{4+}:YAG were all achieved by Nd:YAG laser pumping at 1.064 μm. Figure 12.18 shows the pulsed and CW laser operation results, the slope efficiencies were 22% and 12% respectively [51].

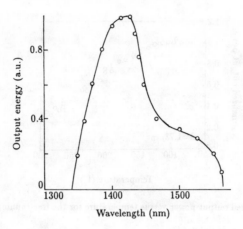

Fig. 12.17. Tuning range of Cr^{4+}:YAG laser.

12.5   Cr:alexandrite Crystals

Alexandrite (Cr^{3+}:$BeAl_2O_4$) is a crystal which belongs to the chrysoberyl group of minerals ($BeAl_2O_4$:iron group ions). Its structure is isomorphous with olivine, the space group is D_{2h}^{16}, orthorhombic, with four molecules per unit cell. The lattice parameters are: a =0.9404 nm, b =0.5476 nm, c=0.4427 nm.

Cr^{3+} ions substitute isomorphically for two types of Al^{3+} ion in the chrysoberyl structure: Al^{3+}(I) in octahedral sites with a plane of symmetry and Al^{3+}(II) in octahedral sites with a center of symmetry. These chromium ions are designated as Cr^{3+}(I) and Cr^{3+}(II).

The first alexandrite tunable laser was demonstrated by J.C. Walling et al. in 1979. The first tunable laser crystal developed in China was also alexandrite single crystal (Cr^{3+}:$BeAl_2O_4$), which was grown by Czochralski pulling technique on a seed oriented along [001] direction, because the c-axis crystals have good optical quality and the continued tunable laser output in wavelength range of 720 nm to 780 nm was obtained with c-axis crystal rods [52]. The laser performances of alexandrite laser crystal pumped by flashlamp depend upon the crystal perfection greatly. We have systematically studied the inclusions, dislocations, impurities in the crystals [53, 54], and the perfection of the crystal was examined by X-ray transmission topography [55]. However, as-grown crystals generally contain high density of dislocation (up to 10^5 /cm^2), besides the defects, such as bubbles, inclusions etc. The grown-in dislocations mainly originate from the seed and propagate along the path normal or nearly perpendicular to the growing interface. The dislocation density and arrangement is closely related to the quality of the seed crystal. Figure

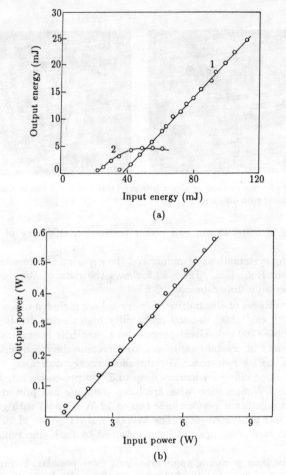

Fig. 12.18. Laser output characteristics of Cr^{4+}:YAG crystal. (a) pulsed, $T = 10\%$; (b) CW, $T = 3\%$.

12.19 shows the interference pattern of an alexandrite crystal rod prepared in SIOFM.

The alexandrite (Cr^{3+}:BeAl$_2$O$_4$) crystal can be tuned in the wavelength range of 730–800 nm pumped by flashlamp. The maximum output energy of crystals grown in SIOFM was about 1 J/pulse with a crystal rod size of ϕ5 mm×7 mm. The E-O efficiency was about 0.54% [56]. Using alexandrite crystal rod of 6.8 mm in diameter and 65 mm in length, the Q-switched tunable laser has been performed rather early [57]. The giant pulse with pulse width of 30 ns and peak power of 10 mW has been achieved. Using a LiNbO$_3$ crystal as frequency doubler, the SHG

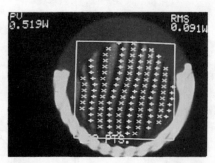

Fig. 12.19. Interference patterns of alexandrite crystal rod with 6.3 mm in diameter and 50 mm in length and optical path difference 0.5 λ.

tunable range of 368–393 nm and power conversion efficiency of 9% have been obtained.

J.C. Walling systematically summarized the experimental results of alexandrite crystal lasers in Ref. [58]. Table 12.5 shows the data of laser performances of alexandrite lasers published before 1985.

The disadvantages of alexandrite laser crystal are included in low emission cross section (6×10^{-21} cm^2), high damage probability, high thermal lensing and short fluorescence lifetime (260 μs). Great progress in alexandrite lasers has been achieved with improvement in crystal quality and to overcome the disadvantage by reasonable design of the laser devices. The threshold energy decreased down to several joules and the slope efficiency increased up to 5% at pulsed operation by flashlamp pumping [59]. CW operation with arc lamp pumping has proved more difficult to achieve, yet significant performance (up to 12 W output) with good transverse mode quality was generated [60]. The average output power of 20 W at 750 nm with repetition rate of 20 pps has been obtained by flashlamp pumping (see Fig. 12.20).

Recently the laser pumping application gets more popular. Pumped by tunable dye laser at the peak wavelength of alexandrite absorption band, the slope efficiency was up to 64%, and pumped by laser diode at 640 nm, the slope efficiency was 28%.

12.6   Cr:LiCaAlF$_6$ and Cr:LiSrAlF$_6$ Crystals

Recently two Cr laser media—Cr^{3+}:LiSrF$_6$(LiSAF) and Cr^{3+}:LiCaAlF$_6$(LiCAF) were discovered [62, 63]. Cr:LiCAF and Cr:LiSAF were grown by a zone melting process, gradient freeze and Vertical Bridgman method. Cr^{3+}:LiCaAlF$_6$ single crystals were also grown. The Vertical Bridgman method was applied. Crystal boules were grown unseededly first and then along [10$\bar{1}$0] at 1 mm/hr in a flowing Ar atmosphere. The defects in the crystal were also examined [64]. The minimum

Table 12.5. Alexandrite laser performance summary [58].

Operational mode	Rod size Diameter (mm)/ Length (mm)	Cr^{3+} conc. (at%)	Wavelength (nm)	Bandwidth (nm)	Pulse energy (J)	Pulse duration (ns)	Pulse repetition rate (Hz)	Average power (W)	Mode character	Diffraction limit	Efficiency (%)
Normal	6.3/10	0.28	755	1.0	7	200 μs	5	35	Multimode	—	2.5
Normal	6.3/10	0.14	750	1.0	4.5	150 μs	20	90	Multimode	30×	1.2
Normal	6.3/11	0.14	750	1.0	0.6	150 μs	20	12	Multimode	5×	0.3
Normal	6.3/11	0.14	750	1.0	1.2	150 μs	125	150	Multimode	15×	—
Q-switched	6.3/2×11	0.14	790	0.01	0.4	<1 μs	125	50	Multimode	15×	0.4
Q-switched	6.3/11	0.14	750	0.1	1.5	28	20	30	Multimode	30×	0.55
Q-switched	6.3/11	0.14	750	0.1	2.0	40	20	40	Multimode	30×	0.19
Q-switched	6.3/11	0.14	740	0.1	0.55	25	10	5.5	Multimode	10×	0.21
Q-switched	6.3/11	0.14	750	0.1	0.55	16	10	5.5	Multimode	12×	0.19
Q-switched	6.3/11	0.14	765	0.1	0.55	26	10	5.5	Multimode	10×	
Passive mode-locked	6.3/11	0.14	750	0.025	0.5 mJ	38 ps	10	—	TEM(00) transform limited	1.5×	—
Active mode-locked and Q-switched	6.3/11	0.14	750	0.02	2 mJ	150 ps	10	—	TEM(00)	—	—
CW pumped	3/10	0.2	755	1	—	—	—	60	Multimode	—	1.21
CW pumped	5/10	0.09	755	1	—	—	—	2	TEM(00)	1.3×	0.31
Oscillator/ Amplifier	6.3/11	0.14	755	1	3	30	10	30	Multimode	7×	—
Oscillator/ Amplifier	6.3/11	0.14	755	1	0.5	30	10	5	Near TEM(00)	1.2×	—
Oscillator/ Amplifier	6.3/11	0.14	755	1	1	10	10	10	Multimode	7×	—

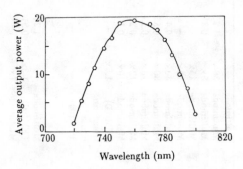

Fig. 12.20. Average output power versus tuning wavelength of alexandrite crystal laser.

Table 12.6. Laser output performances of diode laser pumped Cr:LiCAF and Cr:LiSAF.

Laser	Output power (mW)	1 W-diode pumped slope efficiency (%)	Maximum slope efficiency (%)	Optical conversion efficiency (%)	Overall electrical efficiency (%)
Cr:LiCAF	158	28	50	19	1.7
Cr:LiSAF	176	25	45	21	1.9
Cr,Nd:GSGG	173	23	42	21	1.9

scattering and absorption losses of 0.3–0.5% cm^{-1} can be achieved now. As shown in Fig. 12.21, the slope efficiency of Cr:LiCAF by Kr-laser pumping was 61%, and the extrapolated slope efficiency pumped by flashlamp was 4.1% [65].

The luminescence lifetime is 170 μs in Cr:LiCAF and 67 μs in Cr:LiSAF. These two materials complement each other in that Cr:LiCAF provides a better energy storage lifetime, while Cr:LiSAF has a broader tuning range (see Fig. 12.22), and its higher emission cross section better facilitates extraction of stored energy with short pulses without exceeding optical damage thresholds of the laser element and cavity optics.

Cr:LiCAF or Cr:LiSAF crystal is an attractive alternative to Ti:sapphire since it can be pumped with a diode and exhibits greater energy storage and lower lasing threshold.

Cr:LiCAF and Cr:LiSAF crystals have been successfully diode-pumped. Using 1 W single stripe visible diode laser (0.67 μm wavelength), the laser output performances of Cr:LiCAF and Cr:LiSAF are shown in Table 12.6 [66]. The Cr,Nd:GSGG listed in the Table 12.6 is for comparison.

CW diode-pumped single frequency (resolution limited linewidth<20 MHz)

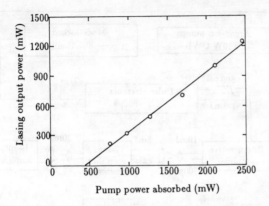

Fig.   12.21.   Lasing output power versus absorbed power of Cr:LiCAF crystal output coupling=4.3%.

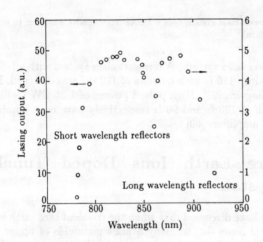

Fig. 12.22. Tuning curve of Cr^{3+}:LiSrAlF$_6$ for two sets of cavity mirrors.

Cr:LiSAF microlaser and diode-pumped sub-100-fs passive mode-locked Cr:LiSAF using an A-FPSA are reported at CLEO'94 Meeting recently.

Due to the broad band absorption, which overlaps well with the emission bands of Xe flashlamp and longer luminescence lifetime in comparison with Ti:sapphire, Cr:LiCaF or Cr:LiSAF should be a suitable high power amplifier medium for flashlamp pumping. P. Beaud *et al.* measured the gain properties of Cr^{3+}:LiSAF. Maximum small-signal gain value of 0.17 cm^{-1} were reported for π-polarized light [67]. Due to the low nonlinear refractive index Cr:LiCAF and Cr:LiSAF can be

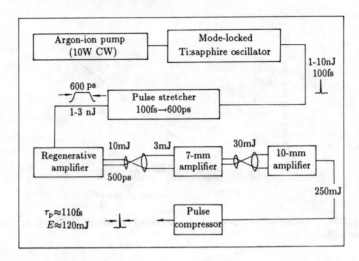

Fig. 12.23. Terawatt-class system uses a LiSAF regenerative amplifier to produce 110-fs pulses with outputs of 120 mJ.

used in high power laser system. A terawatt class system with a LiSAF regenerative amplifier to produce 110 fs with outputs of 120 mJ was designed, Fig. 12.23 shows the schematic diagram [68]. High pulsed powers of 1.25 TW ($\times 10^{12}$W) and 8 TW with pulse width of 120 fs and 90 fs respectively have been obtained using multistage Cr:LiSAF amplifiers [69].

12.7   Rare Earth Ions Doped Tunable Laser Crystals

In Chapter 3 we have discussed that when the trivalent rare earth ions doped in low symmetric crystal hosts due to many Stark's mainfolds of upper and lower levels between radiative transitions, the laser emission band width should be broadened. The laser emission wavelength of Tm^{3+}, Ho^{3+} and Er^{3+} ions can be tuned in the vicinity of 1.65~2.1 μm, 2.7~2.8 μm, 1.52~1.56 μm respectively [70, 71]. This kind of wavelength tuning is discrete, rather than continuous. The continuous wavelength tuning for rare earth ions doped laser crystal can only be found in $5d \rightarrow 4f$ transition of some low valent rare earth ions, such as Ce^{3+}, Sm^{2+}, Dy^{2+}, Ho^{2+}. It is well known that the $5d$ electronic configuration is strongly effected by the crystal lattice field; and the $5d \rightarrow 4f$ transition exhibits vibronic in nature.

The vibronic lasing in Sm^{2+}:CaF_2 was achieved at liquid nitrogen temperature at wavelength between 708.5 nm and 745 nm [73], the electronic transition takes place at $5d \rightarrow {}^7F_1(4f)$.

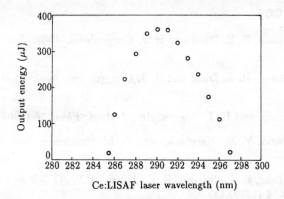

Fig. 12.24. Ce^{3+}:LiSAF laser output energy as a function of output radiation wavelength (pump energy, 12mJ, 264 nm; output coupler T, 26%).

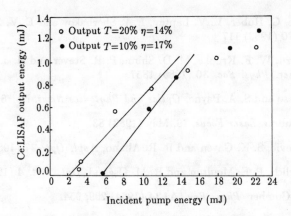

Fig. 12.25. Ce^{3+}:LiSAF laser output energy as a function of incident pump energy (pump wavelength, 266 nm; Ce:LiSAF laser wavelength, 290 nm).

More pronounced tunable laser crystals are Ce^{3+} doped fluorides, the vibronic lasing originates from transition $5d \rightarrow {}^2F_{7/2}$ and ${}^2F_{5/2}(4f)$, the wavelength tuning range is 306~315 nm and 323~328 nm respectively for Ce^{3+}:YLiF$_4$ [74]. As shown in Fig. 12.24, tunable operation of Ce^{3+}:LiSAF laser between 285 and 297 nm was reported by Pinto *et al.* [75] The 0.8% Ce^{3+}:LiSAF laser crystal (4×4×6 mm^3) was oriented with the *c*-axis perpendicular to the resonator axis, and pumped by the fourth harmonic of a 10 Hz Q-switched Nd:YAG laser operating at 266 nm, because the lifetime of upper level is approximately 29 ns. Figure 12.25 shows the laser performance of Ce:LiSAF laser for output coupler transmission values of 26% and 70%. A slope efficiency of 14% was obtained with a 26% T output coupler.

References

1. L. F. Johnson, R. E. Dietz and H. J. Guggenheim, *Phys. Rev. Lett.* **11** (1963) 318.

2. L. F. Johnson, R. E. Dietz and H. J. Guggenheim, *Appl. Phys. Lett.* **5** (1964) 21.

3. L. F. Johnson and H. J. Guggenheim, *J. Appl. Phys.* **38** (1967) 4837.

4. Yu. S. Vagin, V. M. Marchenko and A. M. Prokhorov, *Sov. Phys. JETP* **28** (1969) 902.

5. J. C. Walling, H. P. Jenssen, R. C. Morris, E.W. O'Dell and O. G. Peterson, *Opt. Lett.* **4** (1979) 182.

6. M. L. Shand, J. C. Walling and H. P. Jenssen, *IEEE J. Quant. Electron.* **QE-18** (1982) 167.

7. B. Struve, G. Huber, V. V. Laptev, I. A. Sherbakov and E. V. Zarikov, *Appl. Phys.* **B30** (1983) 117.

8. J. A. Caird, W. F. Krupke, M. D. Shinn, P. R. Staver and H. J. Guggenheim, *Bull. Amer. Phys. Soc.* **30** (1985) 1857.

9. L. L. Chase and S. A. Payne, *Optics and Photonics News*, No. **8** (1990) 18.

10. P. E. Moulton, *Laser Focus*, 19, May (1983) 83.

11. V. Petricevic, S. K. Gayen and R. R. Alfano, *Appl. Opt.* **27** (1988) 4162.

12. D. J. Ehrlich, P. E. Moulton and R. M. Osgood, *Opt. Lett.* **4** (1978) 184.

13. D. E. McCumber, *Phys. Rev.* **A134** (1964) 299; 954.

14. R. F. Moulton, *J. Opt. Soc. Amer.* **B3** (1986) 125.

15. A. Sanchez, *et al.*, *IEEE J. Quant. Electron.* **24** (1988) 995.

16. R. L. Aggarwal, A. Sanchez, *et al.*, *IEEE J. Quant. Electron.* **24** (1988) 995.

17. Chengjiu Wu, *J. Synthetic Crystals* **21** (1992) 31.

18. Peizhen Deng and Yue Chai, *et al.*, *Proc. SPIE* **1338** (1990) 207.

19. Yue Chai and Peizhen Deng, *et al.*, *J. Chinese Silicate Soc.* **4** (1991) 338.

20. Yonzhong Zhou and Peizhen Deng, *et al.*, *Proc. SPIE* **1627** (1992) 230.

21. Qiang Zhang and Peizhen Deng, *et al.*, *J. Appl. Phys.* **67** (1991) 6159.

22. Peizhen Deng, *et al.*, *China and Japan Symposium on Growth and Characterization of Crystals*, 1990, pp. 210.

23. Peizhen Deng, Qiang Zhang, Yang Sun, *et al.*, *Chinese J. Lasers* **19** (1992) 116.

24. *High-Tech Letters (new materials)*, 1992, pp. 107.

25. Reikuen Wu *et al.*, *Chinese J. Lasers* **4** (1991) 208.

26. Kaiyi Yu *et al.*, *Chinese J. Lasers* **4** (1991) 293.

27. *High-Tech Letters (new materials)*, 1992, pp. 107.

28. Peizhen Deng, *Chinese J. Lasers (E.E.)* **1** (1992) 268.

29. Bing Xu, Lihuang Lin, *et al.*, *Acta Optica Sinica* **13** (1993) 425.

30. S. Oga, Y. Segawa, *et al.*, *Japan J. Appl. Phys.* **28** (1989) L1977.

31. P. M. W. French, J. A. R. Williams and J. R. Taylor, *Opt. Lett.* **14** (1989) 686.

32. N. Sarukura, Y. Ishida and H. Nakano, *Opt. Lett.* **16** (1991) 153.

33. N. Sarukura and Y. Ishida, *IEEE J. Quant. Electron.* **28** (1992) 2134.

34. K. F. Wall, R. L. Aggarwal, R. E. Fahey and A. J. Strauss, *IEEE J. Quant. Electron.* **24** (1988) 1016.

35. Peizhen Deng, Qiang Zhang, Fuxi Gan, N. Sarukura, *et al.*, *Techn. Digest of Topic Meeting of Advanced Solid State Lasers*, Salt Lake City, Feb. 1994, p. 15.

36. N. Sarukura and Y. Ishida, *IEEE J. Quant. Electron.* **28** (1992) 2134.

37. N. Sarukura, Y. Segawa and K. Yamagishi, Advanced Solid State Lasers'93, paper ATUD3.

38. T. B. Norris, *Opt. Lett.* **17** (1992) 1009.

39. I. Shimada and J. P. Roberts, *Techn. Digest of CLEO'94* (Anaheim, California), **8** (1994) 136.

40. B. C. Stuart, S. Herman and M. D. Rerry, *Techn. Digest of CLEO'94* (Anaheim, California), **9** (1994) 228.

41. N. Blanchot, C. Rouyer, *et al.*, *Techn. Digest of CLEO'94* (Anaheim, California), **8** (1994) 404.

42. Peicong Pan, Peizhen Deng, *et al.*, *Proc. C-MRS*, Vol. 1, North-Holland Publishers, Amsterdam, 1991, p. 425.

43. Peicong Pan, Shenhui Yan, *et al.*, *J. Cryst. Growth* **121** (1992) 141.

44. Bin Hu, Hongbi Zhu and Peizhen Deng, *J. Cryst. Growth* **128** (1993) 991.

45. Zhi Jui, Hua Lui, Peicong Pan and Hongbi Zhu, *Chinese J. Lasers* **19** (1992) 160.

46. V. Petricevic, S. K. Gayen and R. R. Alfano, *Opt. Lett.* **14** (1989) 612.

47. T. J. Carrig and C. R. Pollock, *Opt. Lett.* **16** (1991) 1662.

48. A. Sugimoto, Y. Segawa, *et al.*, *Japan J. Appl. Phys.* **30** (1991) L495.

49. A. Seas, V. Petricevic and R. R. Alfano, *Opt. Lett.* **16** (1991) 1668.

50. V. Yanovsky, Y. Pang and F. Wise, *Opt. Lett.* **18** (1993) 1541.

51. A. V. Shestakov, N. I. Borodin *et al.*, *Extended Abstracts of CLEO'*91 (Baltimore, Maryland), 1991, pp.592/CPDP11-1.

52. Xingan Guo and Meling Chen, *J. Cryst. Growth* **83** (1987) 311.

53. Peizhen Deng, *Proc. Intern. Conf. on Lasers'*85 (Las Vegas, Nevada, STS Press, 1986), p. 452.

54. Qiang Zhang, Peizhen Deng and Fuxi Gan, *Cryst. Res. Technol.* **25** (1990) 385.

55. Qiang Zhang, Bing Hu, Peizhen Deng and Fuxi Gan, *J. Phys. D: Appl. Phys.* **26** (1993) 1.

56. Guifen Zhang and Xiaoshan Ma, *Chinese J. Lasers* **13** (1986) 707.

57. Bangxing Zhang, Lusen Wu, Meirong Zhao and Haili Zhang, *Acta Optics Sinica* **6** (1986) 408.

58. J. C. Walling, *Tunable Paramagnetic-Ion Solid State Lasers* in *Tunable Lasers*, eds. L. F. Mollenauer and J. C. White (Springer-Verlag, Berlin, 1987), p. 363–395.

59. M. L. Shand, *Proc. Intern. Conf. on Laser'*85 (Las Vegas, Nevada, STS Press, 1986), p. 732.

60. J. C. Walling, *Laser Focus/Electro-Optics* **9** (1983) 213.

61. *Laser Focus World* **28(7)** (1992) 13.

62. S. A. Payne, L. L. Chase, *et al.*, *IEEE J. Quant. Electron.* **24** (1988) 2243.

63. S. A. Payne, L. L. Chase, *et al.*, *J. Appl. Phys.* **66** (1989) 1051.

64. Xiaodong Liu, Peizhen Deng and Bing Hu, *Proc. SPIE* 1863 (1993) 90.

65. S. A. Payne, L. L. Chase, *et al.*, *Proc. SPIE* **1223** (1990) 84.

66. R. Scheps, *Opt. Mater.* **1** (1992) 1.

67. P. Beaud, Y. F. Chen, B. H. T. Chai and M. C. Richardson, *Opt. Lett.* **17** (1992) 1064.

68. M. D. Perry, *Laser Focus World* **27(11)** (1991) 69.

69. P. Beaud, M. Richardson, E. J. Miesak and B. H. T. Chai, *Opt. Lett.* **18** (1993) 1550.

70. H. Saito, *et al.*, *Opt. Lett.* **17** (1992) 189.

71. H. Yanagita, *et al.*, *Electron. Lett.* **26** (1990) 1836.

72. C. Y. Chen, *et al.*, *IEEE Photon. Tech. Lett.* **2** (1990) 18.

73. Y. S. Vagin, V. M. Marchenko, A. M. Prokhorov, *Sov. Phys.—JETP* **28** (1969) 902.

74. D. L. Ehrlich, *et al.*, *Opt. Lett.* **4** (1978) 184.

75. I. F. Pinto, L. Esterowitz, G. H. Rosenbaltt and G. J. Quarles, *Tech. Digest of CLEO'94* (Anaheim, California), **8** (1994) 64.

13. Structure and Laser Emission of Laser Glasses

Laser emission characteristics of laser glasses are greatly dominated by the structure of host glasses and the site state of doping ions in glasses. Up to now only rare earth (RE) ions can perform laser action in glassy hosts. As mentioned in Chapter 3, the nonradiative energy transfer processes between RE ion and host as well as inter-RE-ions are very dependent of the structure of host glass and the site state of RE ion in glass. Therefore, in this chapter we firstly introduce the glass formation and structure of glassy hosts and the site state of RE ions in the glassy hosts. Then, we will summarize the general aspects of laser emission characteristics of RE-doped glasses. The neodymium doped laser glasses are a main kind of RE-doped glasses, we will describe those in detail in the next chapter. The recent developed erbium laser glasses are introduced finally in this chapter.

13.1 Glass Formation and Structure of Host Glasses

The glassy materials are characterized in short range order and long range disorder. Many elements and compounds can form glassy state. For laser glasses most glass hosts belong to different multicomponent glass systems. Oxide glasses, such as silicate, borate, germanate phosphate glasses, and nonoxide glasses, such as halide and chalcogenide glasses. In this section the characteristics of glass formation and structure of different glass systems will be summarized into several basic points.

13.1.1 *Chemical Bond Nature Is the Determinative Factor of Formation and Structure of Inorganic Glasses*

As we have pointed out previously that the nature of chemical bond is a determinative factor for formation and structure of glass [1]. The existence of hybrid in the metallic, ionic and covalent bonds is a prerequisite to glass formation. Ionic and metallic bond materials, such as CsF crystal and Cs metal respectively, exhibit high coordination numbers and non-directional bonds, and are difficult to vitrify. Pure covalent bond molecules, such as F_2, combine each other by Van der Waals forces,

Table 13.1. Tendency of glass formation, electronegativity and chemical bond character of As and Zn atoms with VI and VII group elements.

System	Compound	Electronegativity		Tendency of glass formation	Character of chemical bond
Oxide	As_2O_3	O	3.5		Covalent
Sulfide	As_2S_3	S	2.5		
Selenide	As_2Se_3	Se	2.4		
Telluride	As_2Te_3	Te	2.1		Metallic
Fluoride	ZnF_2	F	4.0		Ionic
Chloride	$ZnCl_2$	Cl	3.0		
Bromide	$ZnBr_2$	Br	2.8		
Iodide	ZnI	I	2.5		Covalent

which are also non-directional. Therefore, pure covalent bond materials are also difficult to form glasses. When ionic and metallic bonds mix with covalent bonds, the chemical bond exhibits direction and saturation through strong polarization. It is then favorable to form low coordinations between the nearest atoms and the glass formation is easy to take place. As an example, Table 13.1 shows the relationship between the tendency of glass formation and the nature of chemical bond of arsenic and zinc compounds. The intermediate compounds, such as As_2S_3 and As_2Se_3, $ZnCl_2$ and $ZnBr_2$, have a mixed chemical bond nature and reach high glass formation ability.

According to chemical bond characteristics the inorganic glass materials can be classified into three groups:

1) Metalloid + non-metallic elements. The metalloid elements are located at IIIA, IVA and VA groups of the periodic table, and the non-metallic elements are located at VIA and VIIA groups.

2) Metalloid + metallic elements. The metallic elements are mainly transition metals and noble elements.

3) Metallic A + metallic B elements. The metallic A elements are located at IIA, IIB, IIIB, IVB groups of the periodic table, and the metallic B elements are noble and rare earth elements.

The chemical bonds in all inorganic materials mentioned above involve sp, spd and $spdf$ hybrid bonds, respectively.

The laser host glasses, including oxide, halide and chalcogenide glasses, belong to the first group.

13.1.2 Calculation of Glass Formation Ability by Chemical Bond Parameters

We propose two chemical bond parameters to characterize the glass formation ability of binary, ternary and multi-component systems [2]. ΔX is the difference of electronegativity and Z/r_c is the ratio of number of valence electron to covalent ra-

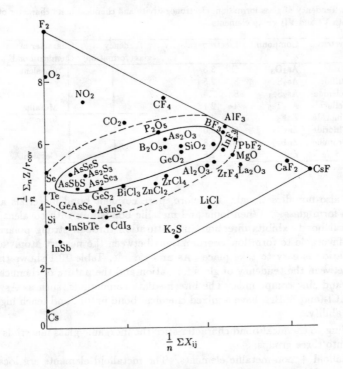

Fig. 13.1. Relationship between chemical bond parameters and glass formation.

dius (in Å). Figure 13.1 shows the chemical bond parameter diagram, which demonstrates the glass formation ability of different compounds. Cs, CsF, F_2 represent the pure metallic, ionic and covalent materials, which are located at the vertex of the triangle diagram. Within the solid line the cooling rate for glass formation is less than 10^{-1} °C/s, and within the dot line the cooling rate is up to 10^3 °C/s.

For multi-component oxide and non-oxide glass systems the average values $\frac{1}{n}\sum_n \Delta X_{ij}$ and $\frac{1}{n}\sum \frac{Z}{r_c}$ can be taken. It can be seen from Fig. 13.1, the critical cooling rate for formation of the most glasses are between 10–10^3 °C/s. Some glass former oxides fluorides and chalcogenides are rather stable, such as BeF_2, As_2Se_3, GeS_2 etc., the cooling rate is less than 10 °C/s.

Wen Yuankai *et al.* proposed a bond parameter criterion in the formation of glass of oxides based on our mixed bond concept [3]. They used two parameters, one is so called Hard and Soft Acids and Bases Value f, which is defined as below [4]:

Table 13.2. Content of IIIA and IVA groups element in As–Se glass system and the character of chemical bond of selenides.

Element	B	Al	Ga	In	Ti	Si	Ge	Sn	Pb
Content	40	3	18	3	50	50	52	10	30
Bond	B–Se	Al–Se	Ga–Se	In–Se	Ti–Se	Si–Se	Ge–S	Sn–Se	Pb–Se
R_σ (a.u.)	0.25	0.49	0.35	0.58	0.50	0.20	0.27	0.48	0.48
R_π (a.u.)	0.31	0.49	0.55	0.58	0.64	0.23	0.45	0.48	0.54

$$f = \frac{z}{r} + 3.0X + 2.2. \qquad (13.1)$$

Here the z/r is the ratio of electric charge and atomic radius, X is the electronegativity. The other factor is F_W,

$$F_W = \sin(\pi \cdot N); \; N = \frac{\Delta X}{3.3}, \qquad (13.2)$$

$\Delta X_{max} = X_F - X_{Cs} = 3.3$. The function F_W gives the following meanings:

1) When N approaches to 1, $F_W \to 0$, it means ionic bond;
2) When N approaches to 0, $F_W \to 0$, it means covalent bond;
3) When N equals 0.5, $F_W \to 1$, this is mixed ionic-covalent bond.

The glass formation value,

$$\phi = f + 12.5 F_W - 11.6. \qquad (13.3)$$

For glass forming oxides, the value $\phi > 0$.

Liang Zhenhua discussed the chalcogenide glass forming tendency in terms of the chemical bond approach [5]. The ionicity R_σ^{AB} and metallicity R_π^{AB} of an A–B bond are defined as follows [6]:

$$R_\sigma^{AB} = (r_s^A + r_p^A) - (r_s^B + r_p^B), \qquad (13.4)$$

$$R_\pi^{AB} = (r_s^A - r_p^A) + (r_s^B - r_p^B). \qquad (13.5)$$

Here the r_s and r_p represent the orbitally dependent ionic radii of s-orbital and p-orbital respectively. The glass forming tendency of IIIA and IVA elements in As–Se glass system and the R_σ^{AB} and R_π^{AB} values of IIIA and IVA elements with Se atoms are shown in Table 13.2. It can be seen that higher content of the IIIA and IV group elements in As–Se glass systems always corresponds with lower ionicity and metallicity.

In mixed anion glasses we found that the glass forming range of mixed anion glass systems can be enlarged with similar electronegativity of anions [7]. According to electronegativity values of VI and VII group elements as shown in Table 13.1, fluoride is easy to form mixed anion glasses with oxide, it is well known that the fluoro-phosphate and fluoro-borate glasses are rather stable. Therefore, chloride

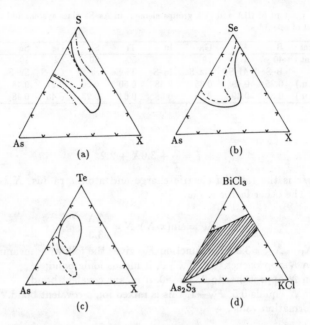

Fig. 13.2. Glass forming regions of As–S–X (a), As–Se–X (b), As–Te–X (c) and As$_2$S$_3$–BiCl$_3$–KCl (d) systems. (a) - - - As–S–I, — · — As–S–Cl, — As–S–Br; (b) - - - As–Se–I, — As–Se–Br; (c) - - - As–Te–I, — As–Te–Br.

with sulfide, bromide with selenide, and iodide with telluride always possess large glass forming regions than those with others. This is obviously due to the similarity of chemical bonds. Figure 13.2 shows the glass forming region of As atoms with VI and VII group elements, and sulfide (As$_2$S$_3$) can continuously form glasses with chlorides (BiCl$_3$+KCl).

13.1.3 Characteristics of Short Range Order of Glassy State and Chemical Bond

The short range order involved in the arrangement of the nearest neighboring atoms can be expressed by three structural parameters: the bond length between the central atom and the nearest neighboring atom, the bond angle and the coordination number [8]. These three structural parameters for different glassy materials are also different from those of the crystalline state, cf. Table 13.3. Comparing the element and the compound (alloy) glasses with the crystalline state, it can be seen clearly that the short range structural similarity depends upon the properties of the chemical bond. These can be classified as three categories: the first are the ionic-covalent bond compounds very stable for forming glass, such as SiO$_2$, B$_2$O$_3$,

Table 13.3. Short range structure parameters of element and compound glasses.

Types	Oxides		Elements				Semiconductors		Halides		Metal alloys	
Compositions	SiO$_2$	B$_2$O$_3$	C	Si	As	Se	III-V II-VI	As$_2$S$_3$	BeF$_2$	ZnCl$_2$	Fe-P	Pd-Si
Central atoms	Si,	B, O					III, V II, VI	As, S	Be, F	Zn, Cl	Fe, P	Pd, Si
Bond angles	o	o	•	•	•	•	• • • •	• o	• •	• •	• •	• •
Bond length	o	o	o	•	o	o	o o o o	• o	o o	• •	• •	o o
Coordination number	o	o	o	•	o	o	o o o o	• o	o o	• •	• •	o o

(o) represents a fixed value; (•) represents a range of value.

BeF$_2$ etc. Their trihedron [BO$_3$] and tetrahedron [SiO$_4$], [BeF$_4$] structures are in agreement with those of the crystalline states (i.e. agreement in bond length, bond angle and coordination number). The second are the elements with covalent bonds as major constituents, such as Si, Ge, Se, As and III-V compound, As$_2$S$_3$ compounds etc. In comparison with the crystalline state, the coordination numbers and the bond lengths are the same, whereas the bond angles have greater difference, which indicates that in the glassy state, it has tetrahedrons whose constitution is similar but not exactly the same with those in crystals (or distorted tetrahedrons). Regardless of the first category or the second category of materials, the connected angles between the triangle and the tetrahedra are not monotonous in glassy state. There is a certain variation range, e.g. the angular variation of Si-O-Si is 120°–180° (144° is in majority), which provides freedom for the vertex angle rotation and reflects the topological disorder of three dimensional structure in the glassy state. The third are metal alloys, e.g. Fe-P, Pd-Si and molten salts etc. High intensity atoms or ions (e.g. Si, P, Zn^{2+} etc) form their own coordination structure in glassy state. They are in agreement in the bond lengths and the coordination number with those of the crystalline states and different in bond angles. However, the short range structures of the low field intensity atoms or ions (e.g. Fe, Pd and Cl$^-$) are not exactly the same as those of the crystalline states. All information of short range order of glassy materials shown above are closely connected with the chemical bond nature.

13.1.4 *The Structural Model of Inorganic Glasses*

It is very difficult to express various kinds of glasses with a single structural model due to the fact that each has its own difference from others in the structure and the structural chemistry in a great variety of glass forming systems. We have introduced three kinds of glass structure models to explain three glass systems with different chemical bond characteristics as classified above, they are random network model, random polyhedra packing model, dense random packing of hard spheres. We believe that the random polyhedra packing model can be summarized as a unified structural model for inorganic glass system.

Consequently there should inevitably be a poly-hedron [AB_n] composed of two atoms (or ions) A, B of unequal radii. This is the most fundamental condition for the short range order in glass as we discussed above, and the most fundamental unit in glass as well. The polyhedron may be a trihedron, tetrahedron, octahedron or a polyhedron with even higher coordination number. The polyhedra can be bonded at the vertex angles, sides or faces, packed disorderly in space to form the spatial arrangement according to the minimum energy principle. Other atoms (or ions) with even greater radii C located at the vacancy in disorder packing of polyhedra form an arrangement of even higher coordination number together with the atoms (or ions) B.

In traditional oxide glasses such as borate, silicate, phosphate etc, B^{3+}, Si^{4+}, P^{5+} cations and $O^=$ anions form trihedra [BO_3], tetrahedra [BO_4], [SiO_4] and [PO_4] in glass forming oxides. These trihedra and tetrahedra are connected at the vertex points and form three dimensional structural networks. Some ions with greater radii, such as Al^{3+}, Ti^{4+} and transition metal ion TM^{3+} etc, can form tetrahedra and enter into the networks, or form octahedra outside the networks, this is random network model, which can be regarded as a special case of the random packing of trihedra, tetrahedra and octahedra. The ions with even greater radii, such as alkali metals and rare earth ions RE^{3+}, locate at the vacancy of the polyhedra structure. BeF_2, $ZnCl_2$-based halide glasses are included in this category, Be^{2+} and Zn^{2+} from tetrahedra [BeF_4], [$ZnCl_4$]. The As- and Ge-based chalcogenide glasses can also be regarded as this structure. Figure 13.3 shows the structure model of an alkali borosilicate glass. The tetrahedra [SiO_4], [BO_4] and triangle [BO_3] form three dimensional random network. The RE ions are located in the interspace of polyhedra. The phosphate glasses exhibit cross-chain structure. P^{5+} ions form tetrahedra with bridging and nonbridging oxygen, the structure model are shown in Fig. 13.4.

In oxide and fluoride glasses with greater ion radii, such as titanate, alumi- nate, tantalate and niobate, zirconium fluoride and thorium fluoride-based fluoride glasses, the cations and the anions form octahedra and the polyhedra with even higher coordination numbers, such as [AlO_6], [TaO_6], [NbO_6], [ZrF_8], [ThF_8] etc. These polyhedra are connected one another at the vertex angles and at sides as well, which form three dimensional random packing. Figures 13.5 and 13.6 show the structural models of K_2O–Ta_2O_5, Nb_2O_5 and BaF_2–ZrF_2Cl_2.

When the metallic bond transits to the hybrid bond (metal-covalent, metal-ionic) the structure of metallic glass tends to be the random polyhedra packing model, the metal or metalloid atoms A with smaller radii and the metal atom B with greater radii form polyhedra. Figure 13.7 shows the Si located at the center of the prismoid composed of Pd in Pd–Si alloy glass, and the prismoids are connected one another at the co-sides.

For making the stability of random packing of polyhedra it have to avoid the close packing of polyhedra in order, and to make more space freedom of the vertex

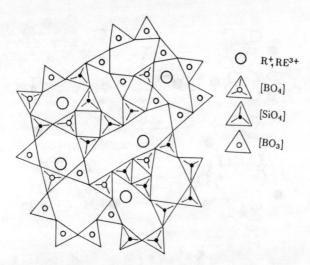

Fig. 13.3. Structural model of $R_2O(RO)$–B_2O_3–SiO_2 glass system ($\frac{Na_2O}{B_2O_3} < 1$).

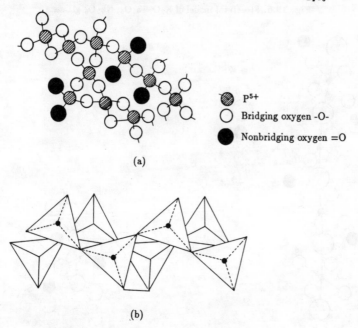

Fig. 13.4. Structural model of phosphate glass. (a) Atomic arrangement; (b) polyhedra packing.

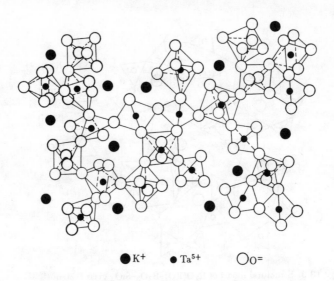

K⁺   Ta⁵⁺   O⁼

Fig. 13.5. Structural model of K_2O–Ta_2O_5, Nb_2O_5 glasses.

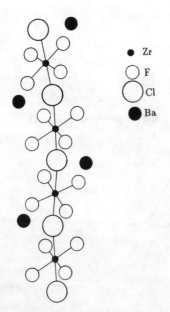

• Zr

○ F

◯ Cl

● Ba

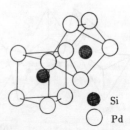

Si

Pd

Fig.   13.6.   Structural model of BaF_2–ZrF_2Cl_2 glasses.

Fig. 13.7. Structure of Pd–Si glass alloy.

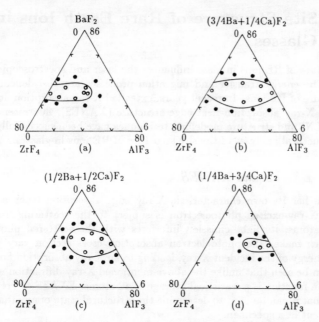

Fig. 13.8. Glass forming regions of ZrF_4–AlF_3–RF_2 (BaF_2, CaF_2) glass systems.

of polyhedra. In this case the atoms (or ions) with greater radii outside the polyhedra play an important role. Take the ZrF_4+AlF_3-based fluoride glasses as an example [9], the alkali earth ion with larger ionic radius, such as Ba^{2+}, makes the "nonbridging" fluorine at the vertex of $[ZrF_8]$ polyhedra and increases the space freedom of connected polyhedra. Therefore, $ZrF_4(ThF_4)$-based fluoride glasses always contain a large amount of BaF_2, and $ThCl_4(BiCl_3)$-based glasses can be doped a lot of KCl and NaCl. It is worth pointing out that for stabilizing the fluoride glass structure, firstly the electric field strength of cation A formed the polyhedron and cation C at the vacancy between the polyhedra should be matched because the cations C are located around the negative charged fluorine polyhedra for charge compensation. The ratio of the electric field strength of cations A and C $(Z_A/r_A^2)/(Z_B/r_B^2)$ is about 5–6. The optimal electric field strength ratio can decrease the structural stress. Therefore, it is easy to form glass when ZrF_4 combines with BaF_2 and AlF_3 with CaF_2. As shown in Fig. 13.8, when CaF_2 replaces the BaF_2, the glass forming regions are gradually moved from ZrF_4 rich area to AlF_3 rich area. Secondly, the content of high field fluoride and low field fluoride should also be matched. The ratio is about 3:2, it means that 20% "non-bridging" fluorines are existing.

Therefore, from chemical bond characteristics we can predict the glass formation ability, classify the inorganic glassy materials and propose the random polyhedra packing model to unify the different structural models for inorganic glass systems.

13.2 Site Structure of Rare Earth Ions in Laser Glasses

The site state of RE ions in glass influences the laser and spectroscopic properties greatly. The energy transfer and migration processes are also dependent on the site structure of RE ions. In recent years, extended X-ray absorption fine structure (EXAFS), X-ray absorption near-edge structure (XANES), fluorescence line narrowing (FLN) and structure modelling techniques have been successfully developed for the structural analyses of the site structure of RE ions in glasses.

13.2.1 *EXAFS and XANES*

Each atom has its own characteristic X-ray edge absorption (such as K, L,...). When an X-ray-excited photoelectron is subject to the scattering from the surrounding atoms, its light emission interferes with the scattered photoelectronic wave, which makes the photoelectron absorption cross section vary periodically with the energy of the incident X-ray, leading to an X-ray absorption fine structure. Thus it can be seen that unlike the above-mentioned X-ray diffraction whose radial distribution function is a statistical average of all atoms. EXAFS can reveal a given atom, so that it can be used to determine the structural state of an atom which has a low content in a specimen.

Figure 13.9 shows the XANES of NdF_3–BeF_2 glass, NdF_3 and Nd_2O_3 crystals. Figure 13.10 are the EXAFS and Fourier transform of NdL_{111} in NdF_3–BeF_2 glass and NdF_3 crystal [10]. It is found that in the glass there is substantial increase in the intensity of Nd_{111} compared with those of crystalline NdF_3. EXAFS analyses show that there is a shrinkage of the innermost Nd–F bond distance and reduction of the nearest neighboring fluorine coordination to $\sim$7 in glass compared with 9 in pure fluoride. It shows the ability of XANES and EXAFS techniques to elucidate the chemical bonding and local structure of a given atomic constitution in glass.

13.2.2 *FLN*

EXAFS and XANES can be used to determine the site state of RE ions in glass, and the results obtained are its statistical average. Actually, the site states of RE ions in glass are not the same, but distributed. The laser-induced fluorescence line narrowing (FLN) can be used to identify different site states. The spectral lines (no matter if they are absorption, fluorescence or paramagnetic resonance spectra) are very broad due to the different site states of excited RE ions, which results in difficulty and inconvenience for analysis. With narrow line laser excitation, only those ions are excited whose sites containing the same energy intercept with that of the laser wavelength (energy), thus the fluorescence line is narrowed. Figure 3.27 shows the FLN of Sm^{3+} in ZBLAN fluoride glass [11]. Sm^{3+} ions were excited via the transition $^6H_{5/2}$–$^4G_{5/2}$, this is the resonant transition. The dashed line in the

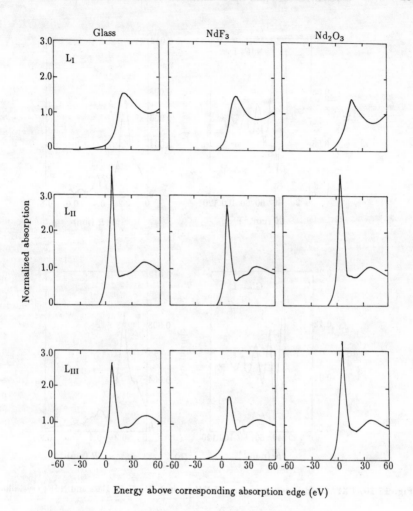

Fig. 13.9. XANES of glass, NdF_3 and Nd_2O_3 crystalline.

figure represents the fluorescence profile measured under non-selective excitation at room temperature. The structures of the profiles mainly contain two parts. The first is the resonant fluorescence at the same wavelength as the excitation laser; this part has a sharp structure and is the most intense in the whole range of fluorescence. This emission was caused by the anti-processes of the excitation transitions, i.e., from the excited levels to the lowest Stark level of the ground state $^6H_{5/2}$. The second part of the fluorescence is in the longer wavelength region which has a larger

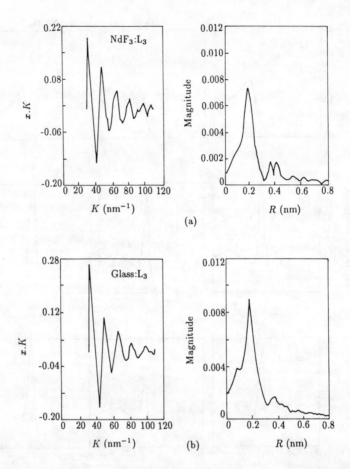

Fig. 13.10. EXFAS and Fourier transform of NdL_{111} in NdF_3–BeF_2 glass and NdF_3 crystalline.

spectral width caused by the transitions from the excited levels to the higher Stark splittings of the ground state. Due to the fact that the upper level is not a single isolated Kramers-degenerate state, accidental coincidences may happen during the excitation; this has little or no influence on the resonant transitions, but the non-resonant emission spectra may be broadened due to the characteristic of ions in several dissimilar environments. It is rather obvious that the inhomogeneous broadening bandwidth consists of several spectral lines with different site states.

The resonant fluorescence at different delay times have also been measured. Pumped at 560 nm, a shortest delay time of 20 μs was performed (see Fig. 3.28).

Table 13.4. Short-range potential parameters used in simulation.

Parameter	F–F	Al–F	Ba–F	Ca–F	In–F	Zn–F	Eu–F
A_{ij}^c (eV)	1127.7	336.73	314.27	669.39	230.44	295.26	1470.72
r_{ij}^c (nm)	0.02753	0.03521	0.04198	0.03286	0.04343	0.03540	0.03057
C_{ij}^c ($\times 10^{-6}$ eV·nm^6)	15.8	0.0	0.0	0.0	0.0	0.0	0.0

From these spectra we concluded that the intensity decreases and the full width at half maximum (FWHM) increases with the increase of the delay time. The minimum FWHM of the resonant fluorescence measured was 5 cm^{-1} and the shape of the spectrum at this time corresponds reasonably to a Lorentzian curve. The gradual broadening of narrowed fluorescence line shows that the energy transfer processes occur between Sm^{3+} ions at different sites.

13.2.3 *Structure Modelling*

In recent years, direct establishment of the site structure of RE ions by means of the Monte Carlo method and the molecular dynamics (MD) method has become very important. According to the given conditions, such as interatomic potential, density and temperature etc. for a given atomic system, the computer can give many kinds of disorder structure in the liquid, and select the models which appear to have high probability, calculate the structure and the thermodynamic properties. To simulate the glassy state, some high temperature liquid models are cooled down to a temperature at which no diffusion will take place, thus forming a series of solidified but disordered models, each representing a structure in a certain microscopical region, then these low temperature models will be built up and the structure of the whole glassy material is formed.

A MD simulation was performed for 40AlF$_3$·20BaF$_2$·40CaF$_2$:Eu (ABCF:Eu) and 40InF$_3$·20BaF$_2$·40ZnF$_2$:Eu (IBZF:Eu) glasses [12]. A constant volume cubic unit cell with periodic boundary condition is employed. The box contains 1178 particles, including 2 Eu ions, corresponding to a doping concentration of 0.6 mol%, for both glass systems. Interactions are applied to the anion-anion and anion-cation, but not to cation-cation as they are assumed to be effectively screened by the intervening anions. The short-range potential is expressed as follows:

$$E_{\parallel} = A_{ij}^c \exp(-\frac{r_{ij}^c}{r_{ij}}) - \frac{C_{ij}^c}{r_{ij}^6}, \tag{13.6}$$

where r_{ij} is the distance between the ith ion and the nearest image of the jth ion. A_{ij}^c, r_{ij}^c, and C_{ij}^c are the constants obtained by the fitting crystal structures and the properties for corresponding crystals (Table 13.4). The initial structure of the sample is heated up to 5000 K, then cooled down to 2500 K, 1000 K and 300 K sequentially. A relaxation of 10,000 timesteps (2×10^{-15} sec/timestep) is allowed for each temperature. The pair distribution functions (PDF) and the coordination number distribution functions (CDF) are averaged over the structure of the last 1000 timesteps at 300 K. The snapshots are taken from the structures at the last timestep.

The bulk structure of ABCF:Eu and IBZF:Eu glasses can be obtained by MD analysis. Figures 13.11(a) and 13.11(b) show the PDF and CDF of the ABCF:Eu glass, respectively. A sharp first Al–F peak at 0.186 nm and a plateau in Al–F CDF with coordination number of 6 indicate that well-defined [AlF$_6$] coordination polyhedra constitute the main framework of the glass. The first Ca–F peak at 0.23 nm is relatively broad, accompanied by a slope in the Ca–F CDF curve, which ranges from coordination number 7 to 9, indicating the distorted [CaF$_7$], [CaF$_8$], and [CaF$_9$] polyhedra. A first broad Ba–F peak and the gradual increase in coordination number in Ba–F CDF indicate that the Ba ions occupy typical modifier sites with a variety of coordination states in the glass. Figures 13.12(a) and 13.12(b) show the PDF and CDF of the IBZF:Eu glass. Interestingly, the first In–F and Zn–F peaks are almost identical, at 0.198 nm and with similar sharpness. CDF also indicate that both In and Zn have a well defined first F shell, with a coordination number of 6, indicating that [InF$_6$] and [ZnF$_6$] together form the main framework of the glass. Ba ions occupy modifier sites in IBZF:Eu glass.

For Eu site structure, Fig. 13.13 shows the PDF and CDF of Eu–F in ABCF:Eu and IBZF:Eu glasses. The Eu–F bond-length is 0.22 nm in both glasses. Quite sharp first Eu–F peaks are observed in both systems, indicating that, due to the high Eu–F bond strength, Eu ions are able to form a well-defined structures in the glasses. The CDF indicate that the [EuF$_7$] and [EuF$_8$] are the principal coordination polyhedra. The site structures of Eu ions are depicted in Fig. 13.14. [EuF$_7$] and [EuF$_8$] polyhedra are found in ABCF:Eu, [EuF$_8$] polyhedra are found in IBZF:Eu.

Figure 13.15 shows the distribution function of Eu^{3+} in BeF$_2$ glass obtained by the Monte Carlo method [13], then the energy levels of Eu^{3+} are calculated in different site states by the point electric charge model. Compare the energy level distribution (histogram) obtained by this calculation and the fluorescence spectrum obtained experimentally to determine the structure of different sites of Eu^{3+} (cf. Fig. 13.16). The coordination number of Eu^{3+} is around 7–8.

13.3   Laser Emission of Rare Earth Ions Doped Glasses

As discussed in Chapter 4 for low-threshold operation it is desirable to have the following requirements for selecting a laser glass:

(i)   Large stimulated emission cross section σ_{32} (narrow fluorescence linewidth $\Delta\nu_L$);
(ii)   Broad (numerous) absorption bands for optical pumping $\Delta\nu_p$;
(iii)   High fluorescence conversion efficiency η_1;

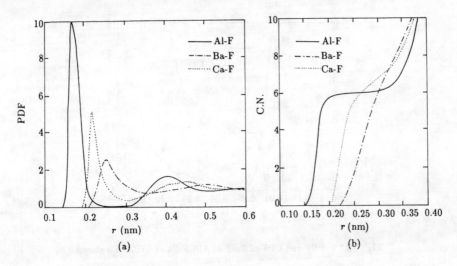

Fig. 13.11. PDFs (a) and CDFs (b) in ABCF:Eu glass.

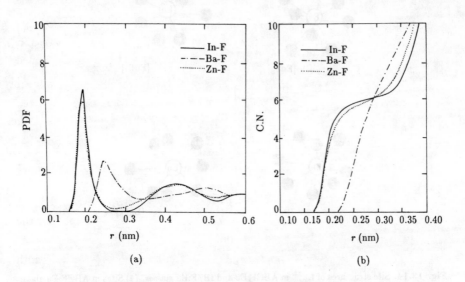

Fig. 13.12. PDFs (a) and CDFs (b) in IBZF:Eu glass.

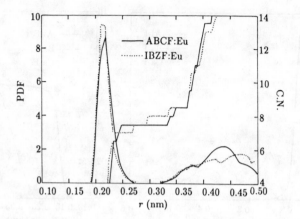

Fig. 13.13. PDF and CDF of Eu-F in ABCF:Eu and IBZF:Eu glasses.

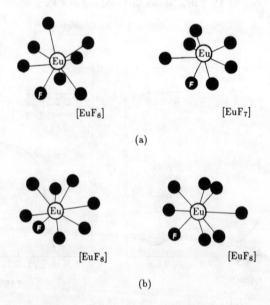

Fig. 13.14. Site structures of Eu^{3+} in ABCF:Eu and IBZF:Eu glasses. (a) Sites in ABCF:Eu glass; (b) sites in IBZF:Eu glass.

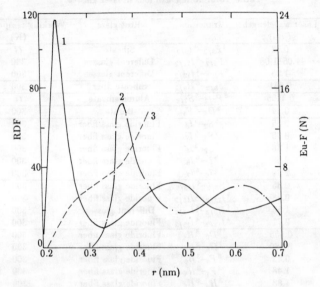

Fig. 13.15. Radial distribution function curve and coordination number variation curve of Eu^{3+} in BeF_2 glass measured by the Monte-Carlo method. 1, Eu–F; 2, Eu–Be; 3, N(Eu–F).

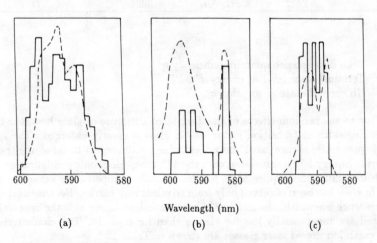

Fig. 13.16. Calculated and measured fluorescence spectra of Eu^{3+} in BeF_2 glass at $^5D_0-^5F_1$, $^7F_1$ transition. (a) Wide band excitation, $^5D_0-^7F_1$; (b) 576.5 nm laser excitation, $^5D_0-^5F_1$; (c) 578.5 nm laser excitation, $^5D_0-^7F_1$.

Table 13.5. Rare earth ions in laser glasses.

Ion	Laser wavelength (μm)	Transition	Host glass	Working temperature (K)
Nd^{3+}	0.93	$^4F_{3/2}-^4I_{9/2}$	Silicate	77
	1.05–1.08	$^4F_{3/2}-^4I_{11/2}$	Different glasses	300
	1.35	$^4F_{3/2}-^4I_{13/2}$	Different glasses	300
Sm^{3+}	0.651	$^4F_{5/2}-^6H_{9/2}$	Silicate fiber	300
Gd^{3+}	0.3125	$^6P_{7/2}-^8S_{7/2}$	Alumo-silicate	77
Tb^{3+}	0.54	$^5D_4-^7F_5$	Borate	300
Ho^{3+}	0.55	$^5S_2-^5I_8$	Fluoride glass fiber	300
	0.75	$^5S_2-^5I_7$	Fluoride glass fiber	300
	1.38	$^5S_2-^5I_5$	Fluoride glass fiber	300
	2.08	$^5I_7-^5I_8$	Fluoride glass fiber	300
	2.90	$^5I_6-^5I_7$	Fluoride glass fiber	300
Er^{3+}	0.85	$^4S_{3/2}-^4I_{13/2}$	Fluoride glass fiber	300
	0.98	$^4I_{11/2}-^4I_{15/2}$	Fluoride glass fiber	300
	1.55	$^4I_{13/2}-^4I_{15/2}$	Different glasses	300
	2.71	$^4I_{11/2}-^4I_{13/2}$	Fluoride glass fiber	300
Tm^{3+}	0.455	$^1D_2-^3H_4$	Fluoride glass fiber	300
	0.480	$^1G_4-^3H_6$	Fluoride glass fiber	300
	0.82	$^3F_4-^3H_6$	Fluoride glass fiber	300
	1.48	$^3F_4-^3H_4$	Fluoride glass fiber	300
	1.88	$^3H_4-^3H_6$	Fluoride glass fiber	300
	2.35	$^3F_4-^3H_5$	Fluoride glass fiber	300
Yb^{3+}	1.01–1.06	$^2F_{5/2}-^2F_{7/2}$	Different glasses	300
Pr^{3+}	1.30	$^1G_4-^3H_5$	Fluoride glass fiber	300
	1.047	$^1G_4-^3H_4$	Silicate	77

(iv) High fluorescence quantum efficiency η;
(v)   Terminal laser level at energy $E \gg kT$;
(vi) No excited-state absorption σ_{ex}.

Due to the random effects of the disordered structure of glass host, the fluorescence linewidth of transition metal ions in glass is greatly enlarged. The spectroscopic properties of rare earth ions are examined with respect to satisfying the above criteria. Energy levels associated with the $4f^n$ group electronic configuration of rare earth ions fulfill the requirements for the optically-pumped laser action. Laser action in glass has been observed only from trivalent rare earths. Because of the larger fluorescence linewidth, the number of rare earth ions lasing and the spectral range covered are substantially less for glasses than for crystals. The characteristics of rare earth ion doped laser glasses are shown in Table 13.5.

Especially Nd^{3+} ions possess a large number of absorption bands from near UV to IR regions, therefore most of laser glasses are activated by Nd^{3+} ions. We will discuss neodymium laser glasses in the detail in next chapter.

Table 13.6. Energy transfer and sensitization schemes for RE ions in glasses [14].

Donor (D)	Transition	Acceptor (A)	Transition	Type in Fig. 13.17	Temperature (K)	Glass
Nd^{3+}	$^4F_{3/2} \rightarrow ^4I_{9/2}$	Nd^{3+}	$^4I_{9/2} \rightarrow ^4F_{3/2}$	d	300	$Ba-K-SiO_2$
					500	
	$^4F_{3/2} \rightarrow ^4I_{13/2,15/2}$		$^4I_{9/2} \rightarrow ^4I_{13/2,15/2}$	c	700	
					900	
Nd^{3+}	$^4F_{3/2} \rightarrow ^4I_{13/2}, ^4I_{15/2}$	Nd^{3+}	$^4I_{9/2} \rightarrow ^4I_{13/2,15/2}$	c	4.2	$Na-SiO_2$
					300	
					450	
					600	
					300	$Na-GeO_2$
					300	$Na-B_2O_3$
					4.2	$Na-P_2O_5$
					300	
	$^4F_{3/2} \rightarrow ^4I_{9/2}$		$^4I_{9/2} \rightarrow ^4F_{3/2}$	d	4.2	$Na-SiO_2$
					300	
					450	
					600	
					300	$Na-GeO_2$
					4.2	$Na-P_2O_5$
Yb^{3+}	$^2F_{5/2} \rightarrow ^2F_{7/2}$	Er^{3+}	$^4I_{15/2} \rightarrow ^4I_{11/2}$	b	300	$K-Ba-Sb-$
					500	SiO_2
					700	
Yb^{3+}	$^2F_{5/2} \rightarrow ^2F_{7/2}$	Yb^{3+}	$^2F_{7/2} \rightarrow ^2F_{5/2}$	d	300	$K-Ba-Sb-$
						SiO_2
Nd^{3+}	$^4F_{3/2} \rightarrow ^4I_{9/2}; ^4I_{11/2}$	Sm^{3+}	$^6H_{5/2} \rightarrow ^6F_{11/2,9/2}$	a	300	
		Nd^{3+}	$^4I_{9/2} - ^4F_{9/2}$	d	300	
		Ho^{3+}	$^5I_8 - ^5I_5$	b	300	
Nd^{3+}	$^4F_{3/2} \rightarrow ^4I_{13/2,15/2}$	Nd^{3+}	$^4I_{9/2} \rightarrow ^4I_{13/2,15/2}$	c	4.2	$Li-La-P_2O_5$
					300	
	$^4F_{3/2} \rightarrow ^4I_{9/2}$		$^4I_{9/2} - ^4F_{3/2}$	d	4.2	
					300	
Eu^{3+}	$^5D_0 \rightarrow ^7F_{0,1,2}$	Cr^{3+}	$^4A_2 \rightarrow ^4T_2$	b	77–700	$La(PO_3)_3$
		Eu^{3+}	$^7F_{0,1} \rightarrow ^5D_0$	d	77–700	

It can be seen from Table 13.5 that the laser action can be easily performed in RE ion doped glass fibers due to high gain length. In Chapter 15 the laser glass fibers will be specially discussed.

For other rare earth ions due to lack of a strong absorption band in the optical pumping wavelength region, codoping with other transition metal ions is always utilized. Table 13.6 shows the energy transfer and sensitization schemes for rare-earths in glasses. There are many types for energy transfer between donor (D) and acceptor (A). Figure 13.17 shows the different energy transfer processes.

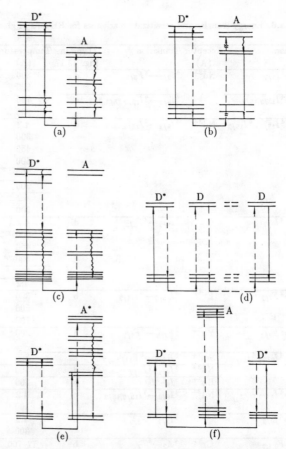

Fig. 13.17. Nonradiative energy transfer processes between donor (D) and acceptor (A). Solid line—radiative transition; dotted line—nonradiative transition between D-A; wavy line—nonradiative transition.

13.4   Erbium Laser Glasses

The most important laser wavelength of erbium laser glasses is at near IR region 1.5–1.6 μm. Due to the weak absorption of Er^{3+} ions at the visible region and the high concentration quenching of radiative transition $^4I_{13/2}$–$^4I_{15/2}$ (1.54 μm) at high Er^{3+} ion concentration (Fig. 13.18) [15], the sensitization of Er^{3+} emission by codoping $Yb^{3+}+Er^{3+}$ is always used for erbium glass lasers because Yb^{3+} ions have a strong and broad absorption band in the vicinity of 800–1100 nm.   Figure 13.19

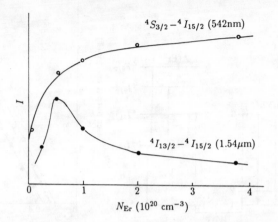

Fig. 13.18. Er^{3+} ion concentration effect of luminescence in phosphate glass.

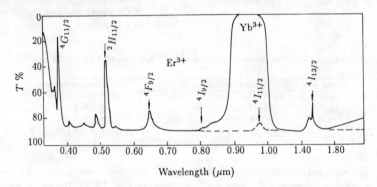

Fig. 13.19. Absorption spectrum of Er+Yb codoped phosphate glass. $N_{Er} = 3\times 10^{19}$ cm^{-3}, $N_{Yb} = 1.3\times 10^{21}$ cm^{-3}, $d = 10$ mm.

Table 13.7. Absorption coefficient of Yb^{3+} ions at 1.055 μm (α_{Yb}), energy transfer rate from Yb^{3+} to Er^{3+} (W_{DA}) and emission lifetime of Er^{3+} (τ_{Er}) at 1.55 μm in different codoped $Yb^{3+}+Er^{3+}$ glasses [16].

Glass system	α_{Yb} at 1.055 μm (10^{-2} cm^{-1})	W_{DA} (10^3 sec^{-1})	τ_{Er} (ms)	N_{Er} (10^{19} cm^{-3})
Alumo–phosphate	~6.0	10.5~11.0	7~8	~5
Na–K–Ba–La silicate	4.5	2~3	12~14	0.33~4.5
Na–K–Ba–Al–germanate	~4.5	~2.9	~6.5	~0.28
Ba–La borate	~9.8	~18	~0.59	~0.15

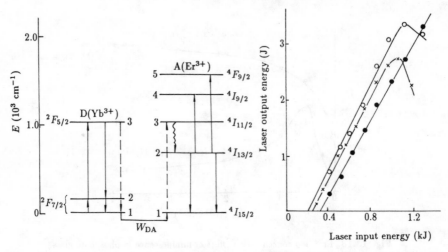

Fig. 13.20. Energy levels and energy transfer scheme of $Yb^{3+} + Er^{3+}$ codoped laser glass.

Fig. 13.21. Relationship between flashlamp input and laser output energy of Cr–Yb–Er codoped phosphate glass at 1.55 μm.

Table 13.8. Sensitization schemes of Er^{3+} luminescence at 1.5–1.6 μm.

Donor ion	Sensitization scheme	Donor ion	Sensitization scheme
Nd^{3+}	Nd^{3+}-Yb^{3+}-Er^{3+}	Mo^{3+}	Mo^{3+}-Yb^{3+}-Er^{3+}
Mn^{2+}	Mn^{2+}-Er^{3+}		Mo^{3+}-Er^{3+}
Cr^{3+}	Cr^{3+}-Er^{3+}	UO_2^{2+}	UO_2^{2+}-Er^{3+}
	Cr^{3+}-Yb^{3+}-Er^{3+}		UO_2^{2+}-Nd^{3+}-Yb^{3+}-Er^{3+}

shows the absorption spectrum of $Er^{3+} + Yb^{3+}$ codoped laser glass. The energy transfer scheme between Er^{3+} and Yb^{3+} is shown in Fig. 13.20. It can be seen from Fig. 13.20 that for high quantum efficiency of Er^{3+} emission, the energy transfer rate W_{DA} and the absorption coefficient α_{Yb} of Yb^{3+} at 1.055 μm should be high and the excited state absorption (transitions from $^4I_{13/2}$ to $^4I_{9/2}$ and $^4F_{9/2}$) should be low. In Table 13.7 the values W_{DA} and α_{Yb} of different $Yb^{3+} + Er^{3+}$ codoped glass are presented [16]. It can be found that the W_{DA} and α_{Yb} values are higher in borate and phosphate glasses than that in silicate glasses. Due to the strong interaction of borate host with Er^{3+} ion the emission lifetime of Er^{3+} in borate glasses are rather low, therefore phosphate glasses are the most suitable hosts for erbium laser glasses. Different sensitization schemes of Er^{3+} emission at 1.5–1.6 μ have been proposed, as shown in Table 13.8.

For example, Fig. 13.21 shows the relationship between the output laser energy and the input flashlamp energy of Cr–Yb–Er codoped phosphate glass at 1.55 μm

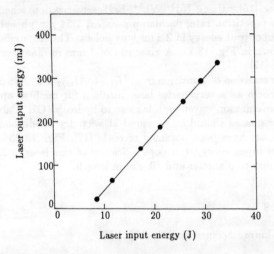

Fig. 13.22. Output vs input energy for Er:glass laser.

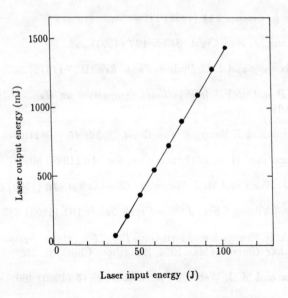

Fig. 13.23. Output vs input energy for Er:YAG laser.

wavelength range [15]. Using $Nd^{3+}-Yb^{3+}-Er^{3+}$ sensitization scheme Hoya Corp. developed a low repetition rate, flashlamp pumped 1.54 μm phosphate glass laser with up to 350 mj output energy in 2.5 ms long pulses. The slope efficiency obtained was 1.4% as shown in Fig. 13.22. A glass rod of 3 mm in diameter and 50 mm in length was used [17].

Erbium laser emission of transition from $^4I_{11/2}$ to $^4I_{13/2}$ is at 2.78 μm wavelength, which is well known as a very useful laser medium for medical applications [18]. Because the laser emission wavelength is close to hydroxy (OH^-) absorption band, the erbium laser glasses should be prepared at very dry condition. High efficient erbium laser glasses have been obtained recently [17], Fig. 13.23 shows the laser output energy vs input energy, the slope efficiency of this laser is 2.1%. The laser glass rod was 6 mm in diameter and 80 mm in length.

References

1. Fuxi Gan, *Chinese Science Bull.* **2** (1963) 18.

2. Fuxi Gan, *Chinese Cer. Soc. Bull.* **3** (1984) 1.

3. Yuankai Wen, Shengzhe Nie and Shuanhu Zhou, *J. Chinese Cer. Soc.* **15** (1987) 560.

4. R. G. Pearson, *Science* **151** (1966) 172.

5. Zhenhua Liang, *J. Non-Cryst. Solids* **127** (1991) 298.

6. J. R. Chelikowsky and J. C. Philips, *Phys. Rev.* **B17** (1978) 2453.

7. Fuxi Gan, *Digest of VII International Symposium on Halide Glasses* (Lorne, 1991) pp. 9.19.

8. R. P. Messmer and J. Wong, *J. Non-Cryst. Solids* **45** (1981) 1.

9. Yuan Cao and Fuxi Gan, *J. Chinese Cer. Soc.* **16** (1988) 50.

10. K. J. Rao, J. Wong and M. J. Weber, *J. Chem. Phys.* **78** (1983) 6228.

11. Fuxi Gan and Yihong Chen, *J. Non-Cryst. Solids* **161** (1993) 282.

12. Yuan Cao, A. N. Cormack and Fuxi Gan, *Proc. 9th Intern. Symp. on Nonoxide Glasses (Halide Glasses)* (May 1994, Hangzhou, China) p. 258.

13. S. A. Brawer and M. J. Weber, *Phys. Rev. Lett.* **45** (1980) 460.

14. M. E. Zabotinskii, *Laser Phosphate Glasses* (Science Press, Moscow, 1980) p. 127–128, 98 (in Russian).

15. Changhong Qi, Xiurong Zhang, Yasi Jiang and Yanyan Jiang, *Chinese J. Lasers* **18** (1991) 16.

16. P. E. Codrashof, Ph.D. Dissertation, State Institute of Optics (GOE), 1976.

17. T. Mochizuki, J. R. Unternahrer *et al.*, *Proc. SPIE* **1021** (1988) 32.

18. S. A. Pollack and M. Robinson, *Electron. Lett.* **24** (1988) 320.

14. Neodymium Laser Glasses

Since Snitzer first reported the Nd^{3+}-doped silicate glass laser in 1962 [1], neodymium laser glasses have been commercialized and put into mass production. But till now the manufacturing techniques are classified and new sorts of neodymium laser glass are still emerging. In this chapter we would like to summarize the development of neodymium laser glasses in general and put more emphasis on research progress of neodymium laser glasses in China.

14.1 Development History of Neodymium Laser Glasses

In 1963, one year after the publication of Snitzer's report, China also succeeded in creating the Nd-doped glass lasers [2]. In 1964, a 100 J output energy and 1% overall efficiency laser was made with a 16 mm diameter, 500 mm long glass rod, and the second harmonic generation and spectral analysis were reported [3]. China also succeeded in the development of borate and phosphate glass lasers early in 1965 [4].

We carried out a series of investigations on new sorts of neodymium laser glasses and their preparation techniques [5]. The composition of the early neodymium laser glass was based on barium crown glass reported by Snitzer [1]. Such glass has poor chemical stability and high transmission loss. We improved the physical and chemical properties of this glass and developed Type I (N_{01}) glass to satisfy China's laser development before 1966. Then we selected a more mature series of optical glasses for neodymium doping and lasing tests. As a result, Type III glass (N_{03}) based on a hard crown glass was developed. Experiments showed that the Type III glass had better chemical stability and mechanical strength and high laser output efficiency. It has become the most developed neodymium silicate glass in China. Large-sized, high-quality samples can be made in platinum or ceramic crucibles.

Various demands were raised from different applications of Nd:glasses, such as improving the laser damage resistance, to increase energy storage in laser amplifiers, to eliminate the thermal distortion and improve the thermal fracture resistance in high repetition rate operation etc. A series of investigations were conducted to modify the glass composition to obtain better physical properties. In the first 10 years we developed a series of Nd-doped silicate laser glasses, the physical, technical

Table 14.1. Physical, technical and laser properties of 10 kinds of silicate laser glasses.

1. Physical and technical properties

| | Physical properties | | | | | | Technical properties | | | |
| | Density $\rho(g/cm^3)$ | Micro hardness $H(kg/mm^2)$ | Mechanical strength $S(kg/mm^2)$ | Elastic modulus $E(10^3kg/cm^2)$ | Coef. of thermal expansion $10^{-7}°C$ 15°C~200°C | α 14°C-T_g | Transformation temperature $T_g(°C)$ | Softening temperature $T_f(°C)$ | Temperature for 100 poise (°C) |
Type									
N_{0112}	2.90	560	9.7	679	90	104	500	590	1400
N_{0212}	2.87	560	9.0	720	83	91	560	630	1380
N_{0312}	2.51	606	11.8	759	80	88	590	660	1430
N_{0412}	2.49	615	9.4	727	52	57	590	670	1680
N_{0612}	2.52	623	10.4	810	87	98	555	610	1270
N_{0712}	2.52	557	9.1	647	80	96	495	560	1470
N_{0812}	2.80	551	9.1	647	107	120	545	600	1440
N_{0912}	2.50	533	10.2	687	87	93	620	680	1450
N_{1024}	2.52	585	8.9	750	89	100	525	585	1420

2. Refractive index and dispersion

| | Refractive index | | | | | | | $n_{1.06}$ | |
Type	Crucible No.	n_C	n_D	n_e	n_F	$n_F - n_C$	ν	Measured	Calculated
N_{0112}	C72-13	1.53965	1.5424	—	1.54898	0.00933	58.14	1.5316	1.5315
N_{0212}	E70-05	1.53860	1.5413	—	1.54768	0.00908	59.5	—	1.5307
N_{0312}	D904	1.51969	1.5224	1.52446	1.52843	0.00874	59.8	1.5122	1.5122
N_{0412}	C7215	1.49969	1.5021	—	1.50774	0.00805	62.5	1.4923	1.4927
N_{0612}	E70-13	1.51725	1.5197	—	1.52549	0.00824	63.1	1.5136	1.5100
N_{0712}	DS65	1.50290	1.5054	1.50746	1.51134	0.00844	59.9	1.4953	1.4955
N_{0812}	C7133	1.53271	1.5354	—	1.54191	0.00920	58.2	1.5248	1.5246
N_{0912}	D926	1.51502	1.5176	1.51972	1.52363	0.00861	60.1	1.5076	1.5075
N_{1024}	B71-07	1.51455	1.5171	—	1.52335	0.00880	58.8	1.5067	1.5068

Table 14.1. (*To be continued*)

Table 14.1. (*Continued*)

3. Thermo-optical properties

Type	$Q(10^{-6}, °C^{-1})$	$W(10^{-6}, °C^{-1})$	$P(10^{-6}, °C^{-1})$	$Q(10^{-6}, °C^{-1})$	Stress optical constants $(10^{-6}, cm^2/kg)$	
					$-C_1$	$-C_2$
N_{01}	-0.86	4	5.0	0.8	0.15	0.38
N_{02}	0.13	4.6	4.9	1.0	0.12	0.42
N_{03}	1.64	5.8	4.0	0.9	0.11	0.36
N_{04}	4.17	6.8	2.7	0.8	0.08	0.37
N_{06}	-0.20	4.3	4.9	1.0	0.10	0.35
N_{07}	0.20	4.5	4.0	1.1	0.11	0.40
N_{08}	-3.20	2.5	5.5	1.0	0.09	0.38
N_{09}	0.12	4.6	4.2	1.0	0.11	0.37
N_{10}	0.80	5.4	4.6	1.0	0.09	0.35

4. Spectroscopical properties

Type	Major composition	Fluorescence branching	$A_{[0.88]}$ (sec^{-1})	$A_{[1.06]}$ (sec^{-1})	$A_{[1.35]}$ (sec^{-1})	$\sum_i A_{ri}$ (sec^{-1})	τ (μs)	$\sum_i A_{nr}$ (sec^{-1})	η $(\%)$	$\lambda^F_{[0.88]}$ (nm)	$\Delta\lambda^F_{[1.06]}$ (nm)	$\sigma^a_{p[0.88]}$ $(\times10^{-20}$ $cm^2)$	$\sigma^F_{p[1.06]}$ $(\times10^{-20}$ $cm^2)$
N_{0112}	SiO_2-K_2O-BaO	40 50 10	410	513	100	1024	600	640	61	22.0	38.0	0.26	1.00
N_{0212}	SiO_2-B_2O_3-K_2O-BaO	41 49 10	427	513	100	1038	620	575	64	23.0	36.0	0.26	1.04
N_{0312}	SiO_2-Na_2O-K_2O-CaO	38 53 9	520	719	124	1364	590	331	81	23.0	39.5	0.32	1.35
N_{0412}	SiO_2-K_2O-CaO	41 45 14	477	524	163	1159	680	312	79	26.0	37.0	0.26	1.05
N_{0612}	SiO_2-B_2O_3-Na_2O-K_2O	45 47 9	405	424	81	904	680	567	62	26.0	36.0	0.22	1.87
N_{0712}	SiO_2-Na_2O-K_2O	46 46 8	330	330	57	725	890	399	65	25.0	33.5	0.19	1.73
N_{0812}	SiO_2-BaO-K_2O	37 52 11	426	600	127	1156	760	160	88	27.0	40.0	0.23	1.11
N_{0912}	SiO_2-CaO-K_2O	42 48 9	405	460	87	959	750	374	72	25.0	36.0	0.23	0.95
N_{1024}	SiO_2-CaO-Na_2O	35 54 12	470	725	161	1355	510	606	69	24.0	37.0	0.29	1.45

Table 14.1. (*To be continued*)

Table 14.1. *(Continued)*

5. Laser properties

Type	Loss at 1.06 μm ($\times 10^4$ cm^{-1})	Laser efficiency of 16 mm diameter $\times$500 mm rod (%)	Laser central wavelength (nm)	Laser spectral linewidth (nm)
N$_{0112}$	0.2	2.4	—	—
N$_{0212}$	0.1	3.0	—	—
N$_{0312}$	0.1	4.0	1062.453	9.4
N$_{0412}$	0.16	3.8	1061.321	9.0
N$_{0612}$	0.29	2.2	1060.427	12.4
N$_{0712}$	0.12	3.5	1058.414	6.2
N$_{0812}$	0.27	2.7	1059.789	11.2
N$_{0912}$	0.10	3.8	1060.861	10.9
N$_{1024}$	0.22	3.5	1061.139	9.1

Table 14.2. Properties of Nd-doped phosphate laser glasses.

No.	Properties	Phosphate glasses	
		N$_{2112}$	N$_{2412}$
1	Fluorescence central wavelength (μm)	1.054	1.054
2	Fluorescence line half-width (nm)	26.5	25.5
3	Fluorescence lifetime (μs)	350	310
4	Stimulated emission cross section ($\times 10^{-20}$ cm^2)	3.5	4.0
5	Fluorescence branching ratio	0.52	0.52
6	Loss at 1.06 μm ($\times 10^{-1}$)	1.5	1.5
7	Laser efficiency (6 mm diameter, 50 mm long)	1.8	1.5
8	Laser central-wavelength (nm)	1054.139	—
9	Laser spectral linewidth (nm)	3.4	—
10	Density (g/cm^3)	3.38	2.95
11	Refractive index	1.574	1.543
12	Abbe number	64.5	66.6
13	Thermal expansion coefficient ($\times 10^{-7}$, °C^{-1})	117	156
14	Temperature coefficient of refractive index ($\times 10^{-7}$, °C^{-1})	−53	—
15	Thermo-optical coefficient ($\times 10^{-7}$, °C^{-1})	7.1	—
16	Stress thermo-optical coefficient ($\times 10^{-7}$, °C^{-1})	7	—
17	Birefringence thermal-optic coefficient ($\times 10^{-7}$, °C^{-1})	4	—
18	Transformation temperature (°C)	510	370
19	Deformation temperature (°C)	535	410
20	Elastic modulus (kg/m^2)	5550	5370
21	Shear modulus (kg/mm^2)	2200	2150
22	Poisson ratio	0.26	0.25
23	Nonlinear refractive index (10^{-13} esu)	1.3	1.2

and laser properties of them are shown in Table 14.1. The production of five kinds of glasses (N_{03}, N_{07}, N_{08}, N_{09}, N_{10}) has been expanded to the 100-liter scale to provide large-sized products. These glasses basically covered the international commercial varieties of neodymium glasses.

In the next 10 years the laser glass was developed with emphasis on the support of high power solid state lasers. The laser induced damage in laser materials caused by high-power short laser pulses are mainly due to nonlinear optical effects such as self-focusing and electrostriction. To improve the damage resistance of laser glasses, the chief task is to reduce the nonlinear refractive index of the glass. As the duration of the optical pumping pulse shortens, the stimulated emission cross section should be increased to improve the gain. Phosphate glass and fluorophosphate glasses satisfy these requirements, which have a lower thermo-optical coefficient than that of silicate glasses. In addition, the thermal distortion caused by the optical pumping can also be improved.

Based on previous work on phosphate and fluorophosphate glasses [6], we have also investigated the spectral and emission characteristics of the glasses, the chemical stability, and the effects of the hydroxyl group on the emission and laser properties of Nd-doped glass. Using the new fabrication technique, two new types of Nd-doped phosphate glasses, N_{21} and N_{24} were developed [7]. Type N_{21} glass was made into 70 mm diameter rods and 300 mm diameter disks. Table 14.2 shows the spectral, lasing, physical and chemical properties of the two Nd-doped phosphate glasses described above.

For comparison the physical, optical and spectral properties of commercial laser glasses produced abroad are listed in Tables 14.3, 14.4 and 14.5 respectively. The properties of these laser glasses are rather similar each other.

14.2 Preparation Technology of Neodymium Laser Glasses

Compared to optical glass, the neodymium glass requires higher purity and homogeneity, and the entire manufacturing process must ensure such quality for the products to be useful. Since 1964 we have been working in the melting technique of the neodymium silicate glass.

We first studied the making of Nd-glass in a platinum crucible using a resistance electric furnace. As the output of Nd-glass laser increased, the working medium often suffered damage. Studies showed that the Pt particles in Nd glass were the principal cause of the glass damage, and Nd glass free of Pt had many times the resistance against laser damage. We improved the technique in the following two areas and obtained glasses without Pt particles.

First we used ceramic crucibles in making Nd glass. The key problem was the selection of refractory materials for the crucible and the stirrer. Such materials must have a low concentration ($< 0.1\%$) of impurities such as iron, nickel, copper,

Table 14.3. Physical and optical properties of commercial silicate laser glasses from different countries.

Country	Sort	ρ (g/cm³)	H	E (×10⁹ Pa)	G (×10⁹ Pa)	μ	α (10⁻⁷ K⁻¹)	C_p (J/kg·K)	K (W/m·K)	ε	$\tan\delta$ (10⁻⁴)	T_g (°C)	T_f (°C)
USA	ED2	2.54	6.12	90.12	35.99	0.24	103	362	1.35	—	—	468	590
Germany	LG57	2.59	4.73	63.06	—	—	96	—	1.00	7.0	17	453	640
	LG630	2.59	4.75	60.61	—	—	97	—	0.85	7.2	20	465	665
Japan	LSG91H	2.77	5.90	84.44	35.21	0.24	105	224	1.03	—	—	465	—
USSR	KГCC7	2.96	—	65.02	26.01	0.25	105	—	—	—	—	—	—
	ГJIC14	2.73	5.50	70.02	27.56	0.23	108	—	—	6.9	15	450	640
France	MG915	2.58	4.04	64.72	—	—	96	—	—	—	—	—	—

Country	Sort	n_λ $\lambda=0.589\,\mu m$	n_λ $\lambda=1.06\,\mu m$	$(n_F - n_C)$ ×10⁻⁵	v	$\frac{dn}{dT}\cdot10^{-7}$ (K⁻¹)	$W\cdot10^{-7}$ (K⁻¹)	$C_{ij}\times10^{-7}$ (cm²/kg) $-C_1$	$-C_2$	$P\times10^{-7}$ (K⁻¹)	$Q\times10^{-7}$ (K⁻¹)	$n_2\cdot10^{-14}$ (esu)
USA	ED2	1.5716	1.5612	994	57.5	29	87	0.3	2.3	74	12	14.1
Germany	LG57	1.517	1.506	—	—	−21	28	—	—	—	—	—
	LG630	1.520	1.509	—	—	−22	28	—	—	—	—	—
Japan	LSG91H	1.559	1.5487	976	57.3	50	73	—	—	—	—	15.9
USSR	KГCC7	1.553	1.542	967	57.2	−36	—	1.4	3.8	18	4	—
	ГJIC14	1.536	1.525	942	56.9	−18	—	—	—	—	—	—
France	MG915	1.5164	1.5070	903	57.2	—	—	—	—	—	—	13.6

Table 14.4. Physical and optical properties of commercial phosphate laser glasses from different countries.

Country	Sort	ρ (g/cm³)	H	E (×10⁹ Pa)	G (×10⁹ Pa)	μ	α (10⁻⁷ K⁻¹)	C_p (J/kg·K)	K (W/m·K)	T_g (°C)	T_f (°C)
USA	EV2	2.71	2.65	35.40	13.73	0.24	155	199	0.60	379	467
	Q88	2.71	4.18	69.82	26.28	0.24	92	324	0.72	367	384
Germany	LG700	2.73	3.20	60.61	24.32	0.24	103	289	0.79	472	543
	LG710	2.96	2.40	43.15	17.16	0.26	132	—	0.54	341	—
Japan	LHG5	2.67	4.97	61.29	25.79	0.18	98	266	0.99	455	486
	LHG7	2.60	3.67	55.31	22.36	0.24	112	273	0.86	510	543
USSR	JITC56	2.70	4.00	66.10	26.48	0.25	128	—	0.56	440	550
	JITCH2	3.39	3.40	51.19	19.81	0.31	111	—	0.42	520	630

Country	Sort	n_λ $\lambda = 0.589\ \mu m$	n_λ $\lambda = 1.06\ \mu m$	$(n_F - n_c)$ ×10⁻⁵	ν	$\frac{dn}{dT}\cdot10^{-7}$ (K⁻¹)	$W\cdot10^{-7}$ (K⁻¹)	$C_{ij}\times10^{-7}$ (cm²/Kg) $-C_1$	$-C_2$	$P\times10^{-7}$ (K⁻¹)	$Q\times10^{-7}$ (K⁻¹)	$n_2\cdot10^{-14}$ (esu)
USA	EV2	1.511	1.5035	727	70.3	−110	−29	—	—	—	—	9.8
	Q88	1.5449	1.5362	841	64.8	−16	34	—	—	—	—	11.4
Germany	LG700	1.5253	1.5173	773	68	−27	—	—	—	—	—	10.8
	LG710	1.5387	1.5298	865	62.3	−72	—	—	—	—	—	12.8
Japan	LHG5	1.5409	1.5308	852	63.5	0	46	—	—	—	—	11.6
	LHG7	1.5131	1.5042	759	67.6	−29	22	—	—	—	—	10.5
USSR	JITC56	1.540	1.532	853	63.3	—	—	1.2	3.2	—	—	—
	JITCH2	1.578	1.568	885	65.3	—	3	2.4	3.7	—	—	—

Table 14.5. Spectral and laser properties of commercial laser glasses.

Company	Sort	$\Delta\lambda^F_{1.06}$ (nm)	$\Delta\lambda^F_{0.88}$ (nm)	λ_L (nm)	$\sigma_p \times 10^{20}$ (cm²)	τ (μs)	J-O ($\times 10^{-20}$ cm²)			β
							Ω_2	Ω_4	Ω_6	
	Silicate									
USA	ED-2	34.43	27.79	1061.0	2.698	359	3.24	4.59	4.80	0.484
Owens-Illinois	ED-3	34.71	28.18	1061.0	2.872	333	3.37	4.99	5.13	0.482
	ED-8	34.00	28.05	1061.0	3.024	316	2.98	4.90	5.23	0.486
Japan, Hoya	LSG91H	34.40	27.41	1061.5	2.422	412	3.49	3.99	4.40	0.489
Germany, Schott	LG650	34.30	23.53	1057.0	1.052	926	3.69	2.24	1.81	0.461
	Phosphate									
Japan	LHG5	26.07	21.96	1053.0	4.148	322	4.64	5.10	5.91	0.500
Hoya	LHG6	26.40	22.21	1053.0	4.125	318	4.10	5.46	5.27	0.492
	LHG7	26.41	22.19	1053.0	3.924	345	5.37	5.07	5.74	0.497
	LHG8	25.91	21.76	1053.0	4.000	338	4.40	5.10	5.60	0.493
USA, Kigre	C88	26.27	21.89	1054.0	3.998	326	3.33	5.12	5.63	0.494
USA	EV2	22.93	19.02	1054.0	4.294	372	3.98	4.57	5.56	0.504
Owens-Illinois	EV3	23.70	19.48	1054.0	4.395	340	4.02	5.25	5.69	0.492
	LC700	26.62	22.33	1053.5	3.673	363	4.22	4.63	5.40	0.500
Germany, Schott	LG710	24.92	20.87	1053.4	3.794	367	3.84	4.52	5.14	0.497
	Fluorophosphate									
	LHG104A	30.69	25.92	1051.0	2.772	465	1.77	4.18	5.03	0.505
Japan	LHG104B	31.25	26.56	1051.5	2.657	464	1.71	4.03	4.83	0.504
Hoya	LHG105A	30.19	25.28	1050.5	2.074	493	1.60	3.96	4.80	0.506
	LHG10	31.25	26.47	1050.5	2.670	461	1.67	4.10	4.86	0.504
	LHG11	31.20	26.44	1051.0	2.651	468	1.51	4.14	4.81	0.502
Germany, Schott	FK50	30.90	25.98	1054.5	2.703	444	2.06	4.20	4.62	0.495
	LG813	30.95	26.05	1051.0	2.668	488	1.86	4.05	4.94	0.506
USA	E111-1	27.69	23.38	1054.0	3.579	368	2.70	4.83	5.54	0.498
	E123-1	30.41	25.58	1052.5	2.946	417	1.91	4.34	5.10	0.502
Owens-Illinois	E133-5	30.75	26.11	1050.5	2.511	507	1.59	3.73	4.55	0.507
	E181-2	30.36	26.02	1050.0	2.563	504	1.54	3.91	4.57	0.503

a high resistance against glass melt corrosion and sufficient load ability at high temperature. We tested various refractory materials and synthesized high purity raw materials. A reasonable melting technique has also been developed [8].

The second was to eliminate the platinum impurity by using a gas atmosphere or using a protective coating so that the Pt-impurity would not enter the glass melt due to oxidation. We have also experimented with glass melting by self-heating [9]. In this method the glass was melted in a 20 MHz high frequency electromagnetic field by dielectric loss and eddy heating. The corrosion of crucible was less in the self-heating method.

With the improvements in the glass making process, the iron content of Nd-doped silicate glass melted in a platinum crucible was about 0.01%, the optical loss at 1.06 μm was about 0.1%cm^{-1}, and the lasing efficiency of a 16 mm diameter, 500 mm long glass rod (N_{330}) was close to 6%. The glass made in a ceramic crucible had an iron content of 0.02%, the optical loss was 0.2%cm^{-1}, and the lasing efficiency of a same-sized rod was 4%. The pulsed output power density was 1.6 GW/cm^2.

Phosphate and fluorophosphate glasses are highly corrosive to ceramic materials and therefore cannot be made in a ceramic system. Moreover, these types of glasses have a tendency to crystallize and have a small viscosity in pouring, so that it is difficult to obtain homogeneous pieces. After years of experimentation, we found that Nd-doped phosphate glass may be melted in a platinum crucible with protective coating and formed by leak-injection [10–12]. The phosphate glass absorbs water easily and the hydroxyl group in the glass can strongly quench the fluorescence of Nd-ions. A special dehydration process is therefore needed in the melting process [13, 14]. We can now produce large size and high quality Nd-doped phosphate glass. After fine annealing the residual stress birefringence was less than 1 nm/cm and the refractive index difference within a 30×15×3 cm^3 elliptical disk or ϕ7×52 cm^3 long glass rod was less than 2×10^{-6} [15]. The lasing efficiency of a 6 mm diameter and 100 mm long glass rod can be 1.3% (free oscillation) and the short pulse small signal gain of the rod is 0.16 cm^{-1}. The gain of a 200 mm diameter disk is 0.048 cm^{-1} after the elimination of parasitic oscillation.

Some of the glass performance characteristics depend on the glass composition, others are decided by the preparation technology used. Taking phosphate laser glass N_{21} for an example, the quality factors are as follows:

(1) *Water content.* The water content can be determined by the absorption coefficient at 3.5 μm due to OH-groups. It is generally less than 10 cm^{-1}; it can even be 6 cm^{-1} for some melts.

(2) *Fluorescence lifetime.* The lifetime of N_{2106} (Nd_2O_3=0.6wt%) is about 340 μs. For some glasses it can be up to 365 μs. The lifetime of N_{2122} is 320–340 μs.

(3) *Stress birefringence.* After special annealing, the birefringence was determined at the points symmetrically located at the edge for a 50×22×5 cm^3 block of glass and an elliptical disk. All the stress birefringence values measured are less than 1 nm/cm. Though it is difficult to measure the magnitude of the stress birefringence for rods of glass, the stress distribution could be observed by placing the rod between two crossed polaroids. A symmetrical distribution of the stress

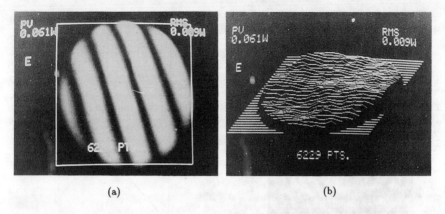

(a)   (b)

Fig. 14.1 Interference pattern (a) and optical path difference diagram (b).

birefringence for a $55 \times 9 \times 8$ cm^3 glass block can be obtained.

(4) *Optical homogeneity.* Using a Zygo interferometer with a He-Ne laser as light source, an elliptical disk of $40 \times 20 \times 5$ cm^3 was used for the test. The interference pattern and the three-dimensional optical path difference diagram (Fig. 14.1 (a) and (b)) show that the peak-valley value of the wave form distortion is 0.061 λ for 5 cm optical paths, and the refractive index variation is estimated to be about 1×10^6. A 10 cm aperture disk amplifier which consists of 6 elliptical disks of 2.5 cm in thickness was tested by a shearing interferometer. The shearing interferogram shown in Fig. 14.2 displays a symmetrical waveform without local distortion.

(5) *Amplification performance of laser.* The performance of the laser as an amplifier depends on the laser configuration, optical pumping, and the properties and quality of the laser glass. As an experimental result, Fig. 14.3 shows the relation between the gain coefficient and the optical pumping density for a 10 cm aperture disk amplifier.

14.3 Laser Performances of Neodymium Laser Glasses [17]

In this section we will review the main laser performances of neodymium glasses and present our experimental results.

14.3.1 *Laser Output Wavelength and Linewidth*

The spectra of laser light depend on the resonator cavity structure and the output energy. At the same laser operation conditions, as shown in Tables 14.6 and 14.7

Fig. 14.2. Shearing interferogram of 10 cm aperture disk amplifier.

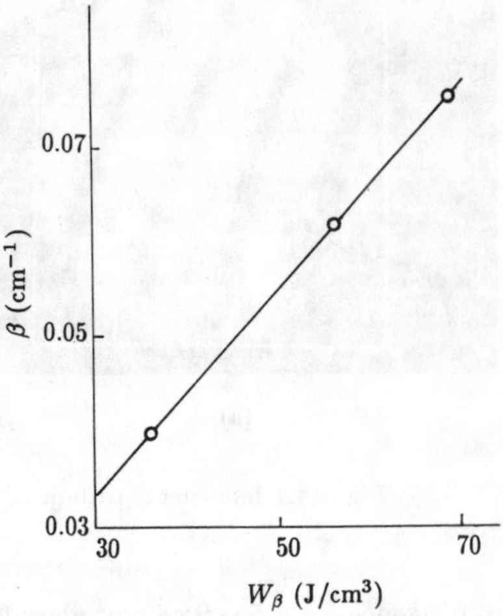

Fig. 14.3. Relationship between the gain coefficient β (cm^{-1}) and pumping energy density W_0 (J/cm^2) for phosphate laser glass amplifier with optical aperture of 20 cm in diameter [16].

Table 14.6. Central wavelength and linewidth of spontaneous and stimulated emission of Nd-doped silicate glasses.

Sort of Nd-glass	SiO_2 content (% mol)	Spontaneous λ_L (nm)	$\Delta\nu_L$ (nm)	Stimulated λ_P (nm)	$\Delta\nu_P$ (nm)
N_{0712}	85	1057	22	1058.4	6.2
N_{0412}	80	1059	27	1061.3	9.0
N_{0312}	75	1061	29	1062.3	9.4

Table 14.7. Central wavelength and linewidth of spontaneous and stimulated emissions of inorganic Nd-doped glasses.

Glass system	Spontaneous λ_L (nm)	$\Delta\nu_L$ (nm)	Stimulated λ_P (nm)	$\Delta\lambda_P$ (nm)
BeF_2-based fluoride	10492	22	1049.5	5.8
Metaphosphate	10541	23	1055.3	6.3
Alkali–silicate	10611	27	1062.2	9.2
BaO–borate	10625	29	1063.6	9.5

Table 14.8. Change of laser central wavelength and linewidth with laser output energy of N_{721} Nd-doped glass *.

Laser output energy (J)	232	565	810	1095
Central wavelength (nm)	1058.15	1058.67	1058.79	1058.74
Linewidth (nm)	3.98	5.25	5.4	6.54

* Nd-glass rod $\phi 30 \times 100$ mm.

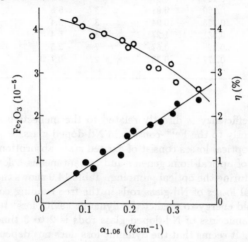

Fig. 14.4. Relationship between Fe_2O_3 content, ineffective absorption coefficient ($\alpha_{1.06}$) and laser output efficiency (η). o Ineffective absorption coefficient with laser efficiency; • Fe_2O_3 content with ineffective absorption coefficient.

the laser central wavelength and linewidth correspond to that of the luminescence spectra, which depend on the chemical composition of Nd-doped glasses. In silicate glasses the central wavelength and linewidth increase with the decreasing SiO_2 content. The laser linewidth is much shorter than that of the luminescence and the laser central wavelength shifts to the longer wavelength side. With increasing laser output energy E_p the laser linewidth $\Delta\nu_p$ broadens while the central wavelength changes not so much, it means that more oscillation modes are generated (Table 14.8). All the measurements were carried out at the free running condition with F-P resonator.

14.3.2 Laser Output Efficiency

The laser output efficiency depends on the static and dynamic optical losses of Nd-doped glass rods or disks in the resonator cavity. The static loss of Nd-doped glass is aroused by macro-inhomogeneities, such as particles, bubbles, striae and stress in the glass, and ineffective absorption at the laser output wavelength. Ferrous ions Fe^{2+} in glass have a strong absorption band around 1.1 μm. As shown in Fig. 14.4,

Table 14.9. Dynamic optical loss of several Nd-glass rods ($\phi 20 \times 500$ mm).

Type of glass	Static loss α (% cm^{-1})	Dynamic loss γ (% cm^{-1})	Ratio γ/α	Thermo-optical coefficient W ($\times 10^{-6}$ °C^{-1})
N_{0330}	0.214	0.67	3	5.8
N_{0312}	0.260	0.91	3.5	5.8
N_{1012}	0.328	0.94	3	5.4
N_{1030}	0.36	1.27	4	5.4
N_{0812}	0.32	0.67	2	2.5
N_{2412} (phosphate)	0.177	0.33	2	0.7

the laser output efficiency is directly related to the ineffective absorption at 1.06 μm and subsequently to the Fe^{2+}-content in Nd-doped glass.

The dynamic optical losses consist of excited state absorption of Nd^{3+}-ions, dynamic distortion of optical homogeneity and instantaneous color center formation by UV radiation during the optical pumping. Table 14.9 shows the measured results of dynamic optical losses of Nd-glass rods in the free running condition. We measured the threshold energy to predict the dynamic cavity loss. It is worth pointing out that the dynamic loss of Nd-doped glass rods is 2 to 3 times larger than the static optical loss. It seems that the dynamic loss does not depend on the Nd^{3+}-ion concentration, but relates to the thermo-optical coefficient W of glass, it means that the thermal distortion of optical homogeneity of the rod is the dominant source.

As shown in Fig. 14.5, due to the dynamic optical loss the laser output efficiency is not proportional to the input pumping energy, and at high pumping energy the laser output efficiency even decreases.

14.3.3   *Laser Beam Divergence*

The static optical homogeneity, of course, shows great influence on the laser beam divergence. Figure 14.6 illustrates the optical quality of two neodymium glass rods measured by different optical methods. The experimental results of divergence angle of the laser output beam at different laser output energy are listed in Table 14.10, and the influence of optical quality on the divergence angle is quite obvious.

Systematic study was made of the careful observation and quantitative measurement of laser beam divergence change caused by thermal distortion of laser glass under optical pumping and interaction between intense laser light and the glass medium.

Figure 14.7 shows the radial distribution of the optical path change $\Delta S(r,t)$ of a Nd-glass N_{0324} rod ($\phi 16 \times 200$ mm) at different pumping duration (optical pumping energy density, 70 J/cm^3) measured by interferometry and high speed photography. The average divergence angle θ caused by thermal distortion can be calculated by

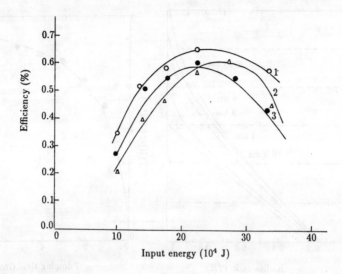

Fig. 14.5. Laser output efficiency curve of several Nd-doped glass rods. Nd-glass rod $\phi 16 \times 500$ mm. 1, D-904; 2, D-912; 3, A7210.

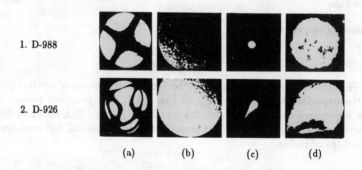

| | (a) | (b) | (c) | (d) |

Fig. 14.6. Comparison of optical quality of neodymium glass D-988 and D-926. 1, D-988; 2, D-926. (a) Stress pattern; (b) optical projection; (c) focusing point; (d) near field laser pattern.

Table 14.10. Laser divergence angle $\Delta\theta$ of two glass rods ($\phi 30 \times 1000$ mm) at different laser output energy E_{out}.

Number of rod	E_{out}(J)	200	500	1000	1500	2000
D-988	$\Delta\theta$	1.26	1.75	2.6	3.6	5.3
D-926	(mrad)	4.77	5.25	5.6	6.3	6.3

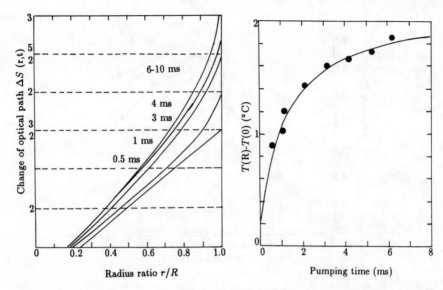

Fig. 14.7. Radial distribution of optical path change during different pumping (pumping density 70 J/cm^3) time of neodymium glass rod N$_{0324}$.

Fig. 14.8. Temperature difference between center and edge of the rod at different pumping duration (pumping density 70 J/cm^3).

the following equation:

$$\theta = \frac{2}{R^2} \int_0^R r \cdot \frac{ds}{dr} dr, \tag{14.1}$$

where R is the radius of rod, ds/dr is the radial distribution of optical path. By using the experimental results shown in Fig. 14.7, in a pumping period of 6–10 ms the divergence angle θ is about 0.3 mrad. Neglecting the effect of Nd^{3+}-ion concentration and thermal stress, the temperature difference ΔT between the center and the edge of rod is

$$\Delta T = T(R) - T(0) = [S(R) - S(0)]/LW, \tag{14.2}$$

where L is the length of rod, W is the thermo-optical coefficient without thermal stress, $S(R)$ and $S(0)$ are the overall optical path change at the center and the edge of the rod, respectively. Figure 14.8 shows the calculated temperature difference between the center and the egde of the rod during optical pumping. Figure 14.9 illustrates the relationship between the central temperature of the rod T and the pumping energy density $E_{\rm p}$.

The dynamic optical distortion induced by the optical pumping depends on the thermo-optical properties and the neodymium-ion concentration of the host glass.

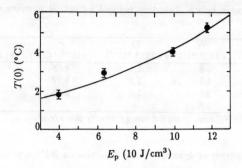

Fig. 14.9. Relationship between the temperature at center of the rod $T(0)$ and the pumping density (E_p).

We measured different thermo-optical properties of Nd-doped glasses, such as the temperature coefficient of refractive index (β), the stress birefringence coefficient (Q), stress thermo-optical coefficient (P) and the thermo-optical coefficient without stress (W). These constants have been discussed in Chapter 5.

The overall optical path difference $\Delta S_{r,\theta}$ and the stress-birefringence change $\Delta S_r \pm \Delta S_\theta$ can be expressed by the aboved-mentioned thermo-optical coefficients,

$$\Delta S_{r,\theta} = l[W\overline{T(R)} + P + \beta(T(r) - \overline{T(R)}) \pm Q(T(r) - \overline{T(r)})], \tag{14.3}$$

$$(\Delta S_r - \Delta S_\theta)/2l = Q[T(r) - \overline{T(r)}], \tag{14.4}$$

$$(\Delta S_r + \Delta S_\theta)/2l = W\overline{T(R)} + \beta + P[T(r) - \overline{T(R)}]. \tag{14.5}$$

From Table 14.11 it can be seen that the overall optical path change or the divergence angle of laser beam actually depends on the thermo-optical coefficients of the host glasses.

The polarizability of Nd^{3+}-ions is different when they are at the ground state or at the excited state. Therefore, the inhomogeneous excitation of Nd^{3+}-ions in the glass rod initiates the instantaneous refractive index gradient. According to Riedel and Baldwin's calculation [18].

$$\Delta n \sim 5 \times 10^{-24} N_3(t), \tag{14.6}$$

where $N_3(t)$ is the number of Nd^{3+}-ions at the excited state. Table 14.12 shows that the overall optical path change ΔS and the beam divergence angle $\Delta\theta$ depend obviously on the Nd^{3+}-concentration.

The optical distortion of laser glass induced by the intense laser pulse ($\sim$ms pulse-duration) is caused by the absorption of laser pulse energy in the glass. It can be clearly observed in the laser amplifiers. Figure 6.16 shows the diffusion process

Table 14.11. Dynamic optical distortion induced by optical pumping of several neodymium glasses *.

Type of Nd-glass	W ($\times 10^{-6}$ °C)	Q ($\times 10^{-6}$ °C^{-1})	P ($\times 10^{-6}$ °C^{-1})	ΔS (at 632.8 nm)	$\Delta\theta$ (mrad)
N$_{0312}$	5.8	1.1	4.68	13.0	1.6
N$_{0712}$	4.5	1.3	5.32	10.0	1.0
N$_{0812}$	2.5	1.0	6.20	7.0	0.4
T-32	0.6	—	—	3.0	0.2

* Glass rod $\phi 16 \times 200$ mm, pumping energy density 200 J/cm^3.

Table 14.12. Dependence of dynamic optical distortion on Nd^{3+}-concentration of Nd-glass rod $\phi 10 \times 220$ mm.

Type of glass	Nd^{3+}-concentration (wt%)	ΔS (at 632.8 nm)	$\Delta\theta$ (mrad)
N$_{0306}$	0.6	2.5	0.5
N$_{0312}$	1.2	3.0	0.5
N$_{0324}$	2.4	4.0	0.7
N$_{0350}$	5.0	6.0	2.0

of the focusing point when an intense free running pulsed laser beam is passing through. The laser divergence angle $\Delta\theta$ changes with time, as shown in Fig 14.10. This distortion of optical homogeneity of glass is a thermal one. It means that the glass medium has absorbed laser energy and formed a temperature gradient, hence a refractive index gradient. Therefore this kind of optical distortion in laser glass depends mainly on the absorption coefficient (α) and the temperature coefficient of refractive index dn/df of the glass. Figure 14.11 illustrates the relationship between the divergence angle $\Delta\theta$ and the material constant K of different glasses,

$$K \equiv \frac{dn}{dT} \cdot P_0 t (1 - e^{-\alpha l}) / \rho C_p n_0. \tag{14.7}$$

where P_0 is the laser power density (W/cm^2), ρ is the glass density, C_p is the specific heat, n_0 is the refractive index, l is the length of glass rod. As we considered that the thermal diffusion process is an adiabatic one, therefore, the divergence angle of the laser output beam increases with increasing laser output energy and also with the laser pulse time delay.

14.3.4 Degradation of Laser Performances

We found the color center formation in neodymium glass, which is induced by UV radiation during the optical pumping very early (1964). The color centers can be accumulated and saturated after optical pumping many times. The experimental results are shown in Fig. 14.12. Because of absorption of pumping energy in the

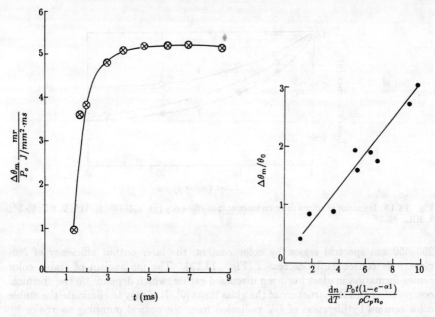

Fig. 14.10. Change of $\Delta\theta_m/P_0$ with time t. Sample: optical glass BaK-7; $P_0 = 0.25$ J/mm²·ms.

Fig. 14.11. Relationship between $\Delta\theta_m/\theta_0$ and dn/dT·$P_0t(1-e^{\alpha l})/\rho C_p n_0$.

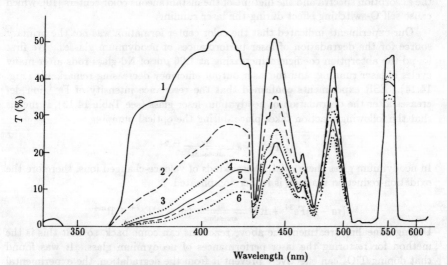

Fig. 14.12. Effect of color center accumulation and saturation. 1, Before optical pumping; 2, 3, 4, 5 and 6 after 2, 4, 6, 8 and 10 times of pumping respectively.

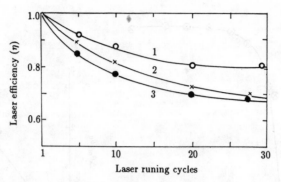

Fig. 14.13. Influence of color center on laser output efficiency (η). 1, Nd30: 65-12$^{\neq}$; 2, HC: 65-3$^{\neq}$; 3, BR$_4$: 65-7$^{\neq}$.

250–450 nm spectral region by color centers, the laser output efficiency of Nd-glass rods consequently decreases (Fig. 14.13). The mechanism of stable color center formation in glass has been discussed earlier, which depends on the chemical composition and the structure of the glass hosts [6]. It is easy to eliminate the stable color centers by filtration of UV radiation from the optical pumping source or by the addition of cerium compounds in the glass. There also exist instantaneous color centers, and their absorption band lies near 1.06 μm. Landry and Snitzer studied the absorption spectra and the lifetime of the instantaneous color centers [19], which cause self Q-switching effect during the laser running.

Our experiments indicated that the color center formation was not the primary source for the degradation of laser performances of neodymium glasses. We first found the absorption coefficient increasing at 1.06 μm of Nd-glass rods after many cycles of laser running, and the laser output efficiency decreasing remarkably (Fig. 14.14). ESR experiments confirmed that the resonance intensity of Fe^{3+}-ion decreases after the degradation of neodymium laser glass (see Table 14.13), it means that the following reaction takes place during the optical pumping.

$$Fe^{3+} + e \underset{\text{annealing}}{\overset{h\nu}{\rightleftharpoons}} Fe^{2+}$$

In neodymium glass there exist several kinds of valence-changed ions, therefore the oxidation-reduction reaction is rather complicated.

$$(m-n)Fe^{3+} + R^{n+} \underset{\text{annealing}}{\overset{h\nu}{\rightleftharpoons}} (m-n)Fe^{2+} + R^{m+}$$

During annealing treatment the above reactions can come back, so that this is the method for restoring the laser performances of neodymium glass. It was found that doping TiO_2 can effectively prevent it from the degradation, the experimental results are shown in Table 14.14. Good results can also be got if the filtration of UV radiation from the pumping source is applied.

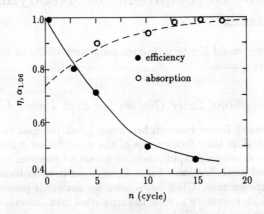

Fig. 14.14. Relationship between laser output efficiency (η), absorption coefficient ($\alpha_{1.06}$)and laser running cycles (n). • η; ○ $\alpha_{1.06}$.

Table 14.13. Change of absorption coefficient $\alpha_{1.06}$, laser output efficiency η_{out} and ESR signal intensity of Fe^{3+} in glass after 40 cycles of laser running and annealing.

No. sample	$\alpha_{1.06}$ (%cm^{-1})			η_{out} (%)			I_{ESR} (Fe^{3+})		
	Before	After 40 cycles	After annealing	Before	After 40 cycles	After annealing	Before	After 40 cycles	After annealing
1	0.21	0.27	0.21	2.0	1.4	2.0	45.6	32.1	43.2
2	0.47	0.65		0.9	0.7		61.5	37.1	
No.2 with $K_2Cr_2O_7$ filtration	0.47	0.47		0.9	0.9		61.5	63.6	

* Nd-glass rod $\phi 16 \times 200$ mm.

Table 14.14. Effect of doping TiO_2 on the degradation of neodymium glass *.

TiO_2 content (wt%)	$\alpha_{1.06}$ (%)		η_{out} (%)	
	before	after	before	after
0	0.47	0.67	0.90	0.70
0.005	0.25	0.25	1.20	1.20
0.1	0.25	0.25	1.00	1.00
0.2	0.25	0.25	1.30	1.30

* Nd-glass rod $\phi 10 \times 100$ mm.

14.4   New Development of Neodymium Laser Glasses

In this section we would like to introduce some new efforts in the development of neodymium laser glasses.

14.4.1   Neodymium Laser Glasses for High Power Laser Systems

Inertial confinement fusion research has made great progress in recent years [20, 21]. High power glass laser facility, one of the main fusion driver candidates, and slab glass lasers developing rapidly require an active medium with excellent laser performances and good thermomechanical properties in order to increase the output power and repetition rate. Glass lasers used for materials processing also require high thermal shock resistance to raise the repetition rate. Several types of new laser glasses with improved thermal shock resistance have been investigated to meet this requirement [22, 23].

It is well known that the glass strength and the chemical durability would increase and the thermal expansion coefficient decrease by introducing Al_2O_3 or B_2O_3 into phosphate glasses. But the fluorescence quenching effect by Al_2O_3 and specially by B_2O_3 is rather pronounced [24]. A wide variety of aluminophosphate and borophosphate laser glasses have been investigated to search glasses with high stimulated emission cross section, low fluorescence quenching effect and excellent thermomechanical properties [25, 26].

Figures 14.15 and 14.16 show the concentration quenching effect of Nd^{3+} in aluminophosphate and borophosphate glasses. Generally, the influence of Al_2O_3 and B_2O_3 on fluorescence lifetime of Nd^{3+} in phosphate glasses is much less than that in silicate glasses for all the glass systems studied, and specially in the glasses containing low charge and large radius cations, such as K^+ and Ba^{2+}. From the experimental results for the spectral and thermal properties, it is estimated that a neodymium-containing glass with a stimulated emission cross section σ_p of 3–4×10^{-20} cm^2, a fluorescence lifetime τ of 170–180 μs at 10^{21} Nd ions/cm^3 and expansion coefficient of 80–100$\times 10^{-7}$°C^{-1} could be obtained from these glass systems. Table 14.15 lists the spectral and physical properties of alumino-silicate and borosilicate glasses.

14.4.2   High Concentration Neodymium Laser Glasses

We found that the concentration quenching of luminescence of Nd-doped glasses depends on the hosts greatly (Fig. 14.17). With increasing P_2O_5 concentration in phosphate glasses, the concentration effect weakens [27]. High concentration Nd-doped phosphate laser glasses have been developed, and the laser performances of these glasses for high repetition rate operation are shown in Table 14.16.   The

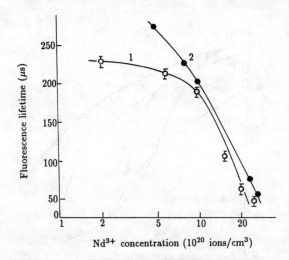

Fig. 14.15. Concentration quenching of Nd^{3+} in (1) $18K_2O \cdot 10Al_2O_3 \cdot 6Ln_2O_3 \cdot 66P_2O_5$; (2) $26Li_2O \cdot 10Ln_2O_3 \cdot 64P_2O_5$.

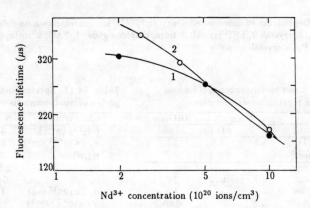

Fig. 14.16. Comparison of Nd^{3+} quenching curves in two borophosphate glasses. The compositions (in mol%) of samples 1 and 2 are as follows: 1, $66P_2O_5$, $10B_2O_3$, $6Ln_2O_3$ (=$La_2O_3 + Nd_2O_3$), $18K_2O$; 2, $68.1P_2O_5$, $11.8Al_2O_3$, $6Ln_2O_3$, $14.1K_2O$.

Table 14.15. Spectral and physical properties of alumino-silicate and borosilicate glasses.

Glass composition	$\tau(\mu s)$	$\sigma_p(10^{-20}$ cm$^2)$	$\alpha_{20/100}$ (K^{-1})	$T_g(°C)$
$18K_2O \cdot 10Al_2O_3 \cdot 6Ln_2O_3 \cdot 66P_2O_5$	182	3.8	80	470
$18K_2O \cdot 10B_2O_3 \cdot 6Ln_2O_3 \cdot 66P_2O_5$	189	4.5	100	445

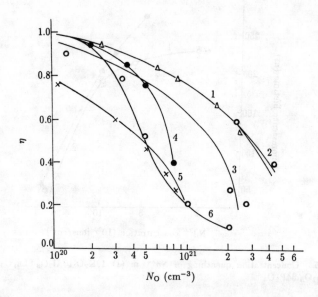

Fig. 14.17. Dependence of quantum efficiency and Nd^{3+} ion concentration on different hosts. 1, $Nd_x La_{(1-x)} P_5 O_4$ crystal; 2, LNP crystal; 3, tetraphosphate glass; 4, YAG; 5, metaphosphate glass; 6, $Nd_x La_{(1-x)} P_3 O_9$ crystal.

Table 14.16. Laser performances of Nd-doped glasses for high repetition rate operation.

Glass type	N_{2135}		NHG_{2390}
Rod size (mm)	$\phi 6 \times 100$	$\phi 4 \times 60$	$\phi 4 \times 60$
Threshold (J)	3.2	1.5	1.5
(at 1 pulse/sec)			
Efficiency (%)			
at 5 pulse/sec	2.0	1.72	1.6
at 10 pulse/sec			1.4
at 20 pulse/sec			0.86

Table 14.17. Spectral and physical properties of neodymium laser glasses.

Glass type	N_{2135}	N_{2390}
Nd_2O_3 (wt%)	3.5	9.0
σ_p ($\times 10^{-20}$ cm^2)	3.5	4.3
τ (μs)	300	160
n_d	1.574	1.566
ν_d	64.8	63.2
n_2 ($\times 10^{-13}$ esu)	1.3	1.2
α ($\times 10^{-7}$ °C^{-1})	120	99

spectral and physical properties of two types of neodymium laser glasses are shown in Table 14.17 [28].

The high concentration neodymium laser glasses have been used for materials processing with high repetition rate. For example, the laser glass rod with size of $\phi 0.8 \times 20$ cm^3 and glass slab with size of $0.5 \times 3 \times 14$ cm^3 can deliver 5–6 J/pulse and 12–15 J/pulse respectively at the repetition rate of 3 pps (pulse per second).

Table 14.18. Spectral and optical properties of different glass systems [24, 29, 30].

Glass	Refractive index	Cross section σ ($\times 10^{-18}$ cm^2)	Wavelength λ_p (nm)	Linewidth $\Delta\lambda_{\text{eff}}$ (nm)	Radiative lifetime T_R (μs)
Oxides					
Silicate	1.46–1.75	0.9–3.6	1057–1088	34–55	170–1090
Germanate	1.61–1.71	1.7–2.5	1060–1063	36–43	300–460
Tellurite	2.0–2.1	3.0–5.1	1056–1063	26–31	140–240
Phosphate	1.49–1.63	2.0–4.8	1052–1057	22–35	280–530
Borate	1.51–1.69	2.1–3.2	1054–1062	34–38	270–450
Halides					
Fluoroberyllate	1.28–1.38	1.6–4.0	1046–1050	19–29	460–1030
Fluorozirconate	1.52–1.56	2.9–3.0	1049	26–27	430–450
Fluorohafnate	1.51	2.6	1048	26	520
Fluoroaluminate	1.41–1.48	2.2–2.9	1049–1051	30–33	420–570
Chloride	1.67–2.06	6.0–6.3	1062–1064	19–20	180–220
Oxyhalides					
Fluorophosphate	1.41–1.56	2.2–4.3	1049–1056	27–34	310–570
Chlorophosphate	1.51–1.55	5.2–5.4	1055	22–23	290–300
Chalcogenides					
Sulfide	2.1–2.5	6.9–8.2	1075–1077	21–22	64–100
Oxysulfide	2.4	4.2	1075	27	92

14.4.3 Neodymium Oxyhalide and Halide Glasses

For high performance lasers the high stimulated emission cross section σ_p and low nonlinear refractive index n_2 are required. In order to reduce the laser self-focusing effect the figure of merit FOM of laser glass disk at Brewster's angle used for high power laser amplifiers is equal to n^2/n_2, where n is the linear refractive index. In Table 14.18 are summarized the spectral and optical properties of different neodymium glass systems.

From Table 14.18 it can be seen that neodymium fluoroberyllate glasses posses high FOM value and suitable σ_p value, but due to poison of beryllium compounds it is difficult to prepare fluoro-beryllate glasses. Thus the fluorophosphate glasses are the second candidate for high power laser glass hosts. Figures 14.18 and 14.19 show the fluorescence lifetime and stimulated emission cross section of Nd^{3+} vs Z/a^2 in fluorophosphate glass system $18MgF_2 \cdot 18CaF_2 \cdot 8SrF_2 \cdot 10BaF_2 \cdot 22.5AlF_3 \cdot 13.5NaPO_3$ (mol%) with Nd^{3+} concentration around 1×10^{20} cm^{-3} [31]. Figure 14.20 shows the concentration quenching of Nd^{3+}-fluorescence in fluorophosphate glass.

Based on the composition effect of spectral properties and other physical properties of glass the practical glass composition and technology of fluorophosphate glass were determined. The properties of applicable fluorophosphate laser glass are listed in Table 14.19. For comparison the physical property data of LHG-10 of Hoya fluorophosphate laser glass are also shown in Table 14.19. It indicated that our

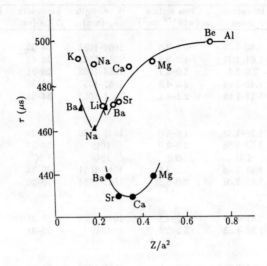

Fig. 14.18. Fluorescence lifetime of Nd^{3+} vs Z/a^2.

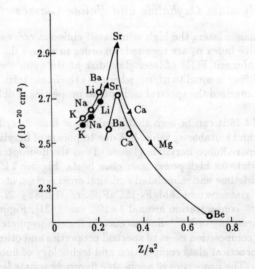

Fig. 14.19. Stimulated emission cross section of Nd^{3+} at 1.05 μm vs Z/a^2.

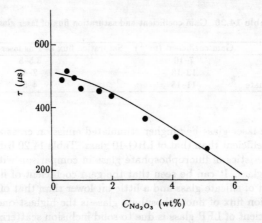

Fig. 14.20. Concentration quenching of Nd^{3+}-fluorescence in fluorophosphate glass.

Table 14.19. Physical properties of fluorophosphate glasses.

Type of glass	LFP	LHG-10
Nd_2O_3 (wt%)	2.0	2.4
Stimulated emission cross section (10^{-20} cm^2)	2.84	2.7
Fluorescence lifetime (μs)	405	384
Principal fluorescence peak (nm)	1053	1051
Principle fluorescence peak linewidth (FWHM) (nm)	26.2	24.1
Attenuation coefficient at lasing wavelength (cm^{-1})	<3	0.15
n_D	1.480	1.467
Abbe value	83.9	87.7
Nonlinear refractive index n_2 (10^{-13} esu)	0.686	0.61
Temperature coefficient of refractive index dn/dt (10^{-70} °C^{-1})	−79	—
Coefficient of linear thermal expansion (10^{-7} °C^{-1})	157	—
Temperature coefficient of optical path length (10^{-7} °C^{-1})	−10	—
Transformation temperature T_g (°C)	420	445
Softening point T_f (°C)	465	470
Density (g/cm^3)	3.52	3.64
Young's modulus E (10^3 N/mm^2)	80.9	72.5
Poisson's ratio μ	0.28	0.3

Table 14.20. Gain coefficient and saturation flux of laser glasses.

Glass	Gain coefficient (m^{-1})	Saturation flux at 1 ns laser pulse (J/cm^2)
Silicate	7–10	3.5–5
Phosphate	13–18	2–4
Fluorophosphate	11–15	4–5

fluorophosphate laser glass has higher stimulated emission cross section and lower thermo-optic coefficient than that of LHG-10 glass. Table 14.20 lists the laser amplification characteristics of fluorophosphate glass in comparison with that of silicate and phosphate glass. It can be seen that the gain coefficient of fluorophosphate is higher than that of silicate glass and a little bit lower than that of phosphate glass, but the saturation flux of fluorophosphate glass is the highest one. The higher attenuation coefficient of LFP glass is due to solid inclusion scattering. The inclusion density is about 10^4–10^5 pieces/cm^3, the laser induced damage threshold is around 5–7 J/cm^2 (1 ns), which is lower than that of silicate laser glasses. This problem has to be solved.

The large stimulated emission cross sections obtainable in chloride and sulfide glasses are of interest for low threshold, high gain laser applications. They are attractive for miniature and fiber lasers.

References

1. E. Snitzer, *Phys. Rev. Lett.* **7** (1961) 444.

2. Fuxi Gan, Zhonghong Jiang and Yingshi Cai, *Science Bulletin* **1** (1964) 54.

3. Yingshi Cai, Xishan Li and Fuxi Gan, *Science Bulletin* **12** (1964) 1112.

4. Fuxi Gan *et al.*, *Science Bulletin* **11** (1965) 1012.

5. Fuxi Gan, *Revue Roumaine de Physique* **33** (1988) 597.

6. Fuxi Gan, *Optical Glasses* (Science Press, Beijing, 1964).

7. Zhonghong Jiang, Xiuyu Song and Junzhou Zhang, *J. Chinese Silicate Soc.* **8** (1980) 1.

8. Collected Research Reports of the Shanghai Institute of Optics and Fine Mechanics, Vol. 2, Neodymium Glasses, 1974.

9. Hongwei Sun, Yasi Jiang and Xingyuan Hu, *Chinese J. Silicate Soc.* **7** (1979) 255.

10. Yasi Jiang, Junzhou Zhang *et al.*, *J. Non-Cryst. Solids* **80** (1986) 623.

11. Yasi Jiang, Junzhou Zhang and Xiongxin Ying, *Glass and Enamel.* **12(2)** (1983) 9.

12. Shibin Jiang, Yasi Jiang and Junzhou Zhang, *Chinese J. Lasers* **18** (1991) 913.

13. Zhongya Li, Zexing Chen and Junzhou Zhang, *Acta Optica Sinica* **4** (1984) 563.

14. Dunshui Zhuo, Wenjuan Xu and Yasi Jiang, *Chinese J. Lasers* **12** (1986) 173.

15. Yasi Jaing and Jie Li, Collected Papers, XIV Intl. Congre. on Glass, New Dehli, 1986, Vol. 2, p. 118.

16. Yuxia Zheng, Wenyan Yu, Dianyuan Fan and Xianghong Tang, *Acta Optica Sinica* **6** (1986) 381.

17. Collected Research Reports of the Shanghai Institute of Optics and Fine Mechanics, Vol. 8, Laser Materials and Components, 1980.

18. E. P. Riedel and G. B. Baldwin, *J. Appl. Phys.* **38** (1967) 2720.

19. R. L. Landry, E. Snitzer, *J. Appl. Phys.* **42** (1971) 3827.

20. S. E. Bodner, *Physics Today* **41** (1988) 65.

21. E. M. Campbell and D. L. Correll, Annual Report of Inertial Confinement Fusion, Lawrence Livermore National Laboratory, 1992, UCRL-LR-105820-92.

22. J. S. Hayden, D. L. Sapak, *et al., Proc. SPIE* **1021** (1988) 36.

23. S. E. Stokowski, Annual Report of Lawrence Livermore National Laboratory, 1987, p. 957.

24. Fuxi Gan, *Optical and Spectroscopic Properties of Glass* (Springer-Verlag Publisher, Berlin, 1992) pp. 148–203.

25. Yasi Jiang, Shibin Jiang and Yanyan Jiang, *J. Non-Cryst. Solids* **112** (1989) 286.

26. Shibin Jiang and Yasi Jiang, *Glastech. Berichte* **64** (1991) 291.

27. Changhong Qi and Fuxi Gan, *J. Lasers* **9** (1982) 691.

28. Yasi Jiang, Sen Mao, *et al., Chinese J. Lasers* **16** (1989) 609.

29. S. E. Stokowski, R. A. Saroyan and M. J. Weber, *Nd-doped laser glass spectroscopy and physical properties*, Lawrence Livermore National Laboratory Reports M-095, 2nd Revision (1981).

30. J. C. Mickel, J. Flahaut, *et al., C. R. Acad. Sc. Paris* **291** (1980) C-21.

31. Yasi Jiang, Junzhou Zhang, *et al., Acta Optica Sinica* **10** (1990) 452.

15. Laser Glass Fibers

In early 1960s Koester and Snitzer observed amplification in a Nd^{3+}-doped silicate glass fiber with 47 dB gain at 1.06 μm wavelength, using a 1 m long fiber coiled around a flashlamp [1]. Due to the high inactive loss of multi-component silicate glass fiber, the laser efficiency was too low, the practical application has not been made. Since 1986 a new technique for fabrication of low loss rare earth (RE) ions doped fibers has led to a prolific generation of novel fiber laser sources. Recently rare-earth-doped glass fiber lasers and amplifiers have drawn great attention as promising contenders for future optical fiber transmission systems. Tunable fiber lasers with low threshold, high efficiency and optical fiber amplifiers with high gain, low noise and large bandwidth have been demonstrated with different pumping sources.

At present, optical fiber communication systems work in the 0.85, 1.3, 1.55 μm wavelength window. Up to now, in the 1.3 and 1.5 μm window, the semiconductor optical amplifiers are available, but they are too sensitive to polarization. Later on, Er-doped silica fiber amplifiers emerged for 1.55 μm window, their performances are very attractive (high gain, high saturation output power, polarization insensitive, etc.). At present one can buy Er-doped silica fibers. The Pr-doped fluoride glass fiber amplifiers are the promising candidates for 1.3 μm window. Infrared glass fiber lasers (1.5–3 μm) are prospective laser sources for medical and remote sensing applications.

In this chapter the status of development of glass fiber for lasers and amplifiers are reviewed and some new experimental results are reported.

15.1 Rare Earth Ions Doped Silica Based Glass Fibers

15.1.1 *Doped Glass Fiber Fabrication*

Silica-based glass fibers were prepared by MCVD method. Soot type preform was used, and RE ions were doped by dipping the soot preform in the RE containing solution. The soot preform was dehydrated in chlorine-oxygen atmosphere, then sintered at high temperature. The characteristics of Nd^{3+}-doped and Er^{3+}/Yb^{3+}-codoped single mode silica glass fibers are shown in Table 15.1 [2].

326

Table 15.1. Characteristics of Nd^{3+}-doped and Er^{3+}/Yb^{3+}-codoped silica glass fibers.

Doped ion	Core composition	Core diam. (μm)	Cladding composition	Outside diam. (μm)	Cutout wavelength (μm)
Nd^{3+}	GeO_2–SiO_2	4–6	SiO_2	120–150	1.3, 1.5
Er^{3+}/Yb^{3+}	GeO_2–SiO_2	4–8	SiO_2	120–160	1.5

Doped ion	Δn (%)	Optical loss (dB/km) at wavelength (μm)				
		800	1000–1100	900	15300	1100–1300
Nd^{3+}	0.8–1.6	10^3–10^5	<10			
Er^{3+}/Yb^{3+}				2×10^3–7×10^4	3×10^2–4×10^3	<10

15.1.2 Laser Performances of Rare Earth Doped Silica Based Glass Fibers

In Table 15.2 the laser output performances and pumping condition of RE-doped silica glass fiber lasers developed up to now have been listed. Among rare earth doped fiber lasers the Nd^{3+} and Er^{3+}-doped glass fiber lasers are the most important for application. We will introduce several Nd^{3+} and Er^{3+}-doped glass fiber lasers developed in China in this section.

A. Laser characteristics of Nd^{3+}-doped fibers pumped by Ar ion laser [20, 21]

Figure 15.1 shows the absorption and luminescence spectra of Nd^{3+}-doped silica fiber excited by Ar ion laser. The luminescent peaks are around 0.9 μm, 1.06 μm and 1.33 μm. Due to the low stimulated emission cross-section and the high excited state absorption it is difficult to get laser action at 1.33 μm. Figure 15.2 shows the experimental arrangement of optical fiber laser. The input mirror had a transmission coefficient of 70–80% in the region 450–550 nm. The output mirror reflectivity ranged from 80% to 99.8%.

There are two absorption bands of Nd-doped fiber in the wavelength region 450–550 nm caused by the transitions $^4I_{9/2}$–$^4G_{9/2}$ and $^4I_{9/2}$–$^4G_{7/2}$ respectively (Fig. 15.1). Ar ion laser also have many components in this region, such as 457.9 nm, 476.5 nm, 488.0 nm, 496.5 nm and 514 nm. Among these, 514.5 nm and 476.5 nm are very close to the peaks of the two absorption bands. So the absorption at these two wavelengths is stronger than that at other wavelengths.

In our experiment, CW fiber laser operations at 1060 nm were achieved when pumped by 476.5 nm, 488.0 nm, 496.5 nm and 514.5 nm light of Ar ion laser. Figure 15.3 shows the laser output characteristics of Nd^{3+}-doped silica fiber laser with different pumping Ar ion laser lines. The most efficient pumping is at 514.5 nm, the next is at 476.5 nm, which tallies with the absorption characteristic. The central laser wavelength is at 9400 cm^{-1} (1064 nm). In the case of 514.5 nm pumping, laser

Table 15.2. Laser output characteristics and puming condition of RE-doped silica fiber lasers.

Ion	Fiber	Pump laser (nm)	Concentration	Output wavelength (nm)	Fiber length (m)	Operation	Threshold (mW)	Ref.
Nd^{3+}	S	Dye, 594	0.1%	1082	0.051	*	6	3
	S	Diode, 820	300 ppm	1088	2	CW	0.1	4
	S	Dye 790–850	0.03%	1050–1075	2–16	Pulse	1.3	5
	S	Diode 820		1088	0.73	**	1.8	6
	S	Ti:Al$_2$O$_3$ 815	3%	1360	0.1	CW	5	7
	S	Diode 806	500 ppm	1550	3.7	CW	10	8
	S	Ti:Al$_2$O$_3$ 980	1100 ppm	1520–1570	5.5, 9.5	CW	40–60	9
Er^{3+}	Yb-codoped, S	Dye 800–845	0.8% (Yb: 1.7%)	1560	0.4–0.7	CW	5	10
	Yb-codoped, S	Nd:YAG 1064	0.8%(Yb: 1.7%)	1560	0.4–0.7	CW	7	11
	Yb-codoped, S	Dye 800–850	0.8%(Yb: 1.7%)	1560	0.4–0.7	CW	4	11
	Yb-codoped, S	Nd:YAG 1064		1560	0.4	CW	13.5	12
	Yb-codoped, S	Nd:YLF 1053			0.9		8	
Ho^{3+}	S	Ar$^+$ 457.9	200 ppm	2040	0.17	CW	46	13
	S	Diode 786	50, 200 ppm	1650–2000	0.45	CW	4.4	14
	S	Nd:YAG 1060	840 ppm	2000	1.75	CW	60	15
Tm^{3+}	S	Dye 800	830 ppm	1880–1960	0.27	CW	30	16
	M	Ti:Al$_2$O$_3$ 790	1000 ppm	1780–2050		CW	50	17
	M	Nd:YAG 1060	840 ppm	1780–2050	0.7	CW	600	17
Sm^{3+}	S	Ar$^+$ 488	1000 ppm	651	4	CW	20	18
Yb^{3+}	S	Diode 822	2500 ppm	1015–1140		CW	8	19
			600 ppm	1028–1064		CW	2	

S, Single-mode; M, multi-mode; CW; *, single-mode, CW; **, mode-locked, Q-switched.

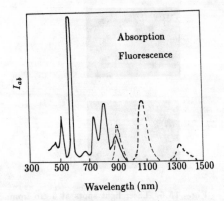

Fig. 15.1. Absorption and luminescence spectra of Nd-doped silica fiber (single mode, 5 μm in diameter).

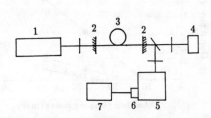

Fig. 15.2. Experimental setup of optical fiber laser. 1, Ar$^+$ laser; 2, resonance cavity; 3, laser fiber; 4, powermeter; 5, monochrometer; 6, detector; 7, recorder.

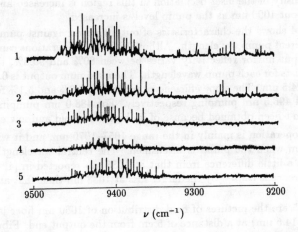

Fig. 15.3. Laser output characteristics of Nd-doped fiber with different pumping laser lines.

No.	1	2	3	4	5
Pumping line (nm)	514.5	514.5	496.5	488.0	476.5
Pumping energy W_p	4 W_{th}	2 W_{th}	2 W_{th}	2 W_{th}	2 W_{th}
Threshold W_{th} (mW)	—	4.8	12	20.6	51
Linewidth (cm^{-1})	130	120	100	90	110

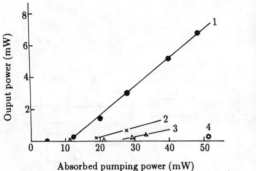

Fig. 15.4. Laser output power of Nd-doped fiber with different pumping laser lines. 1, • 514.5 nm pump; 2, × 476.5 nm pump; 3, △ 496.5 nm pump; 4, ○ 488.0 nm pump.

Fig. 15.5. Laser light spots at 5 cm from the output mirror (a) and through pumping light (514.5 nm) (b), divergence angle 7 mrad, LP_{11} linear polarization.

oscillation occurred in the region of longer wavelength except for that near 1060 nm. The intensity of the laser oscillation in this region is increased and the region extends to about 1090 nm as the pump level is increased.

Figure 15.4 shows the characteristics of output power against pump power absorbed at different pump wavelength. Although the laser operations can be achieved when the output mirror reflectivity varies between 80% and 99.8%, the optimum reflectivity exists for each pump wavelength. The maximum output is 6.8 mW when pumped at 514.5 nm. The slope efficiencies are 18.7%, 6.8% and 5.1% for 514.5 nm, 476.5 nm and 496.5 nm pumping respectively. For 488.0 nm pumping, the slope efficiency hasn't been obtained because of its high threshold and low efficiency.

The laser operation is mainly in the range 1057–1070 nm, and it will extend to about 1090 nm when the pump is intensified. The central wavelength is at 1064 nm. This has a little difference from that at 1088 nm reported in [4, 6], we think that this difference is caused by the larger loss at 1088 nm than that at 1064 nm in our fiber.

Figure 15.5 are the pictures of field distribution of 1060 nm fiber laser and the pump laser (514.5 nm) at a distance of 5 cm from the output end. Fiber laser field is a round spot of 6 cm in diameter. From this we evaluated that the divergence angle was 7°. The light spot of pump laser shows that the pump light travel in LP_{11} mode in the fiber.

The fiber with the core diameter of 5×10^{-3} mm can only allow the LP_{01} mode traveling in at 1060 nm. But for 514.5 nm Ar ion laser, V number of fiber is 4.5, so four modes LP_{01}, LP_{02}, LP_{11} and LP_{21} can exist in this fiber. As observed in the experiment, the mode of pump light can be adjusted by changing the condition of focalization. The best case we have achieved in this experiment is LP_{11} mode.

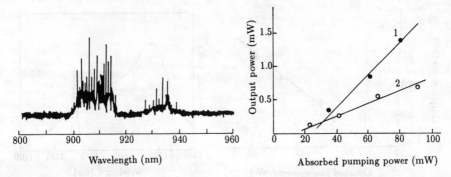

Fig. 15.6. Laser output spectrum of Nd^{3+}-doped silica fiber laser.

Fig. 15.7. Relationship between laser output power and absorbed pumping power. 1, $R = 85\%$; 2, $R = 96\%$.

According to the theory of fiber laser developed by M. J. F. Digonnet [22] the threshold and the slope efficiency can be calculated. Using the data obtained in the experiment, we got that the threshold is 3.7 mW for 514.5 nm pumping and the slope efficiencies are 21.5%, 7.6%, 2.1% and 9.2% for 514.5 nm, 496.5 nm, 488.0 nm and 476.5 nm pumping. These values fit the experiment data reasonably.

Using the same experimental setup and a Nd^{3+}-doped glass fiber, we depressed the simulated emission at 10600 nm and got laser action at 910 nm by applying the cavity back mirror with high transmittance at 10600 nm and high reflectance at 910 nm. The laser output spectrum is shown in Fig. 15.6. There existed two laser emission regions, the 900–915 nm and 925–949 nm respectively. Both the laser linewidths are about 15 nm, and their central wavelengths are located at 910 nm and 935 nm respectively.

The laser output characteristics are demonstrated in Fig. 15.7. The maximum output power and slope efficiency are 1.4 mW, 0.7 mW and 2.5%, 0.97% at absorbed power of 16 mW, 22 mW and at 85%, 95% reflectivity of the output mirror respectively.

B. Laser action at 1.36 μm of Nd:doped silica fiber pumped by Ti:sapphire laser

Due to the low stimulated emission cross section and the high excited state absorption it is difficult to obtain laser action at 1.36 μm. We used a Ti:sapphire laser with laser output wavelength at 0.804 μm, which is the strong absorption wavelength of Nd^{3+} ions in silica glass fiber. For depressing the stimulated emission at 1.06 μm the cavity back mirror with high transmittance at 1.3 μm and high reflectance at 1.06 μm has been used.

Figure 15.8 shows the relationship between the output power at 1.368 μm and

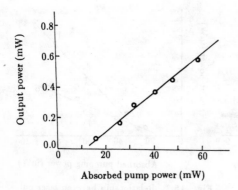

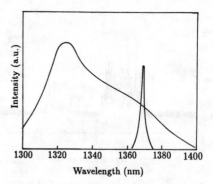

Fig. 15.8. Relationship between the output power at 1.36 μm and absorbed pump power at 0.804 μm of Nd-doped silica fiber.

Fig. 15.9. Fluorescence and laser emission from Nd-doped silica fiber.

the absorbed pumping power at 0.804 μm. The threshold power at reflectivity ($R = 99\%$) of the output mirror was 9.5 mW (absorbed power), the maximum output power at 1368 nm was 0.58 mW and the slope efficiency was 1.3%. The fluorescence and laser emission spectra at 1.3–1.4 μm region are shown in Fig. 15.9. The shift of laser emission peak position to the longer wavelength is due to strong excited absorption at the central fluorescence wavelength (1320 nm) [23].

C. Tunable glass fiber ring lasers [24, 25]

In the glass fiber ring laser the fiber coupler is used not only for coupling the fiber to become a ring cavity, but also as wavelength tuning. In addition, there are not any reflective elements in all-fiber laser, so that the excess loss is very low.

A Nd-doped glass fiber ring laser, end pumped by a Ar$^+$ laser at 514.5 nm wavelength, exhibits the laser wavelength tuning range of 60 nm, the pumping threshold power of 14.5 mW and the slope efficiency of 9.2%.

Using Ar$^+$ laser (514.5 nm) end pumping in a Er/Yb codoped fiber ring cavities of 8 m and 1 m, the slope efficiency of 10.2% and 6.1% and the threshold power of 14.2 mW and 1.9 mW were obtained respectively. The laser wavelength can be tuned continuously from 1554 nm to 1577 nm. With optimum design and fabrication of delicate coupler, the linewidth at central wavelength 1540 nm of less than 0.1 nm for 1 m fiber length and the tuning range of more than 70 nm have been obtained.

D. Laser characteristics of Er^{3+}/Yb^{3+}-codoped fibers pumped by Ar ion laser [26]

Using a 4 m long silica glass fiber doped with high concentration Er^{3+} and Yb^{3+}

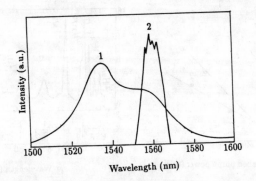

Fig. 15.10. Luminescence and laser spectra of Er^{3+}/Yb^{3+}-codoped silica glass fiber and laser. 1, Luminescence; 2, laser.

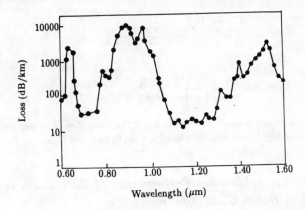

Fig. 15.11. Loss spectra of Er^{3+}/Yb^{3+}-codoped silica glass fiber.

ions and pumped by 514.5 nm line of Ar ion laser, the laser action at 1560 nm wavelength has been obtained at the above mentioned experimental arrangement. Figure 15.10 shows the luminescence and the laser spectra.

The laser central wavelength is not in the peak position of luminescence spectra which due to the higher inactive loss at 1530 nm than that at 1560 nm, which can be seen from Fig. 15.11.

The laser threshold depends on reflectivity of output mirror, the threshold power is 8.7 mW, 12.7 mW and 21.5 mW at reflectivity of the mirror about 96%, 83% and 70% respectively. As shown in Fig. 15.12, the maximum output power is 5.4 mW and the slope efficiency is 9.5% at the mirror reflectivity 70%.

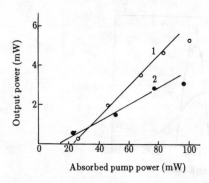

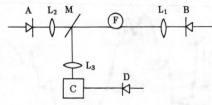

Fig.   15.12.   Output characteristics of Er^{3+}/Yb^{3+}-codoped silica fiber laser.   1, $R = 70\%$; 2, $R = 80\%$.

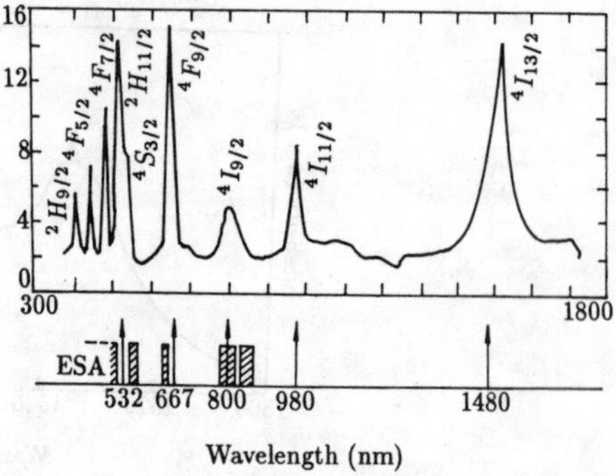

Fig.   15.13.   Absorption spectrum and pump light wavelengths of Er-doped silica glass fibers.

Fig. 15.14. Experimental arrange- ment of Er-doped fiber amplifier: A, pump diode; B, signal diode; C, monochrometer; D, detector; M, mirror; L, lens; F, Er-doped fiber.

15.1.3   *Laser Amplification Performance of Rare Earth Ions Doped Silica Based Glass Fibers*

Er^{3+}-doped or Er^{3+}/Yb^{3+}-codoped silica glass fibers have been widely studied for laser amplifiers at 1.55 μm wavelength.

A. Characteristics of Er^{3+}/Yb^{3+}-codoped fiber amplifier pumped by diode laser [27]

From practical point of view the Er-doped fiber amplifier (EDFA) should be pumped by diode laser. Figure 15.13 shows the absorption spectrum of Er-doped fiber and the suitable pumping light wavelengths. We use 800 nm wavelength light of laser diode as the pumping source. The experimental arrangement for fiber amplifier is shown in Fig. 15.14. The 800 nm pumping light of laser diode A was focused through dichroic mirror M into the fiber and the 1.55 μm signal light of laser diode B is coupled into the fiber from another side of the fiber.   The amplified light was

Table 15.3. Amplification characteristics of Er-doped silica glass fiber.

Signal light intensity (dBm) at pumping power 8.5 mW	47.7	50.2	48.2	47.7	46.2
Gain (dB)		30	5.6	3.8	3.3	0.9
Pumping power (mW) at		20	8.5	7.5	6.6	
Signal intensity 47.7 dBm gain (dB)		30	3.3	1.9	0.7	

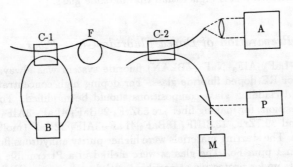

Fig. 15.15. Schematic diagram of all-fiber laser amplifier at 1.55 μm. A, Ar$^+$ laser or diode laser; B, Er-doped fiber signal laser; C-1, C-2, coupler; F, Er-doped fiber; M, monochrometer; P, powermeter.

reflected by mirror M and then focused to detector D through monochrometer C.

Using an Er-doped silica glass fiber with the numerical aperture (N.A.) of 0.22 and 11 m in length, the amplification characteristics of EDFA are listed in Table 15.3. Because of insufficient pump power the amplification gain is not so high.

Due to the excited state absorption (ESA) it makes the 800 nm pump band unattractive. The two last IR bands 980 nm and 1480 nm are free from ESA. Good results have been obtained: a gain coefficient of 5.8 dB/mW for 1.47 μm pump and a 6.1 dB/mW for 970 nm pump. Recently, a 10.2 dB/mW gain coefficient with 980 nm pump was reported by NTT.

B. Er,Yb-doped all-fiber amplifiers [25]

Figure 15.15 shows the schematic diagram of all-fiber laser amplifier working at 1.55 μm wavelength. The pumping sources are Ar$^+$ ion laser working at 514.5 nm or diode laser working at 900 nm. Er,Yb-codoped fiber laser is used as signal source at 1.55 μm. The amplification is achieved about 23 dB for a 1 m long fiber.

15.2   Rare Earth Ions Doped Fluoride Based Glass Fibers

The laser performances and the pumping condition of RE ion doped fluoride glass fiber lasers are listed in Table 15.4. In comparison with Table 15.2 it can be found that for the RE ion doped fluoride glass fiber lasers the useful laser wavelengths are located in IR region, such as 1.33 μm (Nd^{3+}), 1.7 μm and 2.7 μm (Er^{3+}), 2.08 μm and 2.7 μm (Ho^{3+}), 1.88 μm and 2.35 μm (Tm^{3+}). The IR emissions of RE ion are performed by cascade nonradiative energy transfer, the probability between intraionic energy levels is higher than that of oxide glass.

15.2.1   Preparation of Doped Fluoride Glass Fibers

ZrF_4–BaF_2–LaF_3–AlF_3–NaF (ZBLAN) fluoride system was always chosen as the host glass for RE-doped fluoride glass. For doping high concentration of RE ions the core and cladding glass compositions should be modified. For example, the glass compositions for making fiber are $53ZrF_4\cdot20BaF_2\cdot4LaF_3\cdot3AlF_3\cdot20NaF$ (mol%) as a core and $39.7ZrF_4\cdot13.3HfF_4\cdot18BaF_2\cdot4LaF_3\cdot3AlF_3\cdot22NaF$ (mol%) as cladding, respectively. The starting materials were higher-purity anhydrous fluorides containing less than 1 ppm Fe. The glasses were melted in a Pt crucible at temperature of 850–900°C under protective atmosphere with low moisture ($H_2O{<}20$ ppm) in a glove box. The preform was prepared by so called "continuous casting method". The procedure is as follows: the cladding glass melt was cast in a preheated metal mold with a hole in the bottom, the core glass melt was then immediately in the above of the cladding melt, by raising the mold a part of cladding glass melt was flew out of the bottom and a part of core glass melt was filled into the center of the cladding glass. The preform with different core/cladding ratio can be obtained by controlling the times interval. The fiber was prepared by using a preform and a jacketing tube, which was obtained by similar method mentioned above. A set of the preform and the jacketing tube was inserted into a FEP-teflon tube and drawn into a fiber under protective atmosphere. The doping concentration is around 300–1000 ppm. The inactive loss at lasing wavelength is about 100 dB/km.

15.2.2   Laser Performances of Nd^{3+}-doped Fibers

For laser performance measurement the experimental arrangement is shown in Fig. 15.16. An Ar laser was used as pump source. The 514.5 nm laser beam was coupled into the fiber core by a 10× microscopic objective lens. The coupling efficiency was about 15%. Two coated mirrors were placed at both sides of the fiber to form a simple Fabry-Perot cavity. The input mirror possesses a reflectivity of 99% at 1.05 μm and a transmittance more than 92% at 514 nm. The length of the fiber was 95 cm.

Table 15.4. Laser output performances and pumping condition of RE-doped fluoride glass fiber lasers.

Ion	Fiber	Pump laser (nm)	Output wavelength (nm)	Concentration	Fiber length (m)	Operation	Threshold (mW)	Ref.
Nd^{3+}	M	Ar+ 514.5	1060	180 ppm	8	CW	300	28
	M	Ar+ 514.5	1330–1340, 1049	1.5% mol	0.0283	CW	670, 190	29
	M	Ar+ 514.5	1350, 1050	1000 ppm	0.5	CW	84, 33	30
	M	Diode 794	1050, 1350	1000 ppm	0.315	CW	3, 60	31
	Pr-codoped, S	Kr+ 647.1	2700	1000 ppm (Pr. 0, 1000, 2000, 3000 ppm)		CW	40	32
Er^{3+}	M	Ar+ 476.5	2700	1000 ppm	1.5–3	CW	6.9	33
	S	Ar+ 488, 514.5	~1000, ~1560	10000, 500 ppm	0.2–0.6	CW	600–800	34
	S	Kr+ 647.1	980	1200 ppm	0.44	*	19	35
	S	Ar+ 488, 514.5	~1660, 1720	977 ppm	0.4	CW	90	36
	M	Diode 801	850	500 ppm	3	CW	200	37
	M	Diode 802	2700–2800	10% mol		CW	10–60	38
Ho^{3+}	M	Ar+ 488, 457.9	~1380, 2080	993 ppm	0.5	CW	163, 1120	39
	M	Dye 640	~2900	1000 ppm	2.2, 10	CW	76	40
	S	Kr+ 647.1	540, 553	1200 ppm		CW	140	41
	S	Kr+ 647.1	753	1200 ppm		CW	240	41
Tm^{3+}	S	798	~2300	1000 ppm	0.5	Pulse	25 μJ	42
	S	Kr+ 674.6	820, 1480, 1880, 2350	1250 ppm	1.5	CW	30–50	43
Pr^{3+}	S	Nd:YAG 1060	~1300	1000 ppm	6.5	CW		44
	M	Ar+ 476.5	610, 635, 659, 715, 885, 910	1200 ppm	0.6	CW	120–600	45
Yb^{3+}	M	Ar+ 514.5	1000–1050	2 wt%	0.95	CW	50	47

S, Single-mode; M, multi-mode; *, CW or Q-switched.

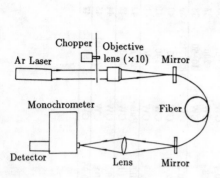

Fig. 15.16. Experimental set-up of optical fiber laser.

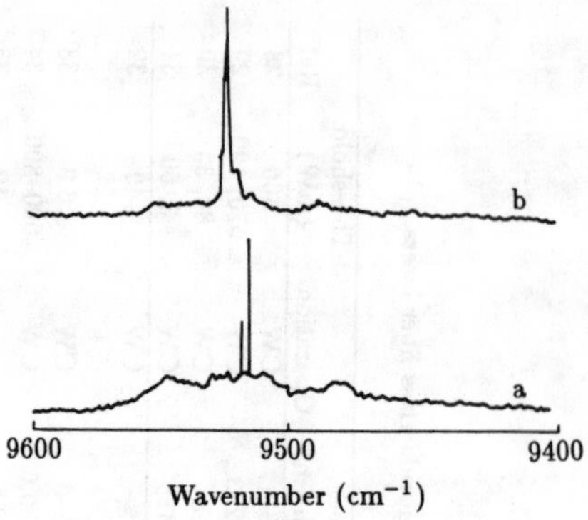

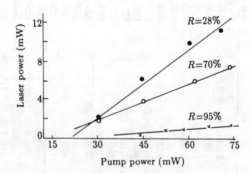

Fig. 15.17. Laser spectra of fiber under various pump power. 1, Near threshold; 2, about 1.4 P_{th}.

Fig. 15.18. Output characteristics of fiber laser.

Figure 15.17 shows the laser spectra of the fiber under various pump power. Two spikes at 1.0502 and 1.0504 μm are observed under the pump power near the threshold. When the pump power increases to 1.4 P_{th}, the laser wavelength moves to 1.0498 μm with a linewidth of 0.2 nm. Figure 15.18 shows the laser output power as a function of the injected pump power using the output mirror with different reflectivity from 95% to 28%. The output power increases with the decrease of the reflectivity of output mirror, and 11 mW CW laser output with a slope efficiency more than 20% was obtained. More detailed result and discussion were presented elsewhere [46].

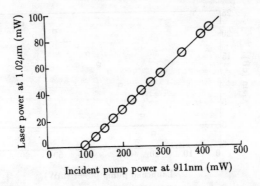

Fig. 15.19. Output power against incident pump power (launch efficiency 50%).

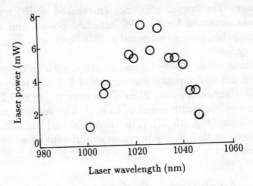

Fig. 15.20. Tuning curve (20 cm, 2.0 wt% Yb, output mirror 43%). Incident pump power 85 mW ($2.0 \times P_{th}$) at 918 nm.

15.2.3 *Laser Performances of Yb^{3+}-doped Glass Fibers*

Trivalent ytterbium is attractive for laser emission owing to its wide pumping range extending from 0.8 to 1.0 μm and to its ability to act as a donor ion in codoped materials. Moreover, it has only one excited level with negligible multiphonon relaxation rate in most materials. The fluorescence spectrum of Yb^{3+} extends from 0.95 to 1.1 μm, a spectral domain where few laser sources are available. The fluorescence lifetime of Yb^{3+} in ZBLAN glass fibers is 1.7 ms for 2.0 wt% Yb. High power laser oscillation at 1.02 μm has been obtained by J. Y. Allain *et al.* [47]. Figure 15.19 shows the laser characteristics of a 3.0 μm core diameter, 12×10^{-3} refractive index difference, 2.0 wt% ytterbium, 20 cm long fiber in an unoptimised cavity (air path, Fresnel output "mirror"). The pump wavelength is 911 nm issued from

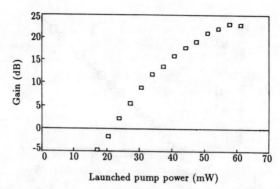

Fig. 15.21. Small-signal gain at 806 nm as a function of launched pump power for the Tm:ZBLAN fiber amplifier. Fiber length, 3 m; pump wavelength, 780 nm.

a Ti:sapphire laser. The launch efficiency (measured at 795 nm) is close to 50%. The threshold was obtained for 50 mW launched power, and the slope efficiency with respect to launched power was equal to 56%, corresponding to 63% differential quantum efficiency.

Inserted a tuning prism inside the laser cavity, with an output mirror of reflectivity 43% at 1.06 μm and a pump wavelength of 918 nm, the pump threshold was lowered to 42 mW (launched power 21 mW). A laser power of 7.2 mW was obtained at 1.02 μm for an incident pump power of 85 mW. The laser emission has been tuned from 1000 to 1050 nm (Fig. 15.20). The threshold and the slope efficiency are decreased as compared to that in the previous experiment as a consequence of higher output mirror reflectivity.

15.2.4 *Laser Amplification Performances at 0.8 μm of Tm^{3+}-doped Glass Fibers*

Thulium-doped fluorozirconate fiber can be used as an efficient amplifier over the wavelength range 800–815 nm in the first telecommunications window. A small-signal gain of 23 dB has been achieved at 805 nm with 50 mW of launched pump power at 780 nm by R. D. T. Lander *et al.* [48]. The fiber used for this demonstration had a core diameter of 3.5 μm, the numerical aperture 0.16 and the thulium ion concentration 800 ppm by weight. The background loss of the fiber was measured to be less than 0.1 dBm^{-1}. These results suggest that a high gain 800 nm amplifier employing an AlGaAs diode-pumped thulium-doped fluorozirconate fiber can now be made.

Figure 15.21 shows how the maximum small-signal gain (< -20 dBm^{-1}), measured at 806 nm, varied with launched pump power at 780 nm. Slightly in excess of 20 mW launched pump was needed to make this 3-level fiber amplifier transparent. For 50 mW launched power a gain of 22 dB was measured.

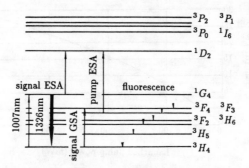

Fig. 15.22. Pr^{3+} energy level diagram.

15.2.5 *Laser Amplification Performances at 1.3 μm of Pr^{3+}-doped Glass Fibers*

Due to a strong application drive, amplification at this wavelength has received greatest attention. Early work centred on the $^4F_{3/2} \rightarrow {^4I_{13/2}}$ transition in Nd^{3+}, however, due to the excited state absorption of signal photons and the competing emission at 1.06 μm, work was reduced on this system in favour of the more promising $^1G_4 \rightarrow {^3H_5}$ transition in Pr^{3+}. The energy level diagram and some of the competing transitions are shown in Fig. 15.22. The metastable $^1G_4$ level is largely depopulated by phonon emission due to the closeness of the lower lying energy levels, this limits the quantum efficiency of the $^1G_4 \rightarrow {^3H_5}$ transition to a few percent. Additionally signal ground state absorption and various excited state absorption processes reduce the efficiency. Thus this system will never be as efficient as the Er^{3+} in silica amplifiers, but is considered sufficiently promising to warrant extensive research.

The most advanced amplifiers have used the relatively well characterised ZrF_4–BaF_2–LaF_3–AlF_3–NaF (ZBLAN) based fibers, where the large Δn and the low background loss have been exploited. Typical Pr^{3+} concentrations are ~1000 ppm, above this level the ion-ion interactions reduce the efficiency. At this concentration τ_s is typically 100 μs. To obtain sufficient gain ~20 m of fiber is needed, and the reported background losses of <0.05 dB/m are low enough to minimize reducing the amplifier performance. Figure 15.23 shows the gain of a 24 m length of fiber as typified above. The fiber was longitudinally pumped with an all solid state source at 1.047 μm into the low energy wing of the $^1G_4$ level. As can be seen, 20 dB gain was achieved for 280 mW launched power. This particular example also included special fluoride silica fiber couplers to allow the amplifier to be easily coupled to standard telecom system fiber and thus be more practical. The gain bandwidth of the amplifier is shown in Fig. 15.24 for a launched pump power of 300 mW. The

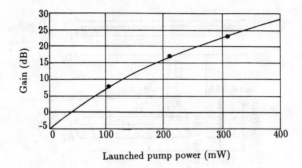

Fig. 15.23. Small signal gain as a function of pump power for Pr³⁺ fiber amplifier.

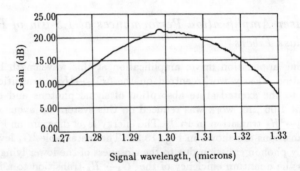

Fig. 15.24. Small signal gain spectrum for Pr³⁺ fiber amplifier for 300 mW launched pump power.

gain peak is situated at 1.30 μm and the 3 dB bandwidth is 20 nm which ties in well with currently used telecom wavelength [49].

Other approaches to improve the amplification gain are to reduce the maximum phonon energy so as to increase the emission quantum efficiency of heat glasses ($\sigma \cdot \tau$). Figure 15.25 shows the summarized result of quantum efficiency for newly developed Pr³⁺-doped glasses. Figure 15.26 shows the calculated efficiency of 0.04 Δn Pr³⁺-doped fiber vs fiber background loss for various glass hosts [50]. It can be seen that for new Pr³⁺-doped glass fibers low background loss and high Δn are necessary to compete with the existing ZBLAN fibers.

15.2.6 *Laser Amplification Performances at 1.5 μm of Er³⁺-doped Glass Fibers*

Up to now the Er-doped fluoride fiber amplifier (EDFFA) suffered poor gain-pump

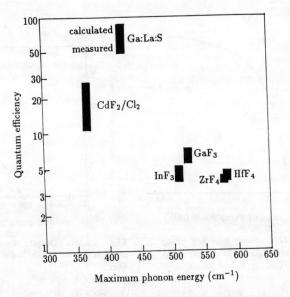

Fig. 15.25. Quantum efficiency of newly developed Pr^{3+}-doped glasses.

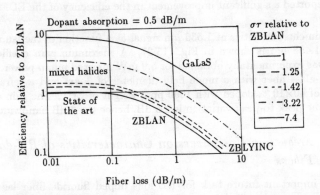

Fig. 15.26. Calculated efficiency of 0.04 Δn Pr^{3+} fiber amplifiers versus fiber background loss for various glass hosts.

power efficiency compared to Er-doped silica fiber amplifier (EDSFA). However, EDFFA have a natural flat bandwidth which is a consequence of the shapes of the Er^{3+} absorption and the emission cross-section spectra in fluoride glasses. The flatness of the bandwidth curve is of interest in wavelength multiplexed digital

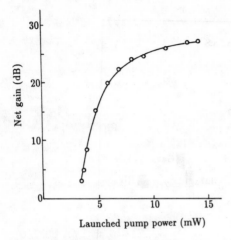

Fig. 15.27. Gain versus launched pump power. Signal wavelength, 1536 nm; input signal power, −38 dBm.

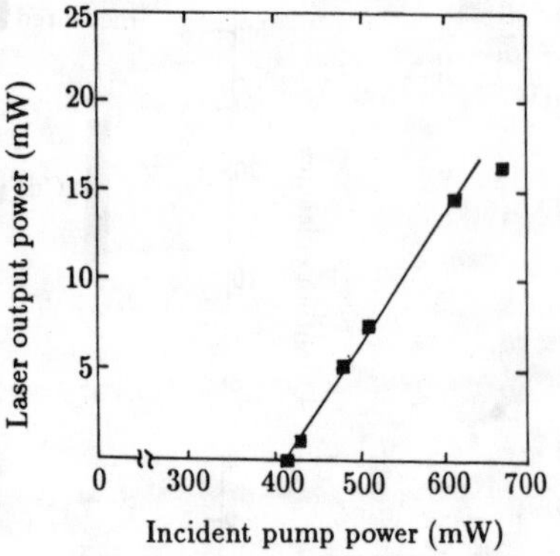

Fig. 15.28. Output power as a function of pump power for Er³⁺-doped upconversion fiber.

systems or in analog distribution systems. For EDSFA it has to use properly designed optical filters.

A lot of work are devoted to improve the gain efficiency of EDFFA [51, 52]. H. Ibrahim reported a significant improvement in the efficiency of the EDFFA recently [53].

The gain characteristics at 1.536 μm signal as a function of the launched pump power at 1.488 μm is shown in Fig. 15.27. A maximum gain coefficient of 3.4 dB/mW was experimentally obtained for 5.9 mW injected pump power. They used a ZBLAN glass fiber with a mode field diameter of 3 μm and a refractive index difference of 40×10^{-3}, doped with 1000 ppm weight of Er³⁺. The gain spectrum is very flat, the peak-to-peak variation is 2.5 dB, hence the 3 dB bandwidth is 36 nm.

15.2.7 Frequency Up-conversion Characteristics of RE-doped Glass Fibers

The most important future task for RE ions doped fluoride fiber laser is to explore the up-conversion fiber lasers. Recently, Allain *et al.* demonstrated CW upconversion lasing action in Ho³⁺-doped fluoride glass fibers at room temperature [54]. According to theoretical modelling of Atkins *et al.*, the threshold power of up-conversion fiber laser operation of Ho³⁺ at 540–553 nm is close to that available from commercial visible semiconductor lasers in the vicinity of 630–670 nm.

Excited by a Ti:sapphire laser at about 800 nm, the up-conversion emission in Nd³⁺:ZBLAN and Er³⁺:ZBLAN fluoride glasses were observed by us [56]. The excitation spectra and excitation intensity dependence of these up-conversions were measured. Our analysis shows that the up-conversion is mainly two photon process with energy transfer characteristics. Continuous-wave, room temperature laser oscillation at 548 nm in Er^{3+}-doped ZBLAN glass fiber pumped at 800 nm was reported by K. Hirao *et al.* [57]. As shown in Fig. 15.28, an output power of 10 mW was obtained with a slope efficiency of 7%.

References

1. C. J. Koester and E. Snitzer, *Appl. Opt.* **3** (1964) 1182.

2. Fangdong Wu and Wen Jiang, Extended Abstracts of SPIE'OE/FIBERS'89.

3. I. M. Jauncey, L. Reekie, J. E. Townsend and D. N. Payne, *Electron. Lett.* **24** (1988) 24.

4. R. J. Mears, L. Reekie, S. B. Poole and D. N. Payne, *Electron. Lett.* **21** (1985) 738.

5. K. Liu, M. Digonnet, K. Fesler, B. Y. Kim and H. J. Shaw, *Electron. Lett.* **24** (1988) 838.

6. I. N. Duliong, L. Goldberg and J. F. Weller, *Electron. Lett.* **24** (1988) 1333.

7. S. G. Grubb, *Electron. Lett.* **26** (1990) 121.

8. R. Wyatt, B. J. Ainslie and S. P. Craig, *Electron. Lett.* **24** (1988) 1362.

9. R. Wyatt, *Electron. Lett.* **25** (1989) 1498.

10. D. C. Hanna, R. M. Percival, I. R. Perry, R. G. Smart and A. C. Tropper, *Electron. Lett.* **24** (1988) 1068.

11. M. E. Fermaun, D. C. Hanna, D. P. Shepherd, P. J. Suni and J. E. Townsend, *Electron. Lett.* **24** (1988) 1135.

12. G. T. Maker and A. I. Ferguson, *Electron. Lett.* **24** (1988) 1160.

13. D. C. Hanna, R. M. Percival, R. G. Smart, J. E. Townsend and A. C. Tropper, *Electron. Lett.* **25** (1989) 593.

14. W. L. Barnes and J. E. Townsend, *Electron. Lett.* **26** (1990) 746.

15. D. C. Hanna, M. J. Mccarthy, I. R. Perry and P. J. Suni, *Electron. Lett.* **25** (1989) 1365.

16. L. Esterowitz, R. Aller and L. Aggarwal, *Electron. Lett.* **24** (1988) 1104.

17. A. Tropper, R. Smart, I. Perry, D. Hanna, J. Lincoln and B. Brocklesby, *Proc. SPIE* **1373** (1990) 152.

18. M. C. Farries, P. R. Morkel and J. E. Townsend, *Electron. Lett.* **24** (1988) 709.

19. D. C. Hanna, R. M. Percival, I. R. Perry, R. G. Smart and P. J. Suni, *Electron. Lett.* **24** (1988) 1111.

20. Yihong Chen, Ruihua Cheng and Fuxi Gan, *Chinese. Sci. Bulletin* **37** (1992) 556.

21. Yihong Chen, Ruihua Cheng and Fuxi Gan, *J. Infrared and Microwave* **11** (1992) 4.

22. M. J. F. Digonnet, *Appl. Opt.* **24** (1985) 333.

23. Ruihua Cheng, Yihong Chen, Fuxi Gan *et al.*, *Chinese J. Lasers* **A20** (1993) 493.

24. Caoyu Yue, Jiangde Peng and Bingqun Zhou, *Electron. Lett.* **25** (1989) 101.

25. Jiangde Peng, Caoyu Yue and Bingqun Zhou, *Acta Optica Sinica* **10** (1990) 922.

26. Yihong Chen, Ruihua Cheng and Fuxi Gan, *Chinese Science Bulletin* **37** (1992) 1224.

27. Zhexin Chen and Yanzhi Hu, *Chinese J. Lasers* **18** (1991) 639.

28. M. C. Brierley and P. W. France, *Electron. Lett.* **23** (1987) 815.

29. W. J. Miniscalco, L. J. Andrews, B. A. Thompson, R. S. Quimby, L. J. B. Vacha and M. G. Drexhage, *Electron. Lett.* **24** (1988) 28.

30. M. C. Brierley and C. A. Millar, *Electron. Lett.* **24** (1988) 438.

31. M. C. Brierley and M. H. Hunt, *Proc. SPIE* **1171** (1989) 157.

32. J. Y. Allain, M. Monerie and H. Poignant, *Electron. Lett.* **27** (1991) 445.

33. J. Y. Allain, M. Monerie and H. Poignant, *Electron. Lett.* **25** (1989) 28.

34. J. Y. Allain, M. Monerie and H. Poignant, *Electron. Lett.* **25** (1989) 318.

35. J. Y. Allain, M. Monerie and H. Poignant, *Electron. Lett.* **25** (1989) 1082.

36. R. G. Smart, J. N. Carter, D. C. Hanna and A. C. Tropper, *Electron. Lett.* **26** (1990) 649.

37. C. A. Miller, M. C. Brierley, M. H. Hunt and S. F. Carter, *Electron. Lett.* **26** (1990) 1871.

38. H. Yanagita, I. Masuda, T. Yammashita and H. Toratani, *Electron. Lett.* **26** (1990) 1837.

39. M. C. Brierley, P. W. France and C. A. Millar, *Electron. Lett.* **24** (1988) 539.

40. L. Wetenkamp, *Electron. Lett.* **26** (1990) 883.

41. J. Y. Allain, M. Monerie and H. Poignant, *Electron. Lett.* **26** (1990) 261.

42. D. C. Hanna, I. M. Jauncey, R. M. Percival, I. R. Perry, R. G. Smart and P. J. Suni, *Electron. Lett.* **24** (1988) 1222.

43. J. Y. Allain, M. Monerie and H. Poignant, *Electron. Lett.* **25** (1989) 1660.

44. Y. Durteste, *Electron Lett.* **27** (1991) 626.

45. J. Y. Allain, M. Monerie and H. Poignant, *Electron. Lett.* **27** (1991) 189.

46. Jie Wang and Fuxi Gan, Extended Abstracts of 9th Intern. Symp. on Non-oxide glasses, Hangzhou, China, 1994, p. 372.

47. J. Y. Allain, M. Monerie, H. Poignant, T. Georges, *J. Non-Cryst. Solids* **167** (1993) 270.

48. R. D. T. Lander, J. N. Carter, *et al.*, *J. Non-Cryst. Solids* **16** (1993) 274.

49. S. T. Davey, D. Szebesta, *et al.*, *J. Non-Cryst. Solids* **161** (1993) 262.

50. B. A. Ainslie, S. T. Davey, *et al.*, Extended Abstract of 9th Intern. Symp. on Nonoxide Glasses, Hangzhou, China, 1994, p. 360.

51. Y. Miyajima, T. Sugawa, T. Tomukai, *Electron. Lett.* **26** (1991) 1527.

52. D. Ronarc'h, M. Guibert, *et al.*, *Electron. Lett.* **27** (1991) 908.

53. H. Ibrahim, D. Ronarc'h *et al.*, *J. Non-Cryst. Solids* **161** (1993) 266.

54. J. Y. Allain, M. Monerie and H. Poignant, *Electron. Lett.* **26** (1990) 262.

55. G. R. Atkins, S. B. Poole and M. G. Sceats, Extended Abstracts of VII Intern. Conf. of Halide Glasses, Lorn, Australia, 1991, p. 513.

56. Fuxi Gan and Yihong Chen, *Opt. Mater.* **2** (1993) 45.

57. K. Hirao, S. Todoroki and N. Soga, *J. Non-Cryst. Solids* **143** (1992) 40.

Subject Index

349

DATE DUE

Demco, Inc. 38-293